AF566922

PROGRESS IN THEORETICAL ORGANIC CHEMISTRY

VOLUME 3

MOLECULAR STRUCTURE AND CONFORMATION

PROGRESS IN THEORETICAL ORGANIC CHEMISTRY

VOLUME 3

MOLECULAR STRUCTURE AND CONFORMATION

Recent Advances

Edited by

I.G. CSIZMADIA

Department of Chemistry
University of Toronto
Lash Miller Chemical Laboratories
80 St. George Street
Toronto, Ontario
Canada M5S 1A1

ELSEVIER SCIENTIFIC PUBLISHING COMPANY
Amsterdam – Oxford – New York 1982

ELSEVIER SCIENTIFIC PUBLISHING COMPANY
Molenwerf 1
P.O. Box 211, Amsterdam, The Netherlands

Distributors for the United States and Canada:

ELSEVIER SCIENCE PUBLISHING COMPANY, INC.
52, Vanderbilt Avenue
New York, N.Y. 10017

Library of Congress Cataloging in Publication Data
Main entry under title:

Molecular structure and conformation.

(Progress in theoretical organic chemistry ; v. 3)
Bibliography: p.
Includes index.
1. Molecular structure. 2. Chemistry, Physical organic. I. Czismadia, I. G. (Imre G.) II. Series.
QD461.M628 547.1'22 82-7422
ISBN 0-444-42089-4 AACR2

ISBN 0-444-42089-4 (Vol. 3)
ISBN 0-444-41720-6 (Series)

Printed in The Netherlands

TABLE OF CONTENTS

IV. CORRELATION ENERGY AS A STABILIZING FACTOR IN MOLECULAR STRUCTURE 156

M.A. Robb

V. ANALYTIC EQUATIONS FOR CONFORMATIONAL ENERGY SURFACES 190

M.R. Peterson and I.G. Csizmadia

VI. QUANTUM CHEMICAL STUDIES ON THE MECHANISM OF ENZYME ACTION 267

G. Náray-Szabó and T. Bleha

PREFACE

Volume 2 of this series, published in 1977, collected a number of papers each of which has been presented at the first symposium on Theoretical Organic Chemistry (TOC) held in 1975 at Teneriff, one of the Canary Islands. TOC, since that time, has developed in a number of frontiers so the present volume, Volume 3, is called upon to show some of these advances.

The first three chapters (I, II and III) are illustrative of the advances that have been made in the field of 'Interpretive TOC' while chapter IV shows some advances in the field of 'computational TOC'.

Chapter V demonstrates how one may reduce a large volume of computed data by fitting analytic equations to them and how such a fitted surface may lead to the determination of critical points as well as to the recognition of critical point topology.

Finally Chapter VI shows some recent advances in the area related to understanding of the mechanism of enzyme action utilizing some methods of TOC.

All in all, this volume demonstrates some of the advances that have been made in diverse directions during the past half decade even though not all advances are represented in a balanced fashion.

CONTRIBUTORS TO VOLUME 3

Fernando Bernardi
Istituto Chimico "G. Ciamician"
Universita di Bologna
Bologna, Italy

Tomás Bleha
Polymer Institute, Slovak Academy of Sciences
809 34 Bratislava
Czechoslovakia

Andrea Bottoni
Istituto di Chimica Organica
Universita di Bologna
Bologna, Italy

Imre G. Csizmadia
Department of Chemistry
University of Toronto
Toronto, Ontario, M5S 1A1
Canada

Gábor Náray-Szabó
CHINOIN Pharmaceutical and Chemical Works
H-1325 Budapest, P.O.B. 110
Hungary

Vladimir I. Minkin
Institute of Physical and Organic Chemistry
Rostov on Don University
344711, Rostov on Don
USSR

Ruslan M. Minyaev
Institute of Physical and Organic Chemistry
Rostov on Don University
344711 Rostov on Don
USSR

Michael R. Peterson
Laboratoire de Chimie Physique Moléculaire
Faculté des Sciences
Université Libre de Bruxelles
Brussels, Belgium

Leo Radom
Research School of Chemistry
Australian National University
Canberra, A.C.T. 2600
Australia

CONTRIBUTORS TO VOLUME 3

Michael A. Robb
Department of Chemistry
Queen Elizabeth College
Campden Hill Road
London W 8 7AH

Molecular Structure and Conformation: Recent Advances,
I.G. Csizmadia (Ed.), *Progress in Theoretical Organic Chemistry*, Volume 3

STRUCTURAL CONSEQUENCES OF HYPERCONJUGATION

LEO RADOM

1. INTRODUCTION

The ability of groups such as CH_3 to interact with double and triple bonds was first recognized nearly half a century ago [1-4]. This phenomenon was termed *hyperconjugation* by Mulliken [3] who also presented a molecular orbital description [3,4]. Since that time, the concept of hyperconjugation has been the subject of vigorous debate and continuing controversy [5-8]. Many of the experimental results ascribed to hyperconjugative interaction have been argued to be explicable equally in other terms [7,8].

During the last ten years, however, *ab initio* theory has been brought to bear on the problem and has yielded unambiguous evidence for the importance of hyperconjugative interactions in certain circumstances. The key to the success of the theoretical approach has been the ability to use theory to examine model systems, i.e. systems artificially constrained in a manner which allows the interactions primarily responsible for particular structural or energetic features of molecules to be separately identified.

In this article, a contemporary theoretical approach to hyperconjugation in its broad sense is presented with particular emphasis on structural consequences of hyperconjugation. The subject spans a very wide area and no attempt is made at comprehensive coverage. Rather, an attempt is made to demonstrate the diverse situations in which hyperconjugative interactions are important through examples heavily flavored by the author's own research interests. The choice of examples has also been dictated by a desire to maintain a uniform and consistent treatment.

Two approaches are used in conjunction in this article to examine hyperconjugative interactions. In the first place, *ab initio molecular orbital theory* [9] is used to provide quantitative data for the systems in question. The second step is to use such data together with *perturbation molecular orbital (PMO) theory* [10,11] to construct a qualitative theory which can explain the results. As a final step, predictions of the qualitative theory may be tested against additional (theoretical or experimental) quantitative data. This overall procedure thus represents a direct exercise in the application of the scientific method [12].

2. THEORETICAL BACKGROUND

2.1 Quantitative Approach

The theoretical results presented in this article all refer to standard *ab initio* molecular orbital calculations [13]. Unless otherwise noted, they were obtained with a recommended combination of split-valence 4-31G (for H,B,C,N,O and F)[14,15] and 5-21G (for Li and Be) [16,17] basis sets. *Ab initio* calculations at this level have been shown to provide a reasonable description of molecular structure [18].

In all of the calculations described in this article, with the exception only of those related to methyl tilts, bond angles have been assigned standard values [19]. Bond lengths have sometimes been assigned standard values [19] or have been optimized where noted. The calculations thus refer to artificial model systems rather than real molecules in many cases. They have been designed as such to probe the factors which influence molecular structure rather than to determine molecular structure *per se*.

2.2 Qualitative Approach

2.2.1 Introduction. Orbital Interaction Diagrams

Although numerical quantum chemical procedures provide quantitative information about systems under examination, they do not always supply accompanying explanations for the observed behavior. In other words, although they generally provide an answer to the question "what?", they frequently do not answer the question "why?". Quantities such as computed molecular orbital coefficients, electron densities and Mulliken charge distributions move partly in this direction but it is desirable to strengthen further the bridge between theoretical chemistry and organic chemistry.

In recent years, qualitative molecular orbital theory has provided a very useful means of rationalizing the quantitative results. If the system in question is considered as a set of interacting fragments, the interaction between them may be taken as a perturbation. This approach is frequently referred to as perturbation molecular orbital (PMO) theory [10,11]. The results of such an approach may be conveniently displayed as orbital interaction diagrams.

Consider, for example, interaction between a donor NH_2 orbital 2p(N) and an acceptor BH_2 orbital 2p(B) as shown in Figure 1. These combine to give a lower energy bonding combination π(N-B) and a higher energy antibonding combination π^*(N-B). The interaction is clearly stabilizing and is referred to as a two-electron stabilization. PMO theory tells us that (i) the magnitude of such a stabilization is inversely proportional to the energy separation of the interacting orbitals, (ii) the stabilization is approximately proportional to the overlap of the interacting orbitals, (iii) the π(N-B) orbital is more concentrated on N while the π^*(N-B) orbital is more concentrated on B,

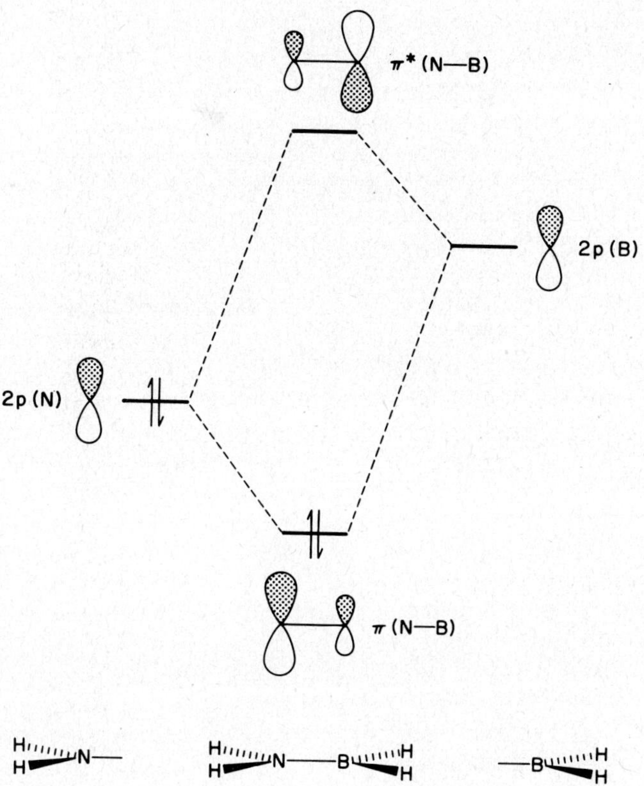

Fig. 1. Orbital interaction diagram showing the interaction of the occupied 2p(N) orbital with the vacant 2p(B) orbital in NH_2BH_2.

and (iv) π(N-B) is lowered slightly less than π*(N-B) is raised. A consequence of (iv) is that if both interacting orbitals are filled (e.g. planar NH_2-NH_2), the interaction is net destabilizing and is referred to as a four-electron destabilization. Such a destabilization is approximately proportional to the overlap and to the mean energy of the interacting orbitals. In general, four-electron interactions are less important than two-electron interactions and indeed, most of the results presented in this article, can be explained in terms of two-electron interactions alone.

Because the two-electron stabilization is inversely proportional to the energy separation of the interacting orbitals, it is generally sufficient to focus on the highest occupied and lowest unoccupied orbitals of the interacting fragments. Factors which influence orbital energies are clearly important in determining interaction energies. Of particular relevance to our discussion here will be the result that electronegative substitution lowers orbital energies while electropositive substitution raises orbital energies. This is discussed in detail for the particular case of substituted methyl groups in Section 2.2.4.

It is not always necessary to draw out in full interaction diagrams such as that displayed in Figure 1. A shorthand representation such as that shown in Figure 2 is used in some cases later in this article.

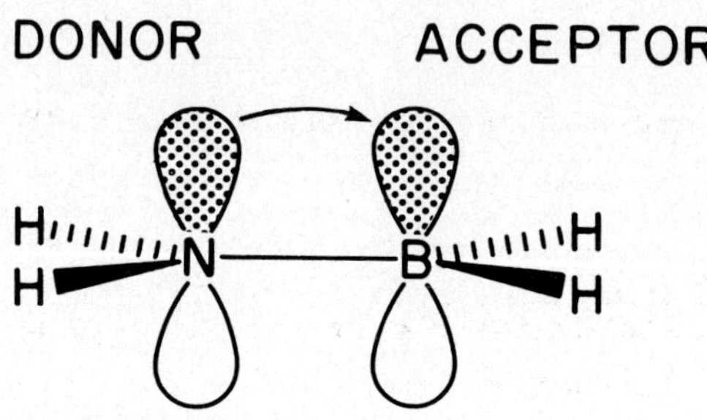

Fig. 2. Shorthand representation of the orbital interaction diagram showing the interaction of the occupied 2p(N) orbital with the vacant 2p(B) orbital in NH_2BH_2.

2.2.2 Classification of Substituents

It is convenient to classify, at this stage, the nature of the substituents with which we will be dealing. The substituents Li, BeH and BH_2 constitute a group of decreasingly powerful σ-donors:

$$Li > BeH > BH_2$$

but increasingly powerful π-acceptors:

$$Li < BeH < BH_2$$

The substituents NH_2, OH and F, on the other hand, constitute a group of increasingly powerful σ-acceptors:

$$NH_2 < OH < F$$

but decreasingly powerful π-donors:

$$NH_2 > OH > F$$

The CH_3 substituent has fairly weak σ-characteristics but normally acts as a σ-donor. It can also act as a hyperconjugative electron donor or acceptor and this is discussed in more detail in the following section.

In most of the examples presented in this article, the BH_2 and NH_2 substituents have been taken to be planar trigonal and examined in artificially constrained "planar" (1) and "perpendicular" (2) orientations with respect to an adjacent interacting orbital.

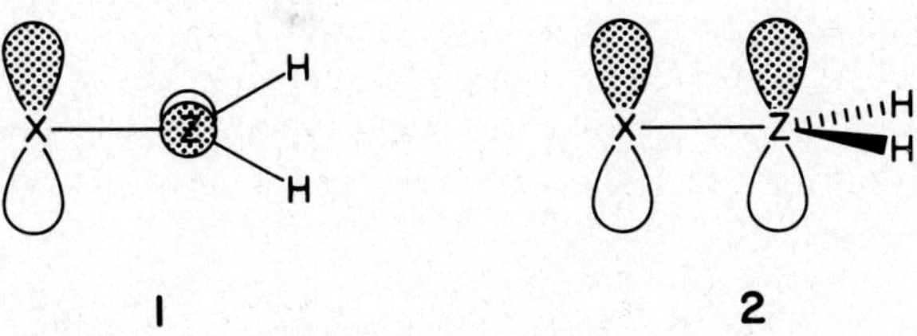

This allows the interaction between the orbitals to be "turned off" (1) or "turned on" (2).

2.2.3 The Orbitals of a Methyl Group

A methyl group may be considered to possess two degenerate pairs of pseudo π-orbitals as shown in Figure 3.

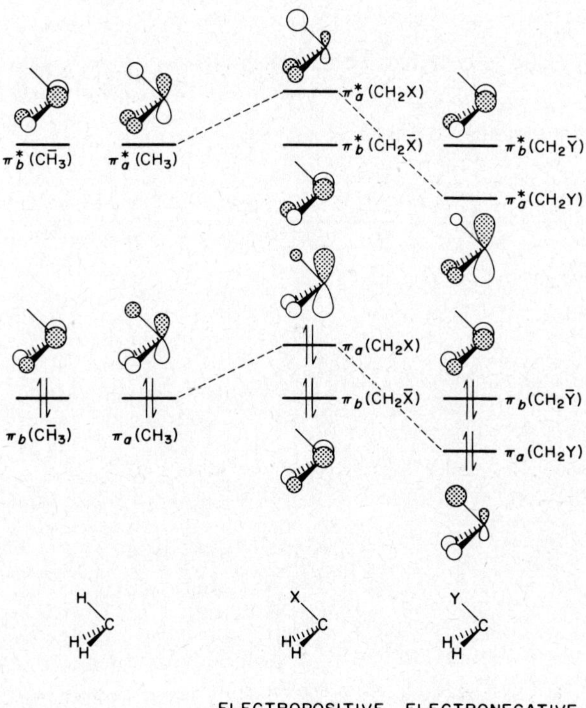

Fig. 3. π-Type orbitals of a methyl group and influence on orbital energies and orbital shapes of electropositive substituents X and electronegative substituents Y.

One pair represents the highest occupied orbitals, $\pi(CH_3)$ and $\pi(C\bar{H}_3)$, while the other, $\pi^*(CH_3)$ and $\pi^*(C\bar{H}_3)$, represents the lowest vacant orbitals. The parenthetic label (CH_3) denotes an orbital which includes a contribution from the "unique" hydrogen. Orbitals lacking this component are labelled ($C\bar{H}_3$). Throughout this article, we will use the convention that pseudo π-orbitals bearing the subscript *a* contain a carbon-centered p-orbital which is symmetric with respect to the plane of the paper. Group orbitals labelled *b* contain a carbon-centered p-orbital that is antisymmetric with respect to reflection in the plane of the paper.

2.2.4 The Orbitals of Substituted Methyl Groups

Substitution in a methyl group by an electropositive group X (e.g. X = Li) has two main effects as demonstrated in Figure 3. Firstly, it *raises* the energies of the $\pi(CH_2X)$ and $\pi^*(CH_2X)$ levels leaving the $\pi(CH_2\bar{X})$ and $\pi^*(CH_2\bar{X})$ levels relatively unaffected. In addition, the electron density in the C-X bond becomes more localized on C in $\pi(CH_2X)$ and more localized on X in $\pi^*(CH_2X)$, in comparison to $\pi(CH_3)$ and $\pi^*(CH_3)$ respectively. Similarly, substitution by an electronegative group Y (e.g. Y = F) *lowers* the energy of the $\pi(CH_2Y)$ and $\pi^*(CH_2Y)$ levels, again leaving $\pi(CH_2\bar{Y})$ and $\pi^*(CH_2\bar{Y})$ relatively unaffected. In this case, $\pi(CH_2Y)$ concentrates electron density on Y and $\pi^*(CH_2Y)$ concentrates the density on C. These points are illustrated in Figure 3 and are of crucial importance in the remainder of this article.

2.2.5 The Non-Equivalence of the Oxygen Lone Pairs in Water

An oxygen atom in, for example, water has two energetically non-equivalent lone pairs (Figure 4) [20].

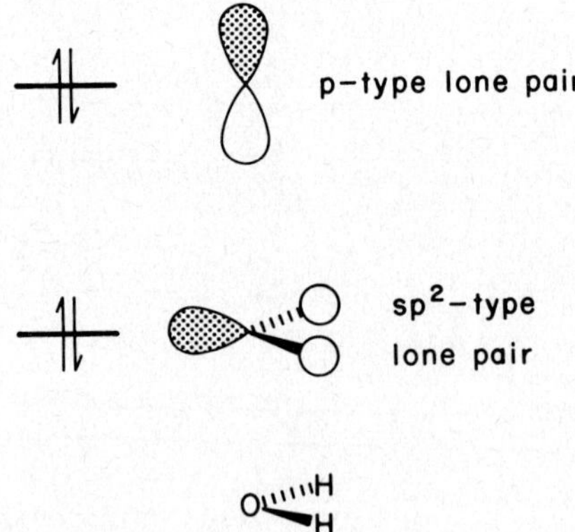

Fig. 4. The energetically non-equivalent lone pairs on oxygen in water.

The higher energy lone pair is a pure p-type orbital while the lower energy sp^2-type lone pair is part of a totally symmetric orbital containing contributions from the hydrogen atoms as well. Electron donation takes place more readily from the p-type lone pair and this is exemplified, for example, in a preference for a planar (**3**) rather than perpendicular (**4**) conformation for phenol.

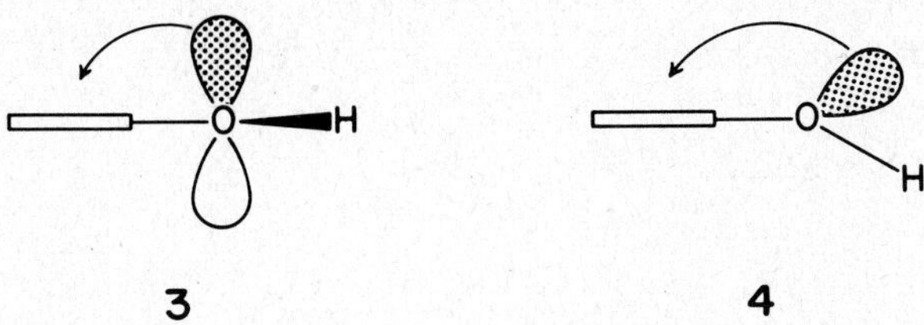

3 **4**

The non-equivalence of the oxygen lone pairs has other conformational consequences which are discussed later in this article.

2.3 Fourier Component Analysis of Rotational Potential Functions

Hyperconjugative interactions are strongly dependent on molecular conformation and hence potential functions for internal rotation in molecules provide useful probes for examining hyperconjugation. Such potential functions, $V(\phi)$, may be readily generated theoretically by carrying out *ab initio* calculations for an appropriate number of rotational angles, ϕ. It is a more difficult task to generate full rotational potential functions from experimental data since only conformations in the vicinity of the potential minima are appreciably populated; information regarding the remainder of the potential function is necessarily indirect.

The full rotational potential $V(\phi)$ is a complicated function and, in order to facilitate its interpretation, it is desirable to break it down into simpler components. A method that has been successfully used to this end is to expand the potential function $V(\phi)$ as a Fourier series [21]:

$$V(\phi) = \tfrac{1}{2} \sum_n V_n(1-\cos n\phi) + \tfrac{1}{2} \sum_n V'_n \sin n\phi \qquad (1)$$

Frequently, the potential is symmetric about $\phi = 0$ so that the sine terms disappear. In addition, the series is generally truncated after three terms. The most commonly used form of the expansion is then:

$$V(\phi) = \tfrac{1}{2} V_1(1-\cos\phi) + \tfrac{1}{2} V_2(1-\cos 2\phi) + \tfrac{1}{2} V_3(1-\cos 3\phi) \qquad (2)$$

or,

$$V(\phi) = V_1(\phi) + V_2(\phi) + V_3(\phi) \qquad (3)$$

The individual components $V_1(\phi)$, $V_2(\phi)$ and $V_3(\phi)$ of the total potential function $V(\phi)$ can be identified with specific physical effects because of their periodic properties. For example, the onefold term, $V_1(\phi) = \frac{1}{2} V_1(1-\cos\phi)$, moves from a maximum value to a minimum value as ϕ changes by 180°. Physical properties which would do this include dipolar or steric interactions (e.g. 5):

5

The twofold term, $V_2(\phi) = \frac{1}{2} V_2(1-\cos 2\phi)$, moves from a maximum value to a minimum value as ϕ changes by 90°. This frequently corresponds to conjugative or hyperconjugative interaction (e.g. 6):

6

Finally, the threefold term, $V_3(\phi) = \frac{1}{2} V_3(1-\cos 3\phi)$, moves from a maximum value to a minimum value as ϕ changes by 60°. This is generally attributed to repulsion between bond pairs of electrons as, for example, in ethane.

Examples of Fourier decomposition will be presented later in this article. Naturally, we shall be most interested in the twofold term and the information that it provides regarding hyperconjugative interaction.

2.4 What is Hyperconjugation?

Hyperconjugation may be readily defined by analogy with conventional conjugation [4]. We consider *conjugation* here as the result of interaction of two "pure" π-systems as, for example, in planar butadiene shown schematically in **7**:

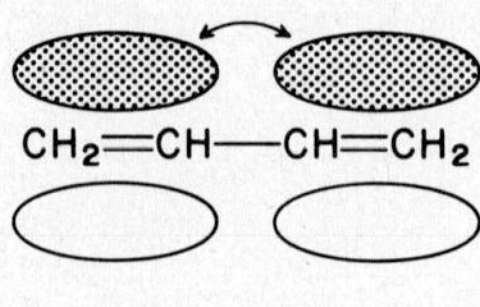

7

First-order hyperconjugation results from the interaction of a pseudo π-system, i.e. a system which has π-symmetry but that incorporates σ-bonds (e.g. CH_3), with a pure π-system as, for example, in propene, shown schematically in **8**:

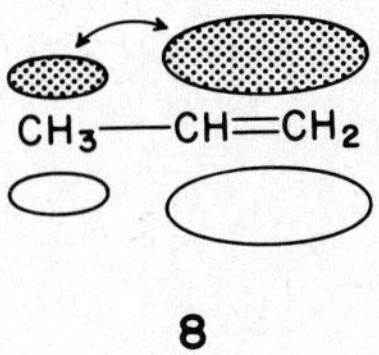

Finally, *second-order hyperconjugation* corresponds to interaction of two pseudo π-systems as, for example, in ethane (9):

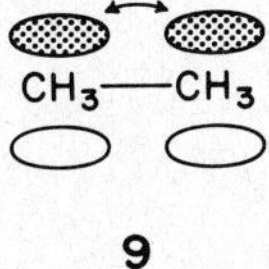

In this article, we shall be examining examples of both first- and second-order hyperconjugation but shall refer to either by the general term hyperconjugation.

3. HYPERCONJUGATION AND MOLECULAR CONFORMATION

3.1 1,3-Butadiene, Propene and Ethane

The close correspondence between conjugation and hyperconjugation is revealed by examining the key interactions which determine the conformations of 1,3-butadiene [22], propene [23] and ethane [24].

An orbital interaction diagram showing the interaction of two ethylenic units in the *cis* conformation of 1,3-butadiene is displayed in Figure 5. All three important interactions shown (i.e. **A**,**B** and **C**) favor the *trans* conformation which is of course well known to be the preferred conformation. Interaction **A** involves four electrons and is therefore destabilizing. It is decreased by decreasing the 1,4-overlap, which may be achieved by rotation to the *trans* conformation. Interactions **B** and **C** involve two electrons and are stabilizing. However, the terminal interacting p orbitals have opposite phases in both cases and therefore have a subtractive influence on the stabilization. Rotation to the *trans* conformation decreases this subtractive influence and hence is favorable.

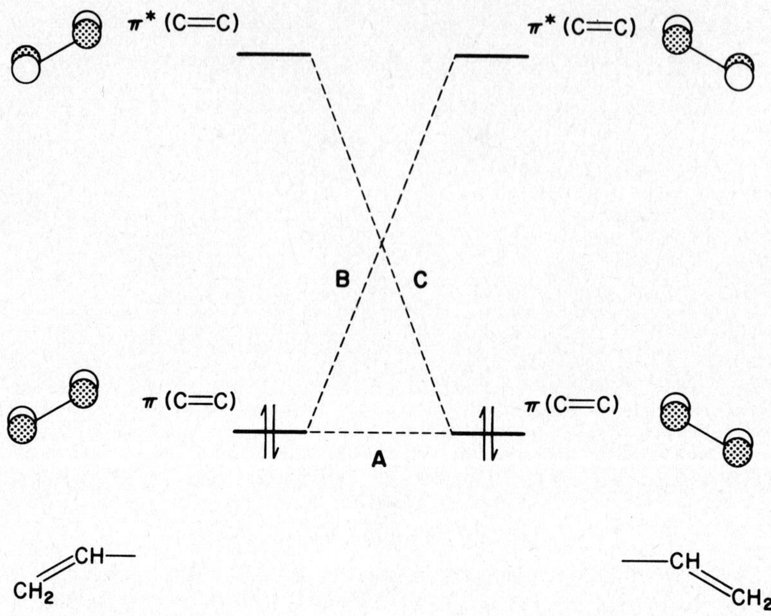

Fig. 5. Orbital interaction diagram showing the key interactions between the ethylenic fragments in the *cis* conformation of 1,3-butadiene.

Almost identical arguments lead to the prediction of a double bond eclipsed conformation for propene and a staggered conformation for ethane. The π and π* orbitals of the double bonds in Figure 5 are simply replaced successively by pseudo π and π* orbitals of a methyl group leading to the interaction diagrams of Figure 6 (propene) and Figure 7 (ethane). As in the case of 1,3-butadiene, interaction **A** is four-electron destabilizing while **B** and **C** are two-electron stabilizing interactions but with negative overlap of the terminal orbitals. All interactions thus favor rotation to the double bond eclipsed conformation for propene and the staggered conformation for ethane.

3.2 Substituted Ethyl Cations $XCH_2\text{-}CH_2^+$

An excellent probe for hyperconjugative interaction is provided by the conformational behavior of β-substituted ethyl cations, $XCH_2\text{-}CH_2^+$ [25,26]. Internal rotation about the C-C bond in the parent ethyl cation involves a six-fold barrier and occurs relatively freely. In particular, the eclipsed (**10**, X=H) and perpendicular (**11**, X=H) conformations of the ethyl cation have very similar energies.

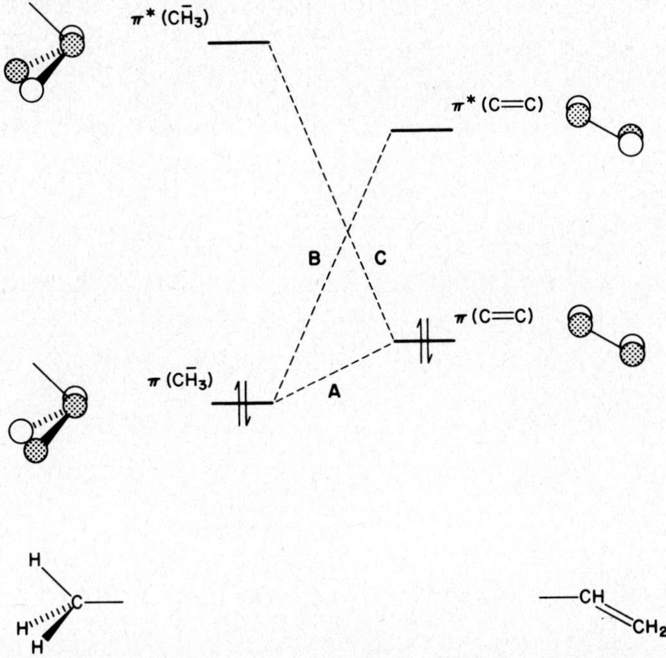

Fig. 6. Orbital interaction diagram showing the key interactions between a methyl group and the double bond in the double bond staggered conformation of propene.

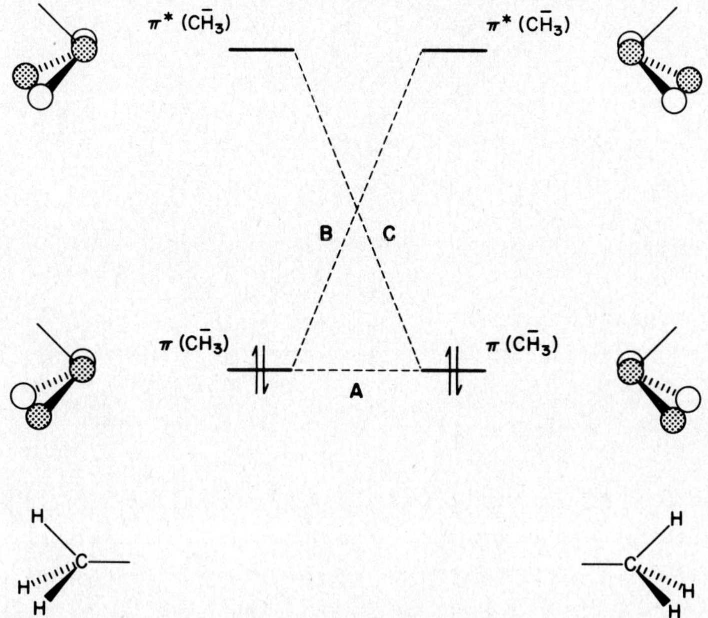

Fig. 7. Orbital interaction diagram showing the key interactions between two methyl groups in the eclipsed conformation of ethane.

10 11

This may be seen from the orbital interaction diagram of Figure 8 which shows that hyperconjugative electron donation from an appropriate $\pi(CH_3)$ orbital into the formally vacant 2p orbital, $2p(C^+)$, at C^+ is very similar for the eclipsed and perpendicular conformations.

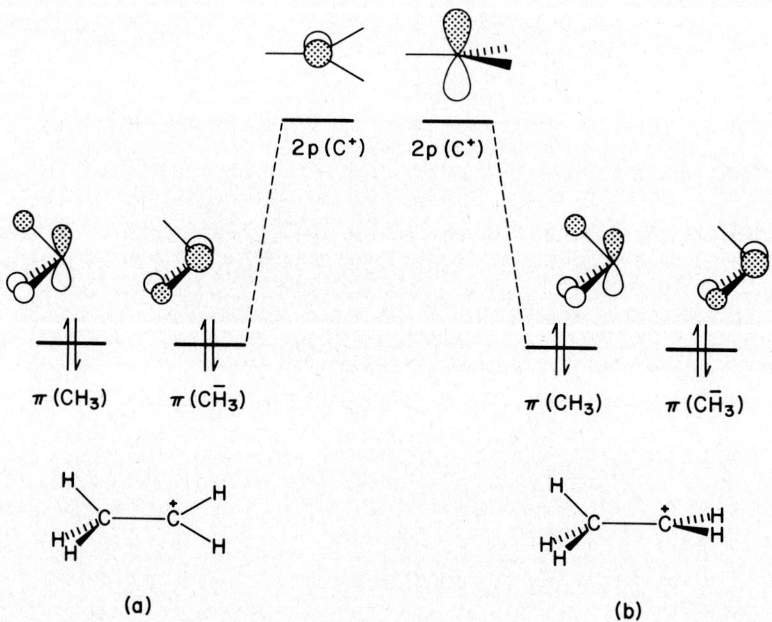

Fig. 8. Comparison of hyperconjugative interactions in (a) eclipsed and (b) perpendicular conformations of the ethyl cation, CH_3-CH_2^+.

The presence of a β-substituent in XCH_2-CH_2^+ breaks the degeneracy of the $\pi(CH_3)$ orbitals leading to unequal interactions in the eclipsed and perpendicular conformations. The situation for an electropositive substituent X is shown in Figure 9. As noted in Section 2.2.4, the $\pi(CH_2\bar{X})$ orbital is relatively insensitive to the nature of X so that the hyperconjugative interaction in the eclipsed conformation of XCH_2-CH_2^+ is very similar to that in the unsubstituted ethyl cation.

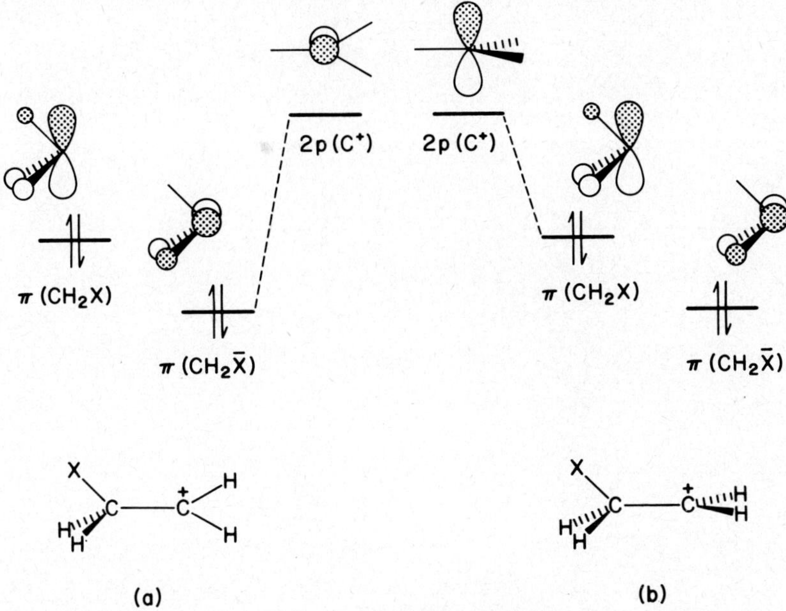

Fig. 9. Comparison of hyperconjugative interactions in (a) eclipsed and (b) perpendicular conformations of substituted ethyl cations, XCH_2-CH_2^+ (X electropositive).

On the other hand, the $\pi(CH_2X)$ orbitital, which can interact with $2p(C^+)$ in the perpendicular conformation, is strongly influenced by X. In particular, an electropositive substituent *raises* the energy of $\pi(CH_2X)$ and makes this orbital more localized on C. The first of these effects decreases the energy gap between the interacting orbitals while the second leads to improved overlap. Both effects therefore result in increased stabilization and improved hyperconjugative interaction in the perpendicular conformation of XCH_2-CH_2^+. The perpendicular conformation is thus expected to be favored over the eclipsed form.

For an electronegative substituent, the reverse predictions apply. Although hyperconjugative interaction in the eclipsed conformation is again relatively unaffected by the substituent, in the perpendicular conformation it is actually *made worse* through an increased energy gap and decreased overlap between the interacting orbitals. The eclipsed conformation is thus expected to be preferred.

In summary then, consideration of hyperconjugative interactions in β-substituted ethyl cations leads to a prediction of a preferred perpendicular conformation for electropositive substituents and a preferred eclipsed conformation for electronegative substituents. These predictions are borne out by *ab initio* calculations [25,26], the results of which are summarized in Table I. There is a strong correlation between the energy difference between eclipsed and perpendicular conformations and the electronegativity of the substituent.

TABLE I

Comparison of Conformational Preferences [E(eclipsed) - E(perpendicular), 4-31G//std, kJ mol^{-1}] in β-Substituted Ethyl Cations, Radicals and Anions

Substituent	Electronegativity	$XCH_2\text{-}CH_2^{+}$ [a]	$XCH_2\text{-}CH_2^{\bullet}$ [b]	$XCH_2\text{-}CH_2^{-}$ [c]
H	2.1	0	0	0.1
Li	1.0	+208.2	+30.2	+162.5
BeH	1.5	+95.4	+13.3	+39.4
BH_2 (plan.)	2.0	+41.4	+0.7	-18.4
BH_2 (perp.)	2.0	+34.7	+9.2	+64.8
CH_3	2.5	+10.3	+0.1	+8.8
NH_2 (plan.)	3.0	-32.3	+1.1	+40.2
NH_2 (perp.)	3.0	-1.6	-3.0	+6.6
OH	3.5	-47.9	-2.9	+60.9
F	4.0	-58.7	-2.7	+47.1

[a]From ref. 25d.
[b]From ref. 31.
[c]From ref. 27.

3.3 Substituted Ethyl Anions $XCH_2\text{-}CH_2^-$

Substituted ethyl anions, $XCH_2\text{-}CH_2^-$, may also be used to probe hyperconjugative interactions [27]. If we consider model systems in which the CH_2^- group is constrained to be planar trigonal, eclipsed (12) and perpendicular (13) conformations may again be distinguished.

The conformational preference predicted on the basis of hyperconjugative interactions may be derived from the orbital interaction diagram of Figure 10 which illustrates the situation for an electronegative substituent X. The key interaction involves electron donation from the doubly occupied $2p(C^-)$ orbital at C^- into a π* orbital of the CH_2X group. As noted in Section 2.2.4, for electronegative X, the π^* (CH_2X) orbital is firstly, lowered in energy and secondly, is concentrated on C leading to a decreased energy gap and increased overlap. Both of these effects lead to improved hyperconjugative interaction in the perpendicular conformation compared to the eclipsed conformation in which the interaction (with π^* $(CH_2\bar{X})$) is relatively unaffected by the substituent. For an electropositive substituent X, the reverse considerations apply. Interaction in the perpendicular conformation is made worse through an increased energy gap and poorer overlap leading to the expectation of a preference for the eclipsed conformation.

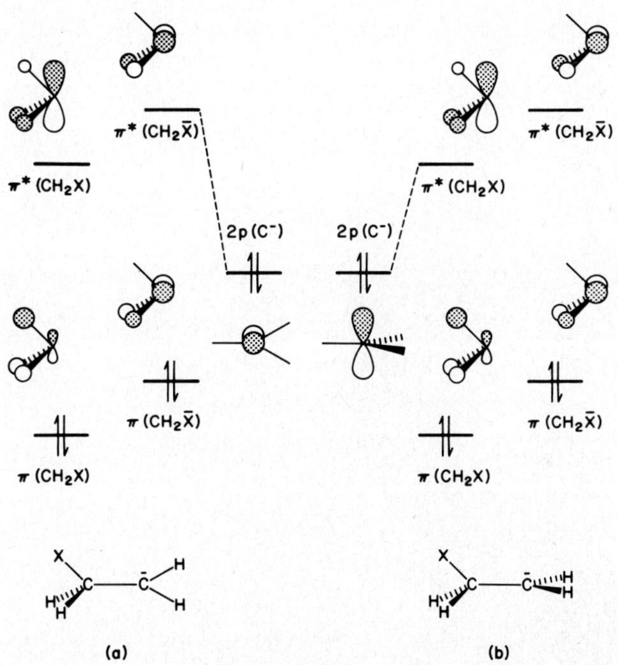

Fig. 10. Comparison of hyperconjugative interactions in (a) eclipsed and (b) perpendicular conformations of substituted ethyl anions, XCH_2-CH_2^- (X electronegative).

Examination of the results of *ab initio* calculations [27] on XCH_2-CH_2^-, included in Table I, show that the predictions for electronegative substituents (NH_2, OH and F) are confirmed with the perpendicular conformation being favored. On the other hand, with the exception of BH_2 (plan.), the electropositive substituents also lead to a preferred perpendicular conformation. This result has been rationalized [27] in terms of direct 1,3-overlap between the $2p(C^-)$ orbital and a vacant orbital, 2p(X), on X for X = Li, BeH and BH_2 (perp.)

(see Figure 11).

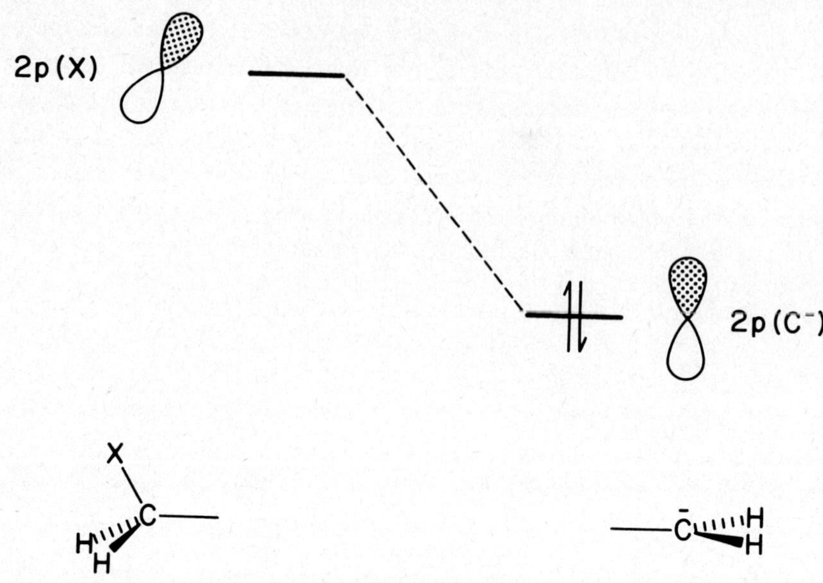

Fig. 11. Direct 1,3-interaction between a formally vacant orbital, 2p(X), on X and the 2p(C^-) orbital in the perpendicular conformation (13) of $XCH_2CH_2^-$.

Indeed, when the BH_2 group is rotated to BH_2 (plan.), this direct 1,3-overlap is no longer possible and the conformational preference reverts back to the hyperconjugatively favored eclipsed conformation (*cf.* 14).

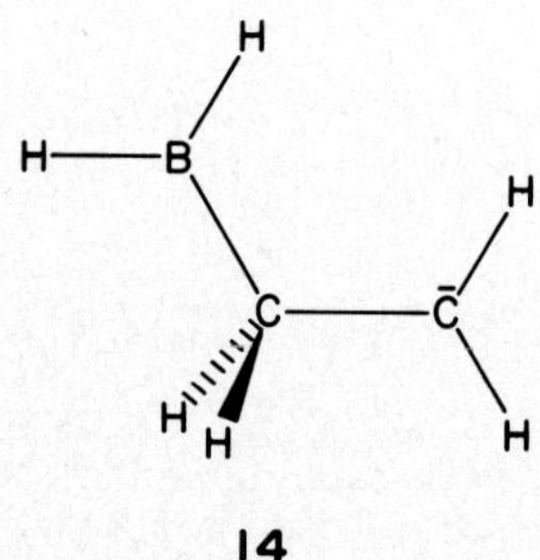

14

Thus hyperconjugation *does* contribute to a preference for a perpendicular conformation for electronegative substituents and an eclipsed conformation for electropositive substituents in β-substituted ethyl anions. In the case of electronegative substituents, this is the dominant contribution. On the other hand, for electropositive substituents with an appropriately oriented vacant orbital, 1,3-direct overlap appears to be dominant in determining the conformational preference.

It is of interest to note that Apeloig and Rappoport [28] have concluded, on the basis of calculations on XCH_2-CH_2^-, that hyperconjugation is the major factor which determines the stereochemistry of nucleophilic vinyl substitution.

3.4 Substituted Ethyl Radicals, XCH_2-$CH_2^{\bullet}$

Although it is well established that β-substituted ethyl cations and anions exhibit strong conformational preferences, both experimental results [29] and theoretical calculations [30,31] for a number of β-substituted ethyl radicals have shown relatively easy rotation about the C_α-C_β bond. Significant conformational preferences in radicals of this type have only been observed experimentally [32] for substituents containing second- or higher-row atoms [e.g. $Si(Et)_3$, $Ge(Et)_3$ etc.] and theoretically [31] for the electropositive substituents Li, BeH and BH_2 (perp.). The calculated conformational energy differences for radicals, included in Table I, pose the question: why is there little conformational preference for X = BH_2 (plan.), CH_3, NH_2, OH and F and yet pronounced conformational preferences for X = Li, BeH and BH_2 (perp.)?

The relevant orbital interaction diagram for eclipsed and perpendicular conformations of XCH_2-$CH_2^{\bullet}$ radicals is shown in Figure 12.

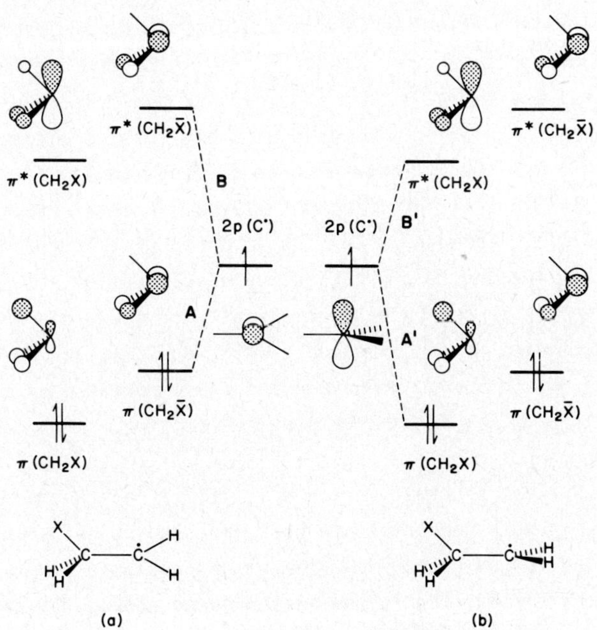

Fig. 12. Comparison of hyperconjugative interactions in (a) eclipsed and (b) perpendicular conformations of substituted ethyl radicals, XCH_2-$CH_2^{\bullet}$ (X electronegative).

In the eclipsed conformation, the $2p(C^{\cdot})$ orbital may interact with the occupied $\pi(CH_2\bar{X})$ orbital (interaction **A**) in a three-electron stabilizing interaction, or with the vacant $\pi^*(CH_2\bar{X})$ orbital (interaction **B**) in a one-electron stabilizing interaction. Interaction **A** involves the hyperconjugative *donation* of charge into the $2p(C^{\cdot})$ orbital (analogous to that in substituted ethyl cations) while interaction **B** involves hyperconjugative *acceptance* from the $2p(C^{\cdot})$ orbital (analogous to that in substituted ethyl anions). Since the $\pi(CH_2\bar{X})$ and $\pi^*(CH_2\bar{X})$ orbitals are relatively insensitive to substitution, the interactions in the eclipsed conformation are likewise quite insensitive to substitution. In the perpendicular conformation, on the other hand, interaction of $2p(C^{\cdot})$ takes place with the $\pi(CH_2X)$ and $\pi^*(CH_2X)$ orbitals whose energies are strongly dependent on the substituent X leading to modified interactions **A'** and **B'**. The changes in the two interactions appear, however, to approximately cancel. Thus, for example, whereas for an electronegative substituent (as shown in Figure 12) there is an improved interaction **B'**, there is also a poorer interaction **A'**. The consequence is that the improvement in one interaction is largely offset by a reduction in the other interaction leading to only a small dependence of the hyperconjugative interaction energy on substituent in the perpendicular conformation, as in the eclipsed conformation. The conformational preference is thus small. This contrasts with the situation for cations $XCH_2\text{-}CH_2^+$ and anions $XCH_2\text{-}CH_2^-$ where there is no compensating effect and large conformational preferences arise.

The arguments presented above suggest that the hyperconjugative contribution to a conformational preference should generally be small for $XCH_2\text{-}CH_2^{\cdot}$ radicals; consequently, the substantial conformational preferences indicated for X = Li, BeH and BH_2 (perp.) must be attributed to an alternative interaction. Evidence has been presented [31] that the dominant alternative interaction is direct 1,3-overlap (Figure 13) between $2p(C^{\cdot})$ and an appropriately oriented formally vacant orbital on X leading to a one-electron stabilization. The importance of such an interaction is demonstrated, for example, by the preference (by 7.0 kJ mol^{-1}) in the perpendicular conformation of $XCH_2\text{-}CH_2^{\cdot}$ for X = BH_2 (perp.) compared with X = BH_2 (plan.). We have noted the importance of such an interaction in anions $XCH_2\text{-}CH_2^-$ in the previous section.

There is strong experimental support [32] for the suggestion that pronounced preferences in $XCH_2\text{-}CH_2^{\cdot}$ generally arise only if 1,3-interactions are important. In all cases where such preferences are observed, the substituent is a group containing a second-row or higher-row atom which possesses low-lying vacant orbitals capable of interacting in the manner noted above. In all cases also, the perpendicular conformation is preferred. Examples are [32] X = $Si(Et)_3$, $Ge(Et)_3$, $Sn(n\text{-}Bu)_3$, PR_2, PR_3^+, AsR_2, and AsR_3^+.

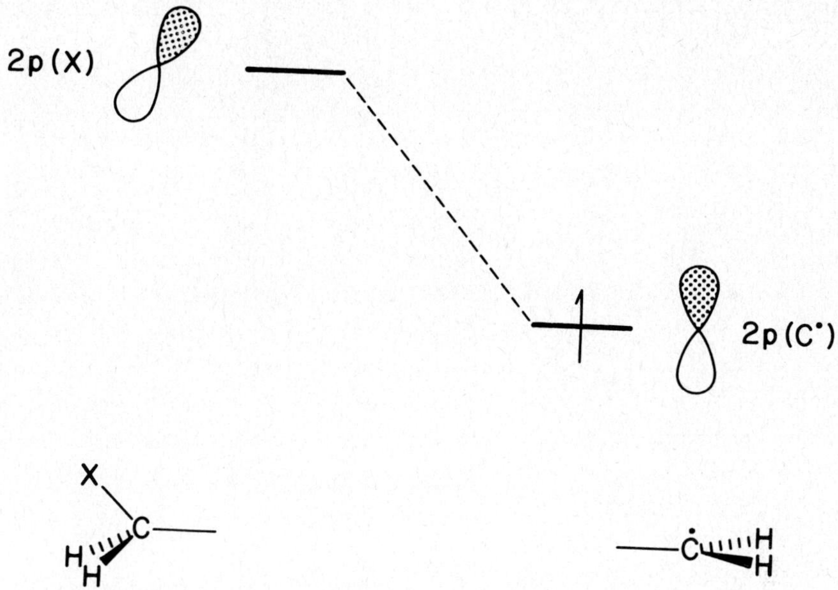

Fig. 13. Direct 1,3-interaction between a formally vacant orbital, 2p(X), on X and the 2p(C•) orbital in the perpendicular conformation of XCH_2-CH_2•.

3.5 Substituted Methanols, XCH_2-OH

Hyperconjugative interaction in methanols XCH_2-OH may take place through electron donation from the p-type lone pair on oxygen, 2p(O), or from the sp^2-type lone pair, into one of the π* orbitals of the XCH_2 group. We have noted in Section 2.2.5 that donation from the p-type lone pair occurs more readily than from the sp^2-type lone pair and it is this factor which leads to strong conformational preferences in some substituted methanols.

In methanol itself, hyperconjugative interaction of 2p(O) with the $\pi^*(CH_3)$ orbitals is conformationally independent because of the degeneracy of the $\pi^*(CH_3)$ pair: interaction with one of the orbitals is increased and with the other decreased during the rotation process. In substituted methanols, on the other hand, the degeneracy is broken and the $\pi^*(CH_2X)$ and $\pi^*(CH_2\bar{X})$ orbitals have different energies. For an electronegative X, this leads to a reduced energy separation and an enhanced interaction in the *perpendicular* conformation where interaction with $\pi^*(CH_2X)$ is possible, as shown in Figure 14. For electro-positive X, interaction of 2p(O) with $\pi^*(CH_2X)$ is made worse because of increased separation between the orbitals. The hyperconjugative donation is then improved through rotation to a periplanar conformation (either *cis* or *trans*) where interaction with the $\pi^*(CH_2\bar{X})$ orbital, which is relatively unaffected by substitution, takes place.

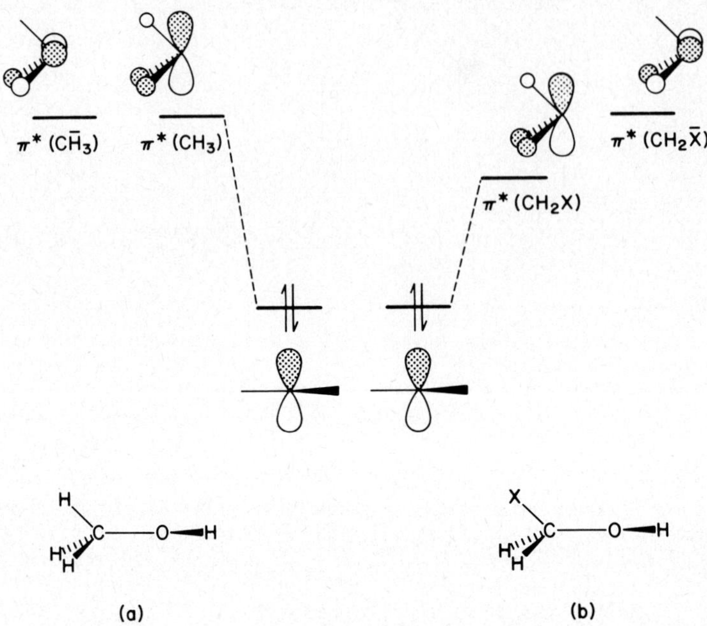

Fig. 14. Hyperconjugative interaction in perpendicular conformations of (a) CH_3-OH and (b) XCH_2-OH (X electronegative).

Hyperconjugative interaction is thus predicted to favor perpendicular conformations for substituted methanols XCH_2-OH with X electronegative, and periplanar conformations when X is electropositive. These predictions bear some similarity to those for substituted ethyl anions (*cf.* Section 3.3).

Since there are factors other than hyperconjugation which contribute to the conformational preferences in substituted methanols, it is not possible to assess the importance of hyperconjugation simply by inspection of the total rotational potential function. Rather, it is necessary to break down the total function into simpler components. Fourier component analysis, described in Section 2.3, provides a convenient method of carrying out the decomposition.

Calculated potential functions $V(\phi)$, and their Fourier components, $V_n(\phi)$, for methanol, ethanol, 2-fluoromethanol and 2-lithiomethanol are displayed in Figures 15-18 respectively [21,33]. Corresponding values of the rotational constants V_n [*cf.* equation (2)] are shown in Table II. Conformationally dependent hyperconjugative interactions of the type described above contribute to the overall potential functions through the twofold or $V_2(\phi)$ term.

TABLE II
Rotational potential constants (V_n, 4-31G//std, kJ mol^{-1}) describing internal rotation in substituted methanols, XCH_2-OH

Molecule	V_1	V_2	V_3
CH_3-OH[a]	0	0	-4.7
CH_3CH_2-OH[a]	-3.9	-0.2	-4.8
FCH_2-OH[a]	+22.0	-9.2	-4.0
$LiCH_2$-OH[b]	-21.3	+8.6	-5.3

[a]From ref. 21.
[b]From ref. 33.

For methanol and ethanol the V_2 term is either zero (methanol) or near-zero (ethanol) and the potential function is dominated by a negative V_3 term leading to minima at *gauche* and *trans* conformations and reflecting a tendency for the staggering of bonds.

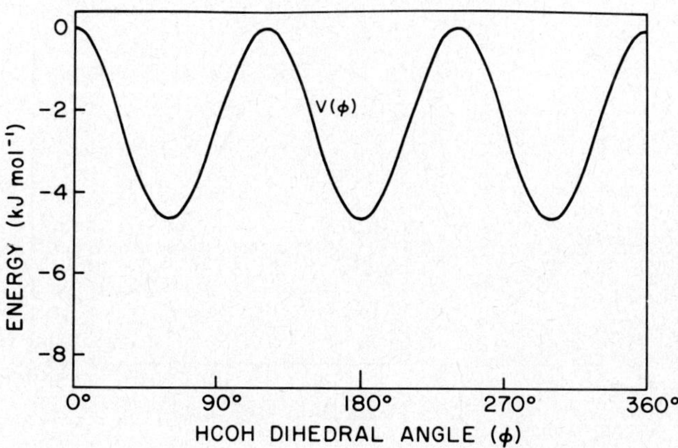

Fig. 15. Potential energy function, V(ϕ), describing internal rotation in methanol, CH_3-OH (4-31G//std).

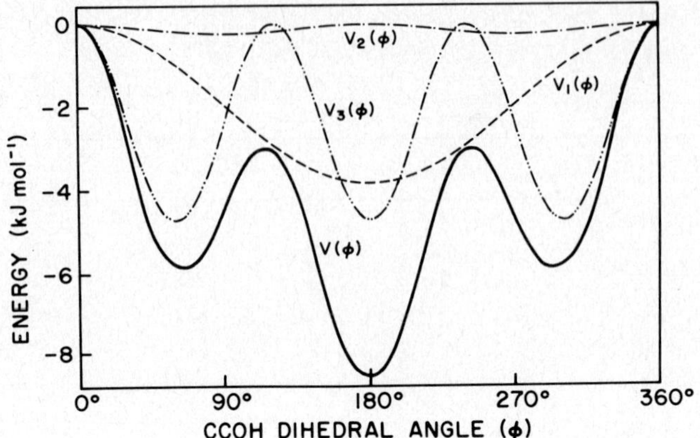

Fig. 16. Potential energy function, V(ϕ), and Fourier components, $V_n(\phi)$, describing internal rotation in ethanol, CH_3CH_2-OH (4-31G//std).

For FCH_2-OH, V_3 is again negative reflecting a preference for staggered conformations but the potential function is now dominated by large V_1 and V_2 terms.

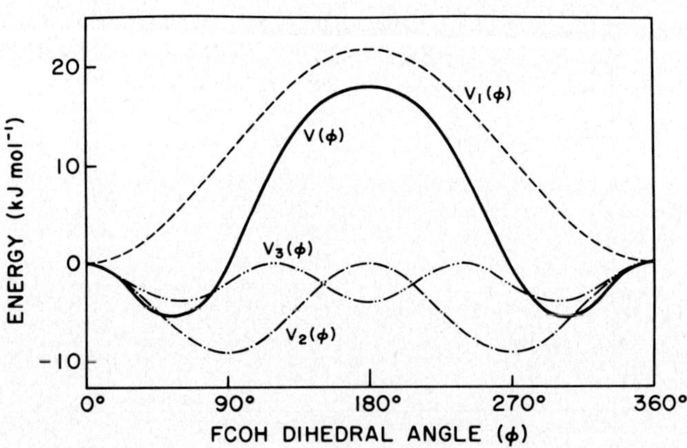

Fig. 17. Potential energy function, V(ϕ), and Fourier components, $V_n(\phi)$, describing internal rotation in fluoromethanol, FCH_2-OH (4-31G//std)

V_2 is negative indicating a preference for perpendicular as opposed to *cis* or *trans* structures, consistent with the orbital interaction diagram of Figure 14 and expressed succinctly as 15:

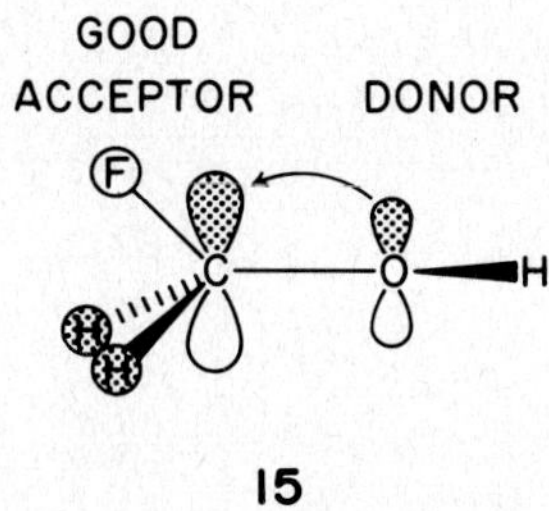

V_1 is positive for FCH_2-OH indicating a preference for *cis* over *trans* structures, a result which is consistent with the dipolar interactions 16 and 17.

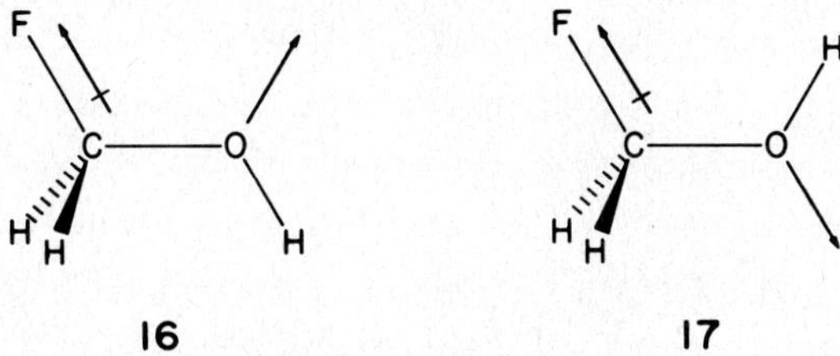

The overall result for fluoromethanol is a potential curve with equivalent *gauche* minima. The staggered *trans* structure is a potential maximum because of the unfavorable V_1 and V_2 terms. The strong preference for a *gauche* rather than a *trans* conformation for fluoromethanol has important implications regarding the experimental observation of the anomeric effect in carbohydrate chemistry (*cf.* Section 3.6) [21,34].

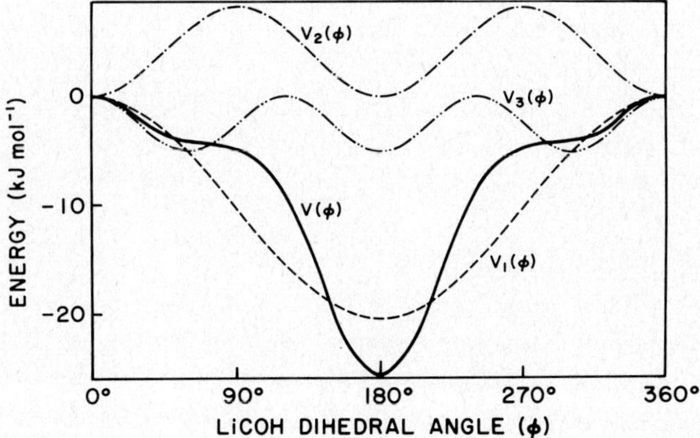

Fig. 18. Potential energy function, $V(\phi)$, and Fourier components, $V_n(\phi)$, describing internal rotation in lithiomethanol, $LiCH_2$-OH (4-31G//std).

For $LiCH_2$-OH, V_3 is again negative but is again dominated by much larger V_1 and V_2 terms. V_2 is positive reflecting relatively unfavorable hyperconjugative interaction in the perpendicular structure, consistent with the qualitative argument above and expressed as 18:

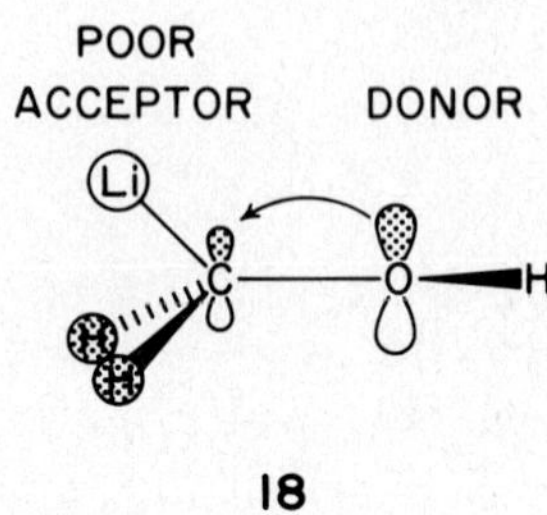

18

V_1 is negative indicating a preference for *trans* over *cis* structures, a result which is expected on the basis of the dipolar interactions 19 and 20.

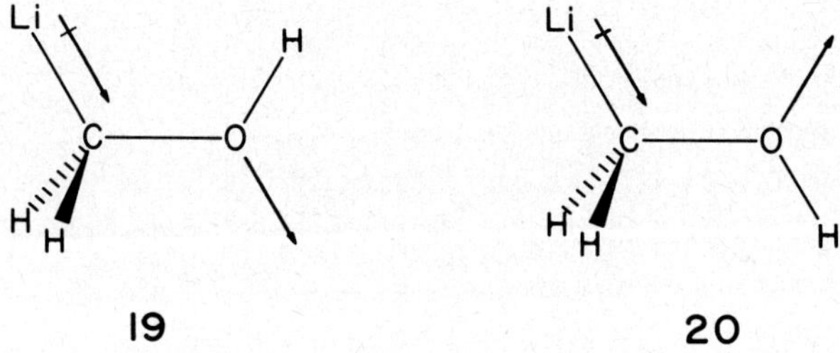

19 20

The total potential function in this case has a single deep minimum corresponding to the *trans* conformation.

We conclude that for substituted methanols, XCH_2-OH, a *gauche* conformation might be generally expected when X is electronegative and a *trans* conformation preferred when X is electropositive.

3.6 The Anomeric Effect

A closely related conformational problem that has received detailed experimental and theoretical attention and which is influenced by hyperconjugative interactions is the *anomeric effect* [35]. This is the name given to the experimentally observed preference for an axial disposition of an electronegative substituent at C_1 in pyranose rings (**21**):

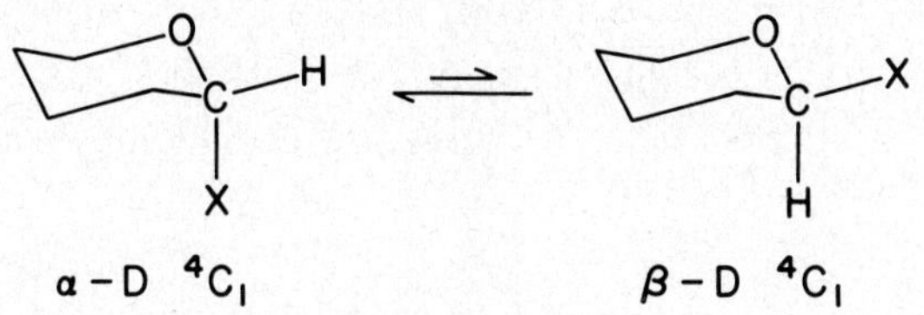

α-D $^4C_1$ β-D $^4C_1$

21

Fluoromethanol (FCH_2OH) is the simplest model for such systems and the predicted *gauche* conformation for this molecule (*cf.* Section 3.5) is consistent with the experimental observation of the anomeric effect [21,34]. Direct STO-3G calculations on 1-chlorotetrahydropyran also confirm the preferred axial orientation of the C-X bond [36].

Methanediol, methoxymethanol and dimethoxymethane have also been used as models to study theoretically the anomeric effect and, in addition, the *exo-anomeric effect* in relation to the observation of these effects in carbohydrate chemistry [37,38]. The exo-anomeric effect refers to the preferred *gauche* conformations about glycosidic bonds in pyranosides (22,23).

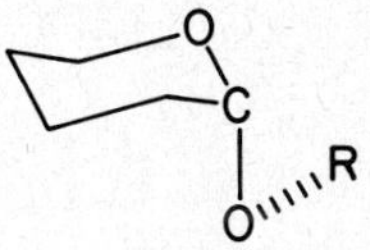

22 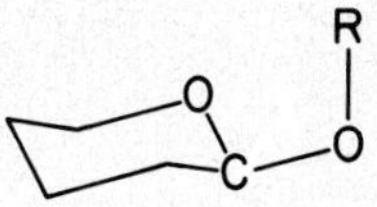 23

When calculations for dimethoxymethane are performed as a function of angles of rotation (θ,ϕ) about the two C-O bonds (24):

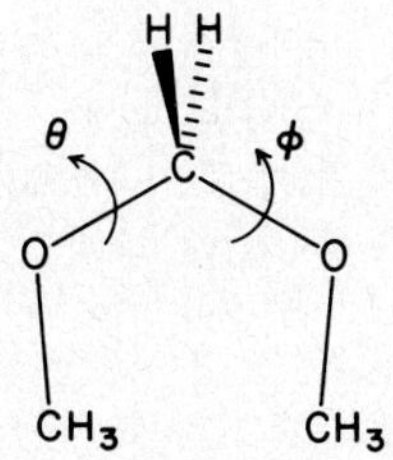

24

The most stable conformation corresponds to $\theta \sim 60°$, $\phi \sim 60°$ (25) [37c].

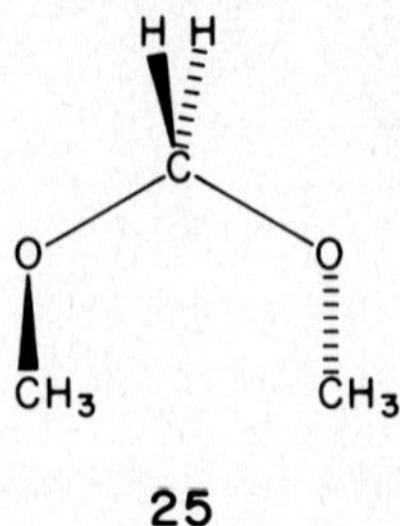

25

This is consistent both with the results of an electron diffraction study of dimethoxymethane [39] and with the anomeric and exo-anomeric effects which, in turn, are supported by a large body of crystal structural data for α-D-pyranosides [37d]. If one of the rotational angles, say θ, is held at 180° in order to model the β-configuration, the potential function in φ shows equivalent minima at $\pm \sim 60°$. This is consistent with the exo-anomeric effect in β-D-pyranosides and the supporting structural data [37d].

A rationalization of the conformational preferences implicit in the anomeric and exo-anomeric effects can be provided along the same lines as for fluoromethanol with contributions from dipolar, hyperconjugative and bond-repulsion terms [21,35-37,40].

3.7 Substituted Acetaldehydes, XCH_2-CHO

Because the carbonyl group represents an electron-deficient center, the hyperconjugative interactions in substituted acetaldehydes, XCH_2-CHO, resemble those of substituted ethyl cations XCH_2-CH_2^+. The key interaction is electron donation from the $\pi(CH_2X)$ orbitals into the $\pi^*(C{=}O)$ orbital. Interaction between $\pi(CH_2X)$ and $\pi^*(C{=}O)$ is favored in the perpendicular conformation, shown in Figure 19, by an electropositive substituent X. When X is electronegative, this interaction is made worse and interaction with $\pi(CH_2\bar{X})$, which takes place in the periplanar (either *cis* or *trans*) conformations, is favored.

In order to test these predictions, we examine the calculated rotational potential functions for XCH_2-CHO with X = H, CH_3, F and Li as displayed in Figures 20-23. Corresponding rotational potential constants are shown in Table III [41].

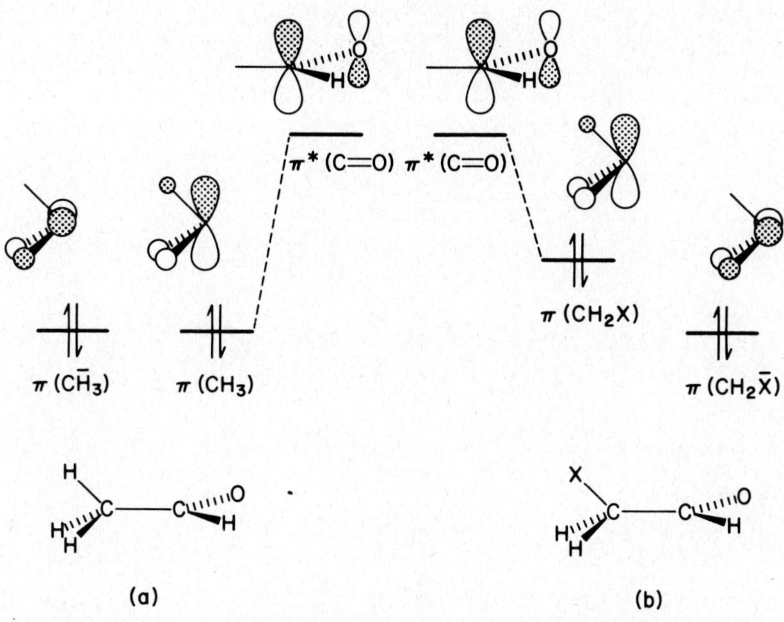

Fig. 19. Hyperconjugative interaction in perpendicular conformations of (a) CH_3-CHO and (b) XCH_2-CHO (X electropositive).

TABLE III

Rotational potential constants (V_n, 4-31G//std, kJ mol^{-1}) describing internal rotation in substituted acetaldehydes, XCH_2-CHO[a]

Molecule	V_1	V_2	V_3
CH_3-CHO	0	0	+3.1
CH_3CH_2-CHO	+2.9	+1.0	+2.5
FCH_2-CHO	-19.7	+15.2	+2.7
$LiCH_2$-CHO	+37.3	-42.3	+5.4

[a]From ref. 41.

For acetaldehyde and propionaldehyde the potential function is dominated by a negative V_3 term, reflecting the tendency for eclipsing of the C=O double bond and leading to minima in the total function for such conformations. The V_2 term, which represents the hyperconjugative interaction described above, is either zero (acetaldehyde) or very small (propionaldehyde).

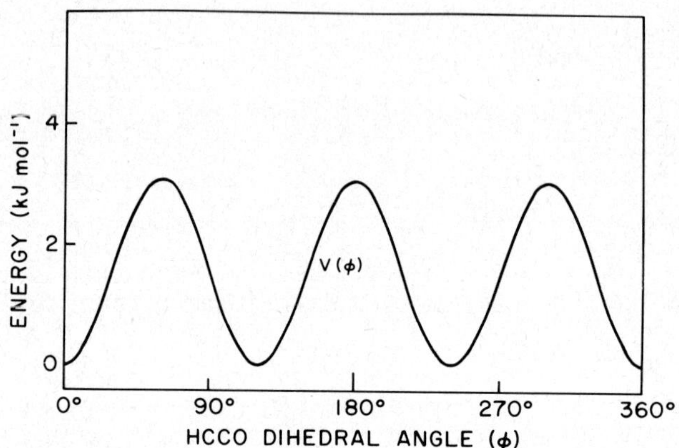

Fig. 20. Potential energy function, V(ϕ), describing internal rotation in acetaldehyde, CH_3-CHO (4-31G//std).

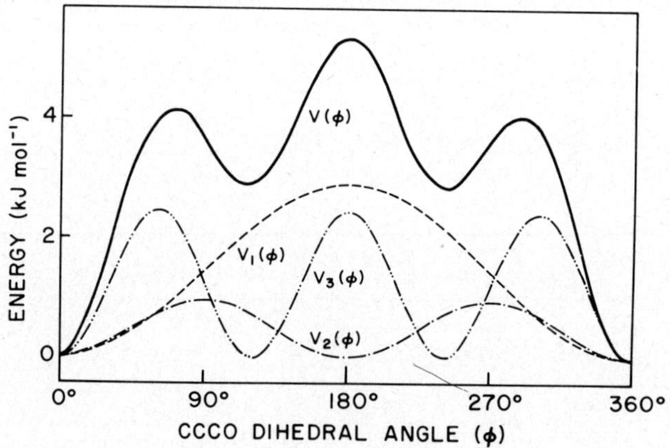

Fig. 21. Potential energy function, V(ϕ), and Fourier components, $V_n(\phi)$, describing internal rotation in propionaldehyde, CH_3CH_2-CHO (4-31G//std)

For FCH_2-CHO, while V_3 remains the same sign and of similar magnitude to its value in acetaldehyde, V_1 and V_2 are much larger. V_1 is large and negative

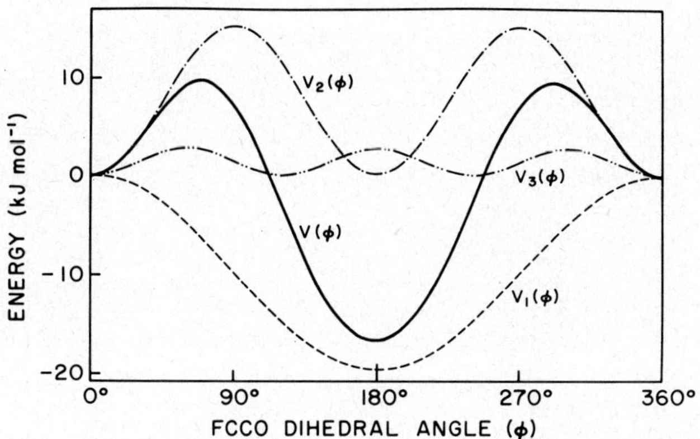

Fig. 22. Potential energy function, V(ϕ), and Fourier components, $V_n(\phi)$, describing internal rotation in fluoroacetaldehyde FCH_2-CHO (4-31G//std).

indicating a preference for a *trans* (ϕ = 180°, **26**) compared with *cis* (ϕ = 0°, **27**) conformation, consistent with the dipolar interactions:

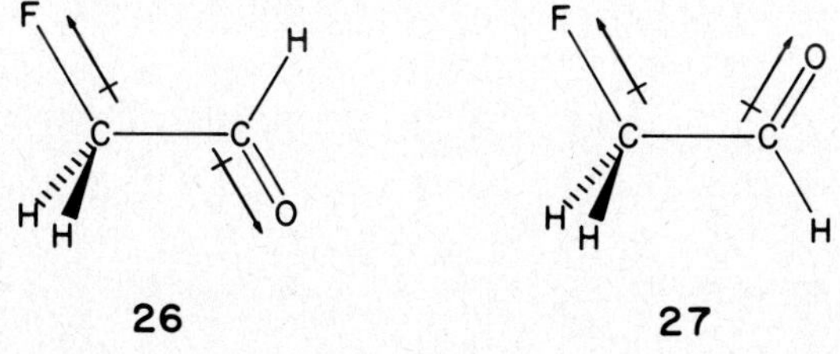

V_2 is positive and reflects relatively unfavorable hyperconjugative interaction (**28**) in the perpendicular structure.

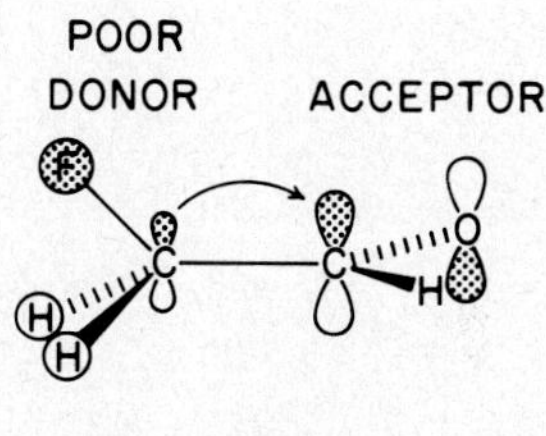

The overall result is a potential function with minima at *cis* and *trans* positions, the latter being favored. These results are consistent with microwave structural studies of the related molecules fluoroacetone and chloroacetaldehyde [42].

For $LiCH_2$-CHO, the V_1 term is positive (*cf*. dipolar interactions 29 and 30) and the V_2 term is negative favoring conformations with $\phi \sim 90°$ (*cf*. hyperconjugative interaction 31).

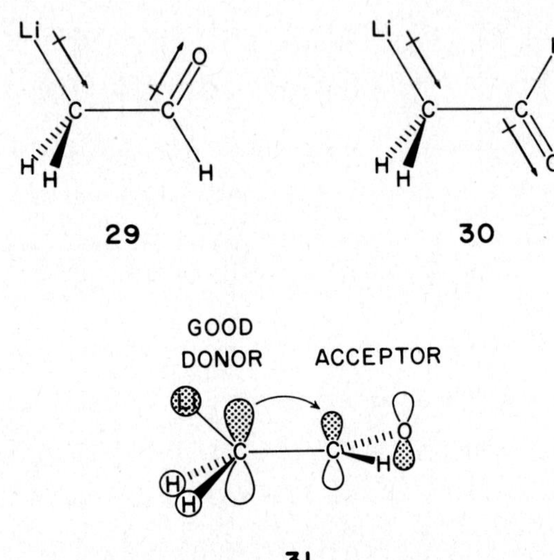

The net result is a potential function with a single minimum with ϕ somewhat less than 90°.

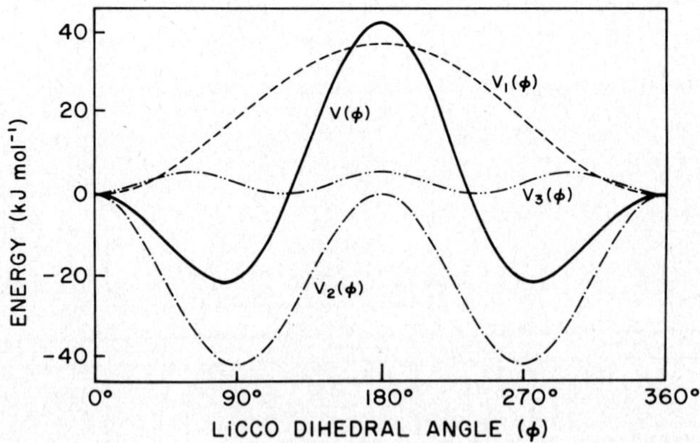

Fig. 23. Potential energy function, $V(\phi)$, and Fourier components, $V_n(\phi)$, describing internal rotation in lithioacetaldehyde $LiCH_2$-CHO (4-31G//std).

3.8 1,2-Disubstituted Ethanes, XCH_2-CH_2Y

When either X or Y in 1,2-disubstituted ethanes XCH_2-CH_2Y is H or CH_3, the rotational potential function is dominated by the V_3 term, reflecting a tendency for staggering of bonds and producing a potential function like that for ethane itself. An example is provided by the potential function for CH_3CH_2-CH_2Li shown in Figure 24 [43].

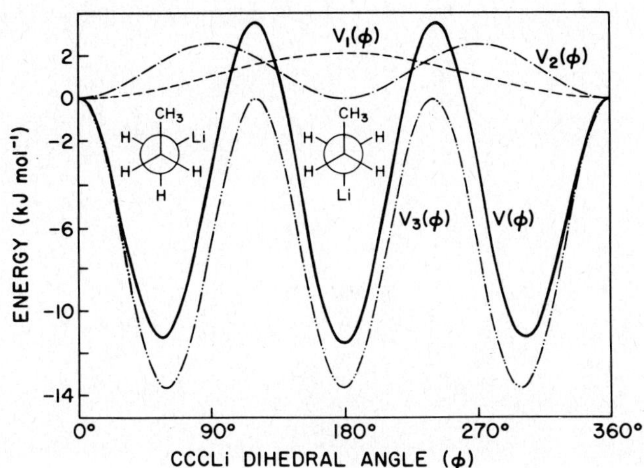

Fig. 24. Potential energy function, $V(\phi)$, and Fourier components, $V_n(\phi)$, describing internal rotation in CH_3CH_2-$CHLi$ (4-31G//std).

When both X and Y are electronegative or electropositive, two additional classes of potential functions arise, depending on the nature of X and Y [43]. In both classes, conformationally dependent hyperconjugative interactions can take place which contribute to the potential function through the V_2 term. These involve interaction between the π and π^* orbitals of the CH_2X and CH_2Y groups.

As an example of one of these classes, we take $LiCH_2$-CH_2F for which the orbital interaction diagram for conformations with ϕ(LiCCF) = 0° and 90° is shown in Figure 25. As detailed in Section 2.2.3, π-type orbitals containing a carbon-centered p-orbital which is symmetric with respect to the plane of the paper are given the subscript *a*. Group orbitals labelled *b* contain a carbon-centered p-orbital that is antisymmetric with respect to reflection in the plane of the paper. By symmetry, orbitals labelled *a* can only interact with other *a* orbitals, and similarly for orbitals labelled *b*. Rotation of the CH_2F group in $LiCH_2$-CH_2F in going from ϕ = 0° to ϕ = 90° interchanges the *a* and *b* labels among the $\pi^*(CH_2F)$ and $\pi^*(CH_2\bar{F})$ orbitals.

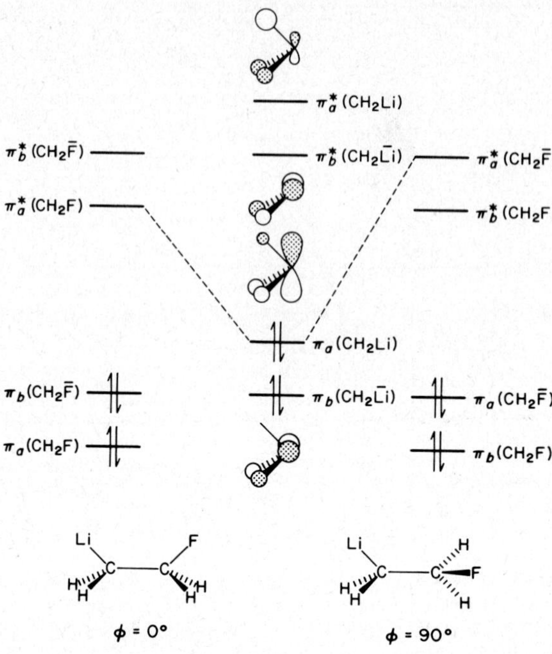

Fig. 25. Orbital interaction scheme showing hyperconjugative tendency of $LiCH_2$-CH_2F to adopt a *cis* conformation, resulting from reduced energy separations between the most strongly interacting orbitals.

This leads to an increased energy gap between the appropriate orbitals when $\phi = 90°$ (compared with $\phi = 0°$) and hence a decrease in the primary stabilizing interaction $\pi_a(CH_2Li)\cdots\pi_a^*(CH_2F)$ for $\phi = 90°$ compared with $\phi = 0°$. Overlap considerations reinforce this effect. Conformations with $\phi = 180°$ are also favored by such hyperconjugative interaction. The conclusion is that hyperconjugation leads to preferences for LiCCF periplanar conformations and a positive V_2 term in the potential function.

The V_1 term in the potential function corresponds to a preference for $\phi = 0°$ rather than 180°. This is consistent with the more energetically favorable alignment of group dipoles in the *cis* conformation (*cf.* **32**) and also with local donor-acceptor interactions, as shown in **33**, which rely on the π-donating ability of F and the π-accepting ability of Li.

π-ACCEPTOR π-DONOR

32 33

The calculated potential function for $LiCH_2CH_2F$ (Figure 26)

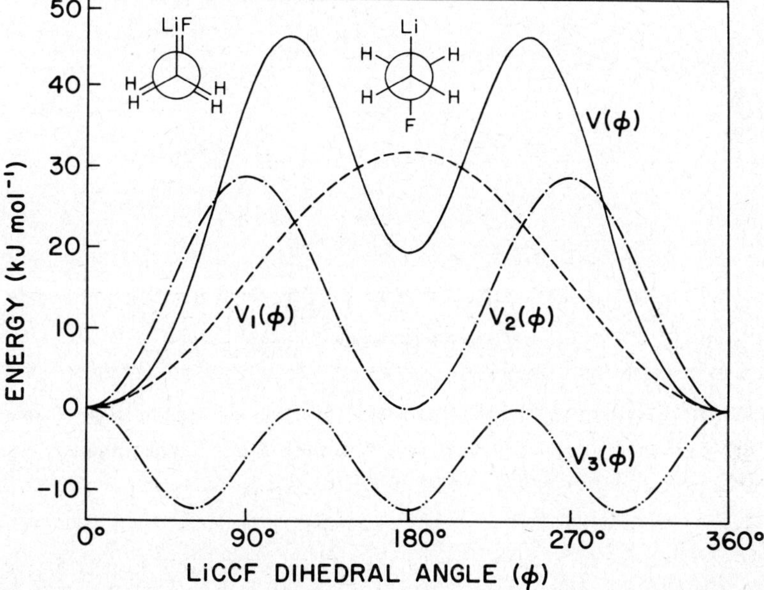

Fig. 26. Potential energy function, $V(\phi)$, and Fourier components, $V_n(\phi)$, describing internal rotation in $LiCH_2$-CH_2F.

is seen to be dominated by the $V_1(\phi)$ and $V_2(\phi)$ terms. The latter favors *cis* and *trans* conformations equally at the expense of orthogonal structures. The former discriminates between the *cis* and *trans* conformations in favor of the *cis* isomer. Although the negative V_3 term reflects a stabilization of the *gauche* conformations with respect to the *cis* conformation, it is too small compared to the V_1 and V_2 terms to be effective here. The net result is the remarkable observation of a preferred eclipsed *cis* conformation for $LiCH_2$-CH_2F, i.e. the stable conformations of this 1,2-disubstituted ethane are *cis* and *trans* rather than *gauche* and *trans* [44].

A third class of potential function is typified by V(ϕ) for $LiCH_2$-CH_2Li (Figure 27).

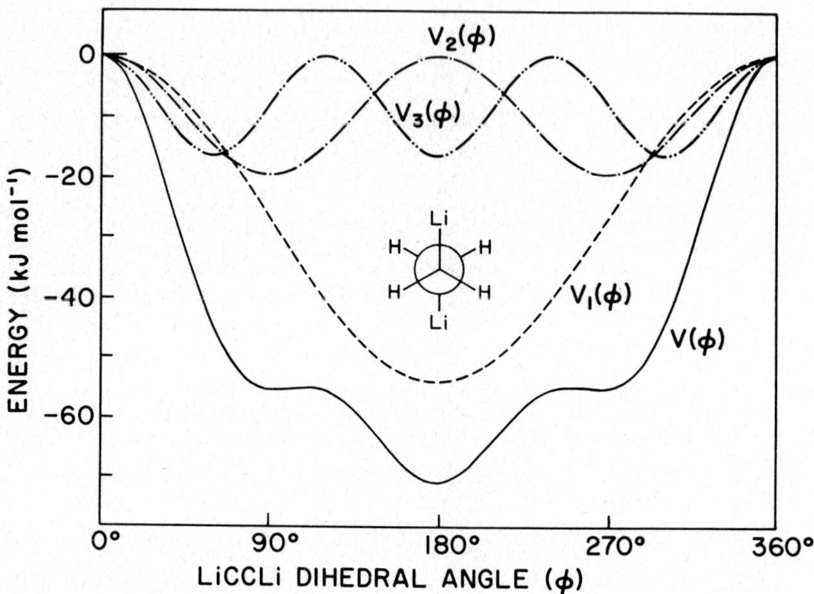

Fig. 27. Potential energy function, V(ϕ), and Fourier components, $V_n(\phi)$, describing internal rotation in $LiCH_2$-CH_2Li.

In contrast to the situation for $LiCH_2$-CH_2F, the V_3 component here is dominated by large *negative* V_1 and V_2 components. The principal term $V_1(\phi)$ produces a strong preference for a *trans* isomer. The possible existence of a *gauche* isomer is only hinted at by a flattening in the potential function in the region ϕ = 95-110°.

The negative V_2 term arises from hyperconjugative interaction which may be described by the orbital interaction diagram of Figure 28. The primary stabilizing interaction, $\pi_a(CH_2Li) \cdots \pi_a^*(CH_2Li)$, is clearly more favorable in the ϕ = 90° than the ϕ = 0° conformation. In addition, the destabilizing interaction $\pi_a(CH_2Li) \cdots \pi_a(CH_2Li)$ is expected to be greater for ϕ = 0° than for ϕ = 90° because of greater overlap. The net effect is a preference for orthogonal rather than coplanar conformations and hence a negative V_2 constant.

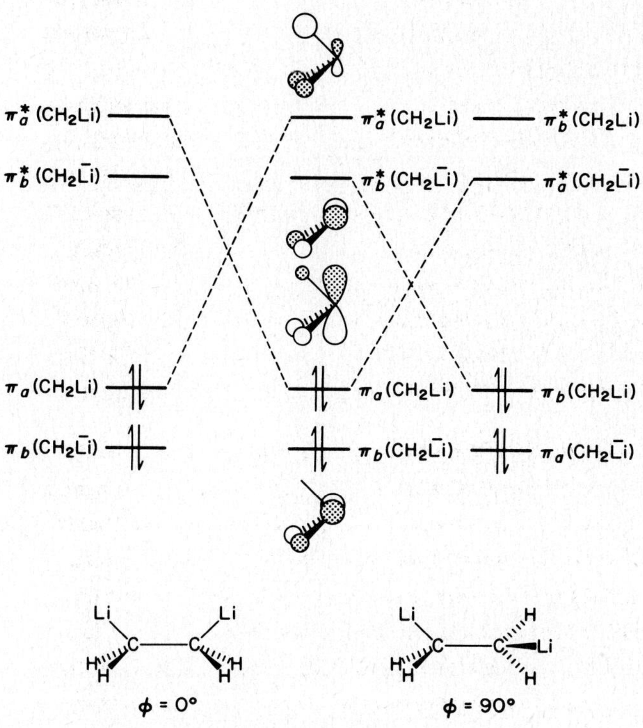

Fig. 28. Orbital interaction scheme showing hyperconjugative tendency of $LiCH_2$-CH_2Li to adopt a conformation with ϕ = 90°, resulting from reduced energy separations between the most strongly interacting orbitals.

The onefold component of $V(\phi)$ for $LiCH_2$-CH_2Li is consistent with the interaction associated with the $LiCH_2$ group dipole moments. This dipolar interaction clearly favors the *trans* conformation (**34**).

Li H H C C H H Li

34

Similar considerations apply to the 1,2-difluoroethane molecule [45]. Hyperconjugative interactions favor the ϕ = 90° conformation (and hence a negative V_2) and dipolar forces again favor a *trans* structure (and hence a negative V_1) for FCH_2-CH_2F. The calculated potential functions are shown in Figure 29.

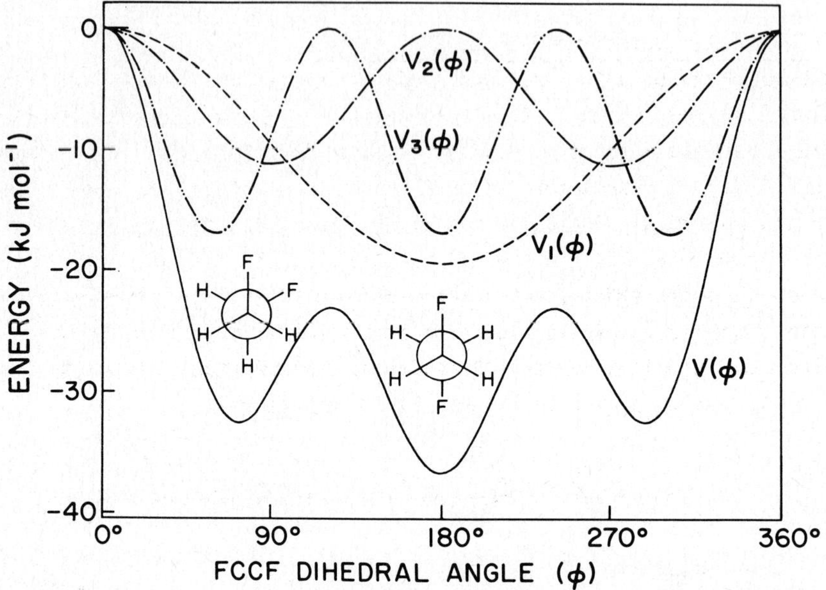

Fig. 29. Potential energy function, $V(\phi)$, and Fourier components, $V_n(\phi)$, describing internal rotation in FCH_2CH_2F.

The overall function predicts a *trans* structure whereas there is strong experimental evidence that the *gauche* form is more stable [46]. This deficiency in the theoretical prediction may be attributed to an exaggeration of the V_1 term by the 4-31G basis set which tends to overestimate dipolar interactions [14]. Our qualitative predictions concerning the V_1 and V_2 terms of the potential function are, however, confirmed.

The analysis presented in this section distinguishes three classes of conformational behavior in 1,2-disubstituted ethanes, XCH_2-CH_2Y [43].
(i) When at least one of X and Y is approximately electroneutral, the potential function is dominated by the threefold term and the stable rotational isomers correspond to staggered *gauche* ($\phi \sim 60°$) and *trans* (ϕ = 180°) structures.
(ii) If X and Y are both electropositive (e.g. $LiCH_2$-CH_2Li) or electronegative (e.g. FCH_2-CH_2F), the potential function is dominated by large *negative* V_1 and V_2 terms. The stable rotational isomers still correspond to *gauche* and *trans* structures, although in these cases the *gauche* isomer may be displaced

towards a substantially larger dihedral angle (i.e. $\phi > 60°$).
(iii) If X is electropositive and Y is electronegative (e.g. $LiCH_2$-CH_2F), the potential function is dominated by large *positive* V_1 and V_2 components. The *trans* structure remains a stable rotational isomer but the other stable isomer may be *cis* and eclipsed rather than *gauche* and staggered.

3.9 1,4-Disubstituted But-2-ynes, $XCH_2C{\equiv}CCH_2Y$

Internal rotation about the carbon-carbon triple bond in but-2-yne (dimethylacetylene) has been shown, both experimentally [47] and theoretically [48], to require a very small barrier, namely less than 0.05 kJ mol^{-1}. It is tempting to conclude from this result that internal rotation about the C≡C bond in substituted but-2-ynes might also occur relatively freely but this turns out not to be the case. For molecules $XCH_2C{\equiv}CCH_2Y$ in which both X and Y are either electropositive or electronegative there are marked conformational preferences resulting from strong hyperconjugative and dipolar interactions [49].

Calculated rotational potential functions for $FCH_2C{\equiv}CCH_2F$, $LiCH_2C{\equiv}CCH_2Li$ and $LiCH_2C{\equiv}CCH_2F$ are displayed in Figures 30-32 respectively.

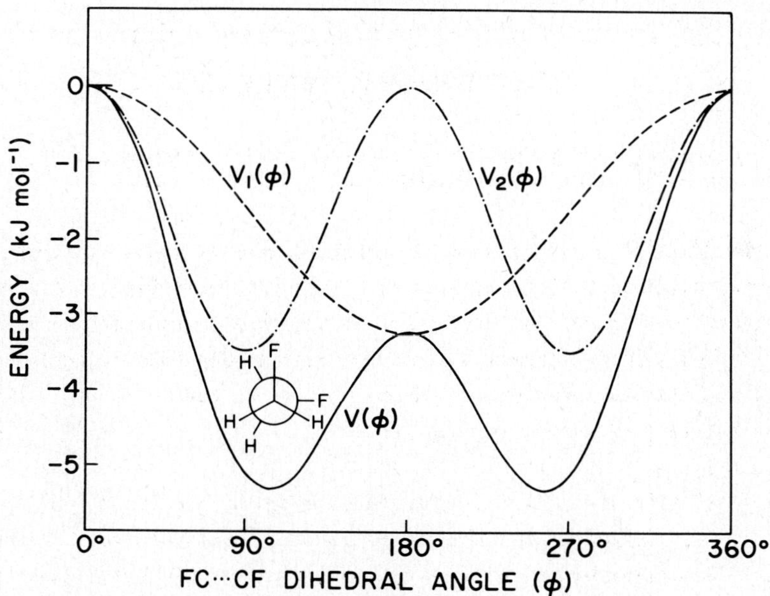

Fig. 30. Potential energy function, $V(\phi)$, and Fourier components, $V_n(\phi)$, describing internal rotation in $FCH_2C{\equiv}CCH_2F$.

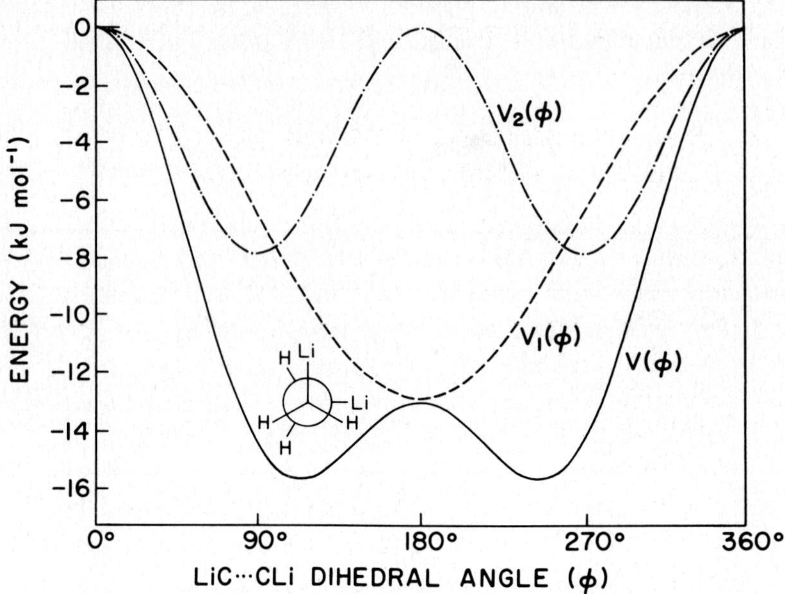

Fig. 31. Potential energy function, $V(\phi)$, and Fourier components, $V_n(\phi)$, describing internal rotation in $LiCH_2C{\equiv}CCH_2Li$.

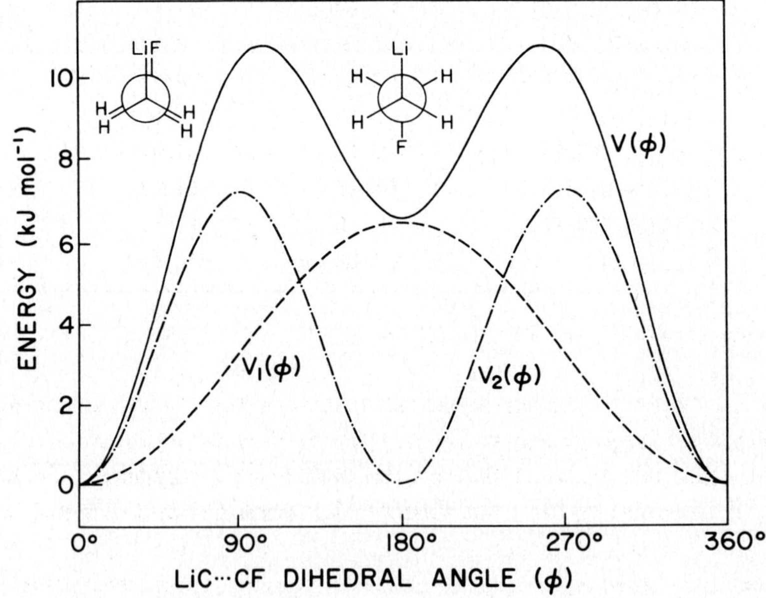

Fig. 32. Potential energy function, $V(\phi)$, and Fourier components, $V_n(\phi)$, describing internal rotation in $LiCH_2C{\equiv}CCH_2F$.

Rotational potential constants for these molecules, the corresponding substituted ethanes, and for some related molecules are shown in Table IV [49].

TABLE IV
Rotational potential constants (V_n, 4-31G//std, kJ mol^{-1}) describing rotation in substitued but-2-ynes and corresponding substituted ethanes[a]

Molecule	V_1	V_2	V_3
$CH_3C{\equiv}CCH_3$	0	0	+0.03
CH_3-CH_3	0	0	-13.63
$CH_3C{\equiv}CCH_2CH_3$	0	0	+0.04
$CH_3CH_2CH_3$	0	0	-15.46
$CH_3C{\equiv}CCH_2F$	0	0	+0.04
CH_3-CH_2F	0	0	-15.18
$CH_3C{\equiv}CCH_2Li$	0	0	+0.03
CH_3-CH_2Li	0	0	-13.78
$CH_3CH_2C{\equiv}CCH_2CH_3$	-0.07	-0.02	+0.04
CH_3CH_2-CH_2CH_3	-13.34	-5.99	-16.15
$CH_3CH_2C{\equiv}CCH_2F$	+0.31	+0.03	+0.03
CH_3CH_2-CH_2F	-2.01	-2.54	-19.54
$CH_3CH_2C{\equiv}CCH_2Li$	-0.29	+0.17	+0.08
CH_3CH_2-CH_2Li	+2.12	+2.61	-13.72
$FCH_2C{\equiv}CCH_2F$	-3.25	-3.50	+0.02
FCH_2-CH_2F	-19.59	-11.39	-17.12
$LiCH_2C{\equiv}CCH_2Li$	-13.01	-7.95	-0.12
$LiCH_2$-CH_2Li	-54.47	-19.89	-16.63
$LiCH_2C{\equiv}CCH_2F$	+6.40	+7.15	+0.10
$LiCH_2$-CH_2F	+31.88	+28.84	-12.47

[a]As summarized in ref. 49.

It can be seen that the potential functions for all three molecules are dominated by large V_1 and V_2 terms. The V_3 term is very small (*cf.* Table IV) and is therefore not included in the potential function diagrams. This indicates that, in contrast to the situation for 1,2-disubstituted ethanes, bond-bond repulsion at such long distances as are operative in the but-2-ynes is very small. The small V_3 term in the substituted but-2-ynes is also consistent with the very low barrier in but-2-yne itself where the V_3 term is the leading term in the Fourier expansion.

Electrostatic dipolar interaction between the CH_2X and CH_2Y groups accounts straightforwardly for the $V_1(\phi)$ potential terms. Thus, for example, in $FCH_2C{\equiv}CCH_2F$ (and $LiCH_2C{\equiv}CCH_2Li$) a *trans* conformation is favored over a *cis* conformation (*cf.* 35) on this basis leading to a negative V_1 coefficient. On the other hand, for $LiCH_2C{\equiv}CCH_2F$, dipolar forces would favor the *cis* over the *trans* conformation (*cf.* 36) leading to a positive V_1 coefficient.

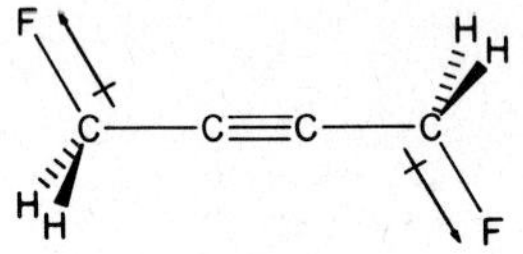

35 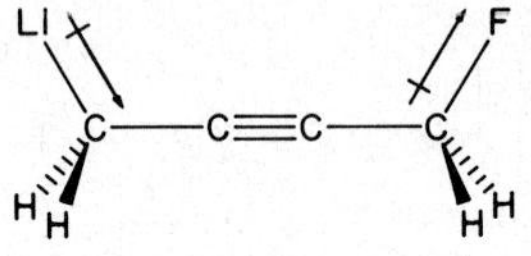 **36**

Note that direct overlap, which is important in the corresponding ethanes XCH_2CH_2Y (*cf.* 33) is unlikely to play a significant role in the but-2-ynes in which the XCH_2 and CH_2Y groups are more widely separated.

The $V_2(\phi)$ contribution to the potential functions may be rationalized in terms of hyperconjugative interaction. It is necessary first to note the orbitals of the $XCH_2C{\equiv}C$ group as shown in Figure 33, derived readily by interaction of XCH_2 and $C{\equiv}C$ group orbitals. The parenthetic labels X and $\bar{X}$ are used to denote respectively $XCH_2C{\equiv}C$ orbitals involving and not involving the substituent X. When X = H, the orbitals exist as degenerate pairs as shown in Figure 33. Substitution by an electropositive or electronegative X leads respectively to a raising or lowering of the $\pi(X)$ orbitals while leaving the $\pi(\bar{X})$ orbitals relatively unaffected.

We can proceed now to the orbital interaction diagram for $FCH_2C{\equiv}CCH_2F$ displayed in Figure 34. This shows that the splittings caused by the F substituent of the otherwise degenerate π_a and π_b orbitals results in a pronounced preference for conformations in which the local symmetry planes of the FCH_2 groups are orthogonal rather than coplanar. This arises through reduced energy separations between the interacting orbitals and hence an increased stabilization. The large negative value of V_2 is consistent with this preference. Large negative values of V_2 can also be expected for but-2-ynes $YCH_2C{\equiv}CCH_2Y'$ in which the substituents are not identical. Analogous arguments can be used to interpret the large negative V_2 coefficient for $LiCH_2C{\equiv}CCH_2Li$.

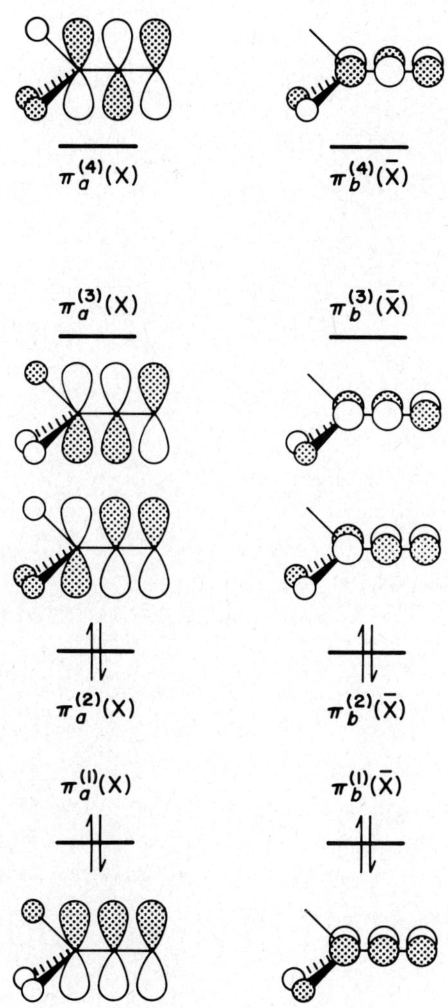

Fig. 33. π-Type orbitals of the $XCH_2C{\equiv}C$ group (X = H).

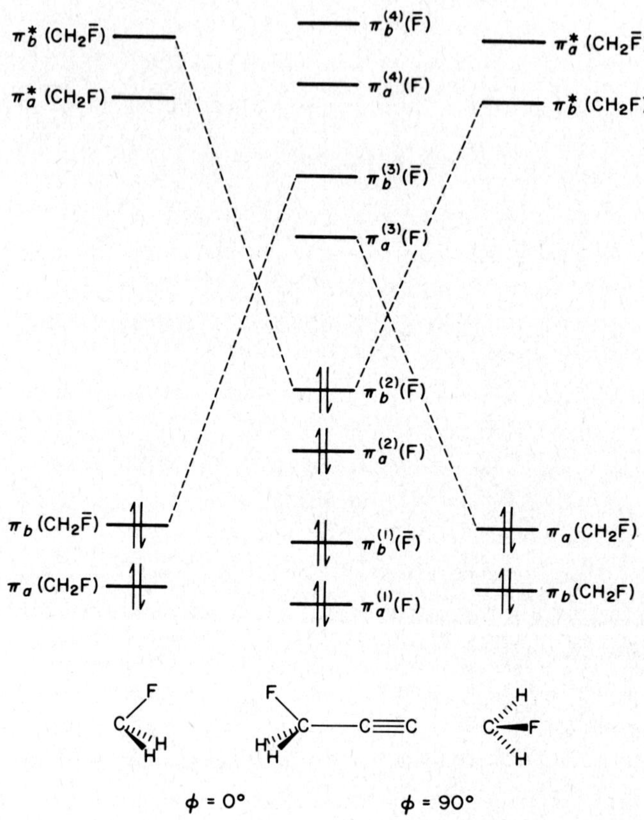

Fig. 34. Orbital interaction scheme showing hyperconjugative tendency of $FCH_2C{\equiv}CCH_2F$ to adopt a conformation with ϕ = 90° resulting from reduced energy separations between the most strongly interacting orbitals.

The situation in which one substituent is electropositive and the other electronegative is exemplified by $LiCH_2C{\equiv}CCH_2F$. The hyperconjugative interactions in this case (Figure 35) contribute to a preference for coplanar rather than orthogonal structures.

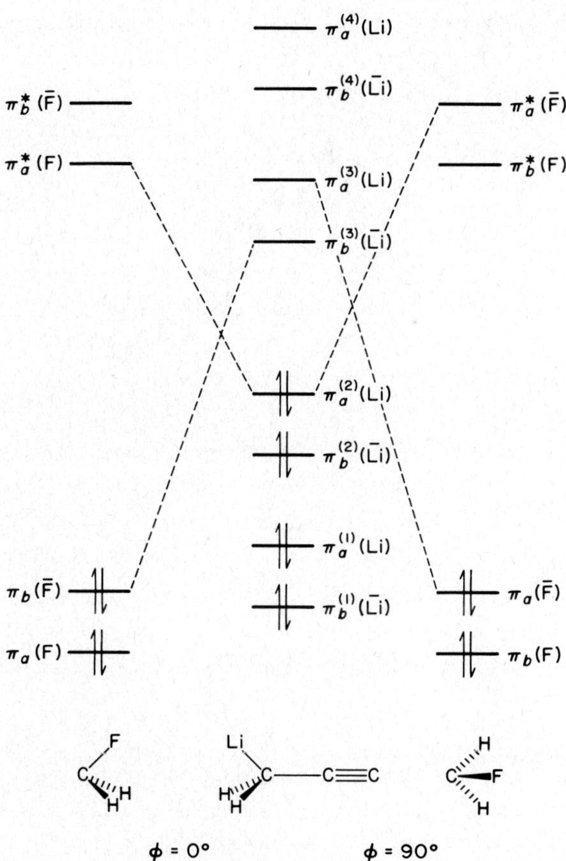

Fig. 35. Orbital interaction scheme showing hyperconjugative tendency of $LiCH_2C{\equiv}CCH_2F$ to adopt a conformation with ϕ = 0° resulting from reduced energy separations between the most strongly interacting orbitals.

The overall results for the 1,4-disubstituted but-2-ynes may be summarized as follows.

(i) When at least one substituent is electroneutral, the rotational barriers in monosubstituted but-2-ynes and 1,4-disubstituted but-2-ynes are very small ($< \sim 0.4$ kJ mol^{-1}).

(ii) There are *pronounced* conformational preferences in 1,4-disubstituted but-2-ynes when both substituents are electronegative ($FCH_2C{\equiv}CCH_2F$) or electropositive ($LiCH_2C{\equiv}CCH_2Li$). The large negative twofold coefficients of such molecules reflect a hyperconjugative tendency to adopt conformations in which the local symmetry planes of the terminal methylene groups are orthogonal. Through-space dipolar interactions between the terminal substituents lead to similarly large and negative V_1 coefficients and to a preference for conformations with $\phi > 90°$ and $\phi < 270°$. The qualitative and quantitative predictions are consistent with a recent electron diffraction study [50] of 1,4-dibromobut-2-yne, $BrCH_2C{\equiv}CCH_2Br$, which revealed a potential barrier of roughly 3 kJ mol^{-1} at the *trans* conformation with respect to the preferred conformation in which the dihedral angle BrC···CBr lies between 90° and 100°.

(iii) But-2-ynes such as $LiCH_2C{\equiv}CCH_2F$ in which one substituent is strongly electropositive and the other is electronegative have stable *cis* and *trans* isomers and reveal a pronounced preference for the eclipsed *cis* conformation.

4. HYPERCONJUGATION AND MOLECULAR BOND LENGTHS AND BOND ANGLES

In addition to influencing molecular conformation, as described in Section 3, hyperconjugation also has a significant influence on bond lengths and bond angles. Conversely, bond lengths and bond angles can sometimes be used as a sensitive probe for hyperconjugative interaction.

4.1 Monosubstituted Methanes, CH_3X

We begin by examining the effect of a substituent X on the C-H lengths of a monosubstituted methane CH_3X [51]. Calculated results for representative substituents are shown in Figure 36. Two effects appear to be operative. Firstly, there are electrostatic effects whereby electron withdrawal by the substituent X leads to a reduction in the C-H bond lengths while electron donation leads to an increase in the C-H lengths [52]. Secondly, there are hyperconjugative interactions involving electron donation from the appropriately oriented $\pi(CH_3)$ orbital into an empty orbital on X (**37**), or electron donation from an orbital on X into the appropriate $\pi^*(CH_3)$ orbital (**38**).

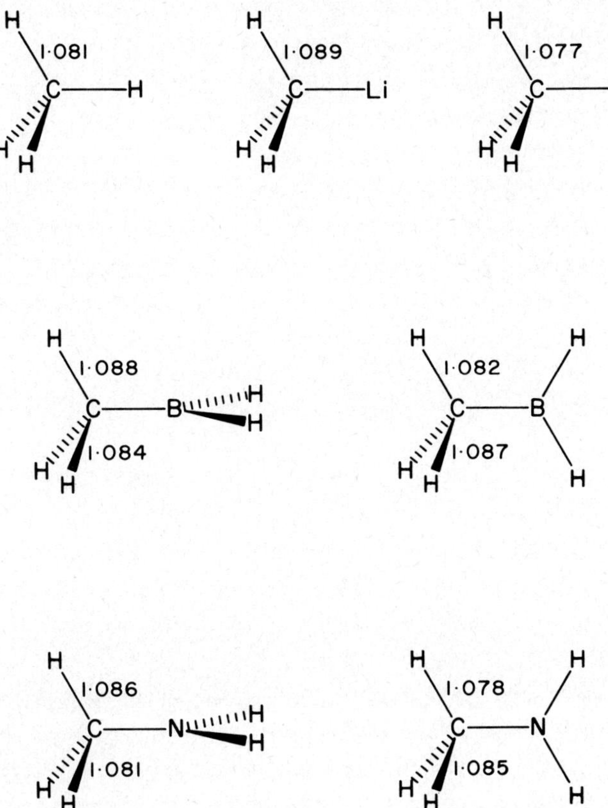

Fig. 36. Calculated C-H bond lengths (4-31G, Ångstroms) for monosubstituted methanes, CH_3X.

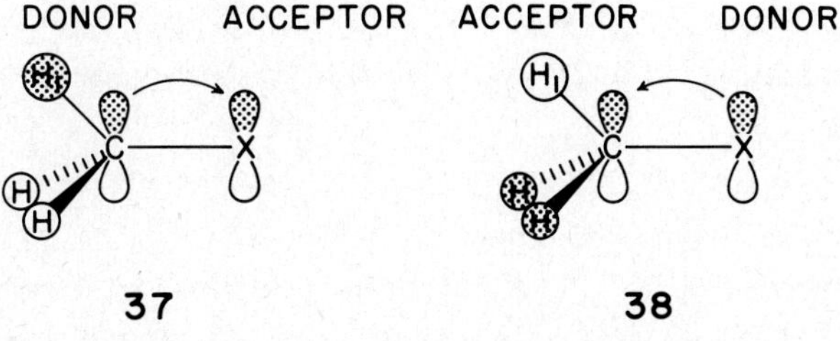

In 37, this leads to withdrawal of bonding electron density from the C-H_1 bond and to increased π-bonding in the C-X bond resulting in lengthened and shortened bonds respectively. In 38 the C-H_1 bond is again weakened and lengthened, this time resulting from increased antibonding character associated with the hyperconjugative interaction. Increased π-bonding in the C-X bond of **38** would again be expected to lead to C-X bond shortening.

Whereas the electrostatic interactions are conformationally independent, the hyperconjugative interactions depend on the mutual orientations of the interacting orbitals and this provides a mechanism for discriminating among the two effects. For example, in the perpendicular conformation of CH_3NH_2, interaction of the 2p(N) orbital with the $\pi^*(CH_3)$ orbital (**39**) leads to enhanced antibonding character in the $C\text{-}H_1$ bond and hence a weaker and longer bond than in CH_4 (1.086 vs 1.081 Å).

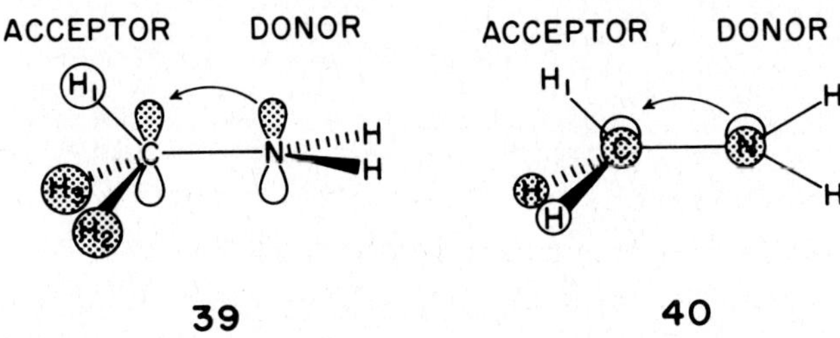

The effect of this hyperconjugative interaction (which would tend to lengthen the $C\text{-}H_1$ bond) is therefore greater than the electrostatic interaction (which would tend to shorten the bond). The $C\text{-}H_2$ and $C\text{-}H_3$ bonds, which are capable of undergoing only a fraction of the hyperconjugative interaction (because of reduced overlap), have the same lengths as in methane indicating a balance of hyperconjugative bond-lengthening and electrostatic bond-shortening effects in these cases.

Consistent with this interpretation, the in-plane $C\text{-}H_1$ bond in the eclipsed conformation **40**, which is oriented so that no hyperconjugative interaction may occur and only the electrostatic bond-shortening effect is possible, is actually shorter than the C-H bonds in methane (1.078 *vs*. 1.081 Å). In contrast, the two out-of-plane C-H bonds are lengthened due to compensating $2p(N) \rightarrow \pi^*(C\bar{H}_3)$ hyperconjugation.

Application of similar arguments to the CH_3BH_2 structures predicts that all C-H bonds should be lengthened compared to those in CH_4 (as is indeed found), reflecting the fact that in this case, both hyperconjugative and electrostatic effects act in the same direction, i.e. are bond lengthening. The longest C-H bond is $C\text{-}H_1$ in **41** where hyperconjugation between $\pi(CH_3)$ and the formally vacant 2p(B) orbital is a maximum while the shortest is $C\text{-}H_1$ in **42** where hyperconjugative interaction with 2p(B) is not possible. Note that in these systems hyperconjugative lengthening results from removal of bonding electron density in the C-H bonds.

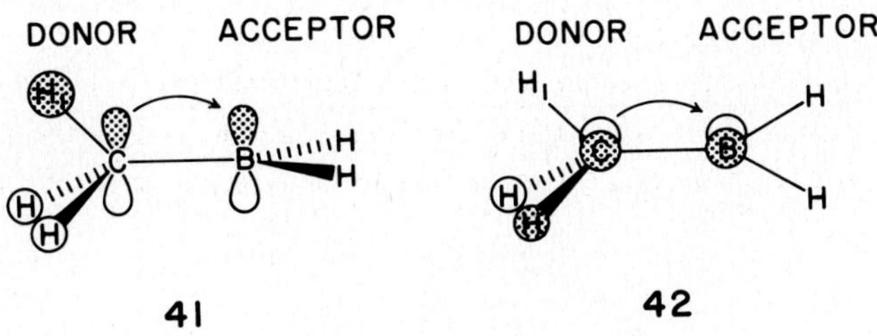

Inspection of C-H bond lengths in CH_3Li and CH_3F, where the hyperconjugative interactions cannot be turned off, shows C-H bond elongation in CH_3Li and C-H bond shortening in CH_3F. The bond elongation in CH_3Li may be attributed to reinforcing hyperconjugative and electrostatic effects. The C-H bond shortening in CH_3F suggests that the electrostatic effect dominates the hyperconjugative effect in this molecule.

Thus it appears that hyperconjugative interactions and electrostatic effects are both important contributors to determining C-H bond lengths in mono-substituted methanes [51]. For a substituent such as F which is a powerful σ-acceptor but only a weak π-donor, the electrostatic term dominates. For NH_2 which is a powerful π-donor, but only a weak σ-acceptor, hyperconjugative effects appear to dominate for the appropriately oriented C-H bonds.

The changes in bond strengths resulting from hyperconjugative interactions lead to β-deuterium isotope effects which have been examined both theoretically and experimentally by Hehre *et al.* [53].

4.2 Disubstituted Methanes, XCH_2Y

There have been a number of theoretical studies of bond lengths changes resulting from geminal interactions in XCH_2Y molecules [35a,37,54,55]. These have attempted to explain *inter alia* the shortening of the carbon-halogen bond lengths in polyhalogenated methanes accompanying successive halogen substitution [56], the systematic variation of C-O bond lengths in carbohydrates [40,57], reactivity patterns of tetrahedral species resulting from nucleophilic addition to a carbonyl group [58], and anomalously low nuclear quadrupole resonance frequencies of ^{35}Cl in chlorofluoromethane and in chlorine-containing ethers, sulphides, fluorides, alkenes and alkynes [59].

Calculated bond lengths for a number of representative disubstituted methanes are included in Table V.

TABLE V

Calculated bond lengths (4-31G, Ångstroms) for mono- and di-substituted methanes (XCH_2Y)[a]

Y	X	C - Y	C - X	C - H
H	H	1.081	1.081	1.081
H	Li	1.089	1.997	1.089
	BH_2 (plan.)	1.082	1.579	1.087
	BH_2 (perp.)	1.083	1.579	1.084
	NH_2 (plan.)	1.078	1.450	1.085
	NH_2 (perp.)	1.086	1.450	1.081
	F	1.077	1.406	1.077
Li	Li	1.968	1.968	1.097
	BH_2 (plan.)	1.999	1.577	1.095
	BH_2 (perp.)	1.999	1.513	1.092
	NH_2 (plan.)	2.017	1.475	1.092
	NH_2 (perp.)	2.032	1.489	1.087
	F	2.027	1.454	1.085
BH_2 (plan.)	BH_2 (plan.)	1.593	1.593	1.092
	BH_2 (perp.)	1.590	1.569	1.090
	NH_2 (plan.)	1.575	1.459	1.091
	NH_2 (perp.)	1.583	1.468	1.087
	F	1.569	1.422	1.083
BH_2 (perp.)	BH_2 (perp.)	1.577	1.577	1.087
	NH_2 (plan.)	1.591	1.462	1.087
	NH_2 (perp.)	1.588	1.463	1.084
	F	1.605	1.421	1.079
NH_2 (plan.)	NH_2 (plan.)	1.445	1.445	1.088
	NH_2 (perp.)	1.454	1.433	1.084
	F	1.431	1.396	1.080
NH_2 (perp.)	NH_2 (perp.)	1.449	1.449	1.080
	F	1.402	1.421	1.076
F	F	1.375	1.375	1.072

[a]From ref. 51

Changes in the C-X and C-Y bond lengths in disubstituted methanes XCH_2Y, from their values in the monosubstituted methanes, CH_3X and CH_3Y, may be readily understood in terms of the hyperconjugative and electrostatic effects described above [51]. Hyperconjugative arguments are based on the fact that for electropositive X (X = Li and BH_2), $\pi(CH_2X)$ orbitals are *more effective* hyperconjugative donors than $\pi(CH_3)$ orbitals while $\pi^*(CH_2X)$ orbitals are *less effective* hyperconjugative acceptors than $\pi^*(CH_3)$ orbitals. Conversely, for electronegative X (X = NH_2 and F), $\pi(CH_2X)$ orbitals are *less effective* hyperconjugative donors than $\pi(CH_3)$ orbitals while $\pi^*(CH_2X)$ orbitals are *more effective* hyperconjugative acceptors than $\pi^*(CH_3)$ orbitals.

The factors which will lead to variation in the C-X bond length in XCH_2Y compared with its value in CH_2X include (a) the electrostatic effect of Y, which is C-X bond lengthening for electropositive Y and C-X bond shortening for electronegative Y; (b) the hyperconjugative donation by $\pi(CH_2X)$ into a vacant p orbital (if present) on Y, which is bond lengthening; (c) the hyperconjugative acceptance by $\pi^*(CH_2X)$ from an occupied lone pair orbital (if present) on Y, which is bond lengthening; (d) the hyperconjugative donation by $\pi(CH_2Y)$ into a vacant p orbital (if present) on X, which is bond lengthening for electronegative Y and bond shortening for electropositive Y; and (e) the hyperconjugative acceptance by $\pi^*(CH_2Y)$ from a lone pair orbital (if present) on X, which is bond lengthening for electropositive Y and bond shortening for electronegative Y.

These principles may now be illustrated with a few examples for XCH_2Y molecules taken from Table V.

For X = Y = NH_2 (plan.) [43], hyperconjugative interactions are likely to be similar to those in CH_3NH_2 since neither of the 2p(N) lone pair orbitals can interact with an adjacent $\pi(CH_2N)$ orbital. Consequently, only the electrostatic effect operates and, due to the electronegative nature of the NH_2 group, a C-N bond shortening is expected and observed: 1.445 Å in **43** *vs*. 1.450 Å in CH_3NH_2.

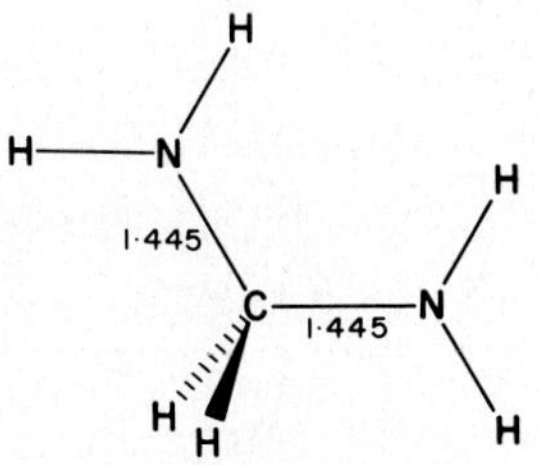

43 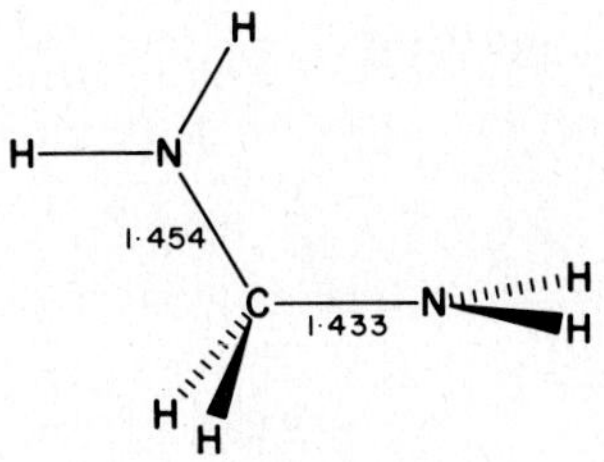 **44**

For the conformation (**44**) in which one NH_2 group is perpendicular and the other remains planar, the C-N (perp.) bond shortens significantly while the C-N (plan.) bond lengthens. This may be attributed to hyperconjugative donation from the 2p(N) (perp.) lone pair orbital into the $\pi^*(CH_2N$ [plan.]) orbital, which tends to strengthen the C-N (perp.) bond but to weaken the C-N (plan.) bond. In the case of **44**, such an orbital interaction term for C-N (plan.) appears to dominate the bond-shortening electrostatic effect.

Substitution of NH_2 (perp.) by F, which is a weaker π-donor but a more powerful σ-acceptor, changes the pattern of bond lengths (45).

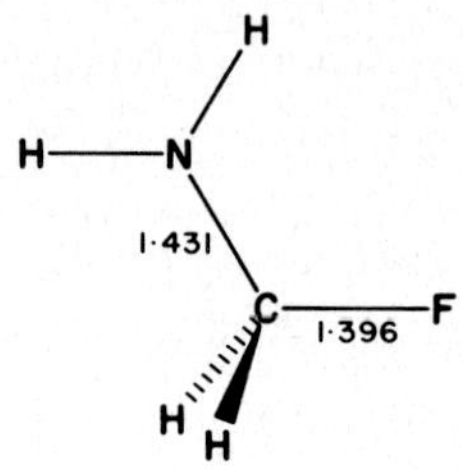

H
H—N
1·431
C—F
1·396
H H

45

The C-N bond length (1.431 Å) is smaller than that in CH_3NH_2 (1.450 Å) suggesting that, in this case, the bond shortening resulting from the electrostatic effect of F dominates the bond lengthening resulting from hyperconjugative donation from the 2p(F) lone pair orbital into $\pi^*(CH_2N)$.

This result [51] strengthens the view that the C-F bond shortening in CH_2F_2 compared to CH_3F is due primarily to electrostatic effects [35,52], rather than to hyperconjugative interactions.

The interpretation of the XCH_2NH_2 results above is reinforced by an examination of the XCH_2BH_2 structures.

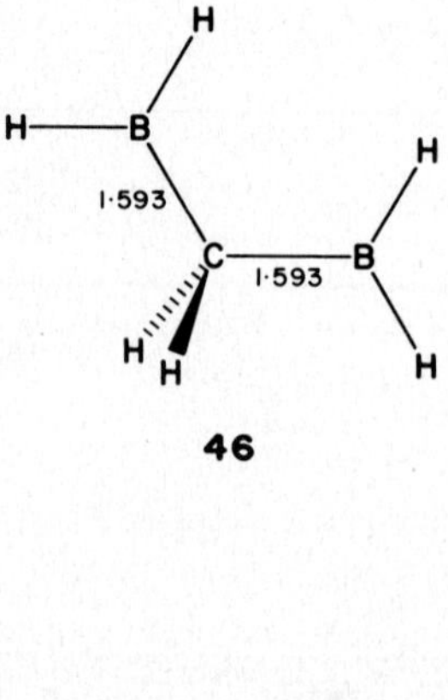

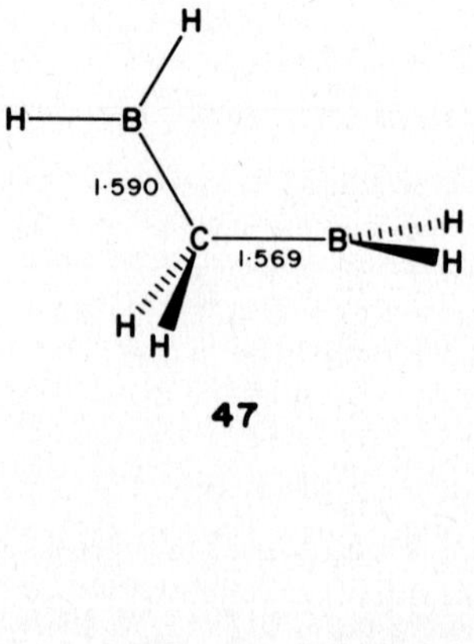

H
H—B
1·593
C—B
1·593
H H H H

46

H
H—B
1·590
C—B
1·569
H H H H

47

For X = Y = BH_2 (plan.) (**46**), the C-B bond (1.593 Å) is longer than in CH_3BH_2 (1.579 Å). This appears to be due primarily to the electrostatic effect of the electropositive BH_2 groups. For the conformer (**47**) in which one of the BH_2 groups is perpendicular and the other remains planar, the C-B (plan.) bond is extended (1.590 Å) while the C-B (perp.) bond is shortened (1.569 Å) compared to CH_3BH_2. This may be attributed to hyperconjugative donation from the $\pi(CH_2B$ [plan.]) orbital into the vacant 2p(B) (perp.) orbital.

Substitution of BH_2 (plan.) in **47** by F, which is a much less effective donor, as in **48** leads to a lengthening of the C-B bond compared to its value in CH_3BH_2.

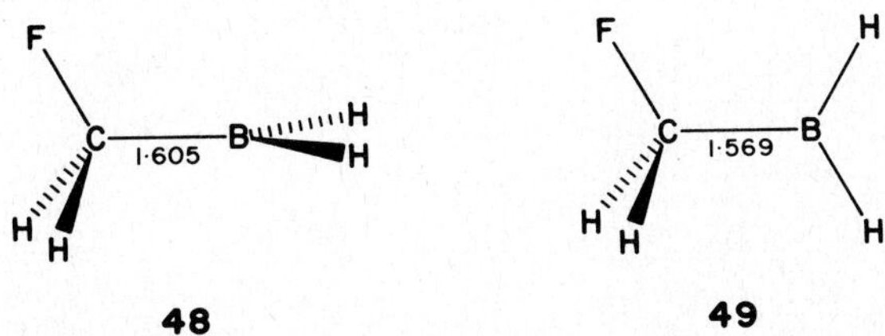

In this case, the markedly different hyperconjugative abilities of $\pi(CH_2F)$ and $\pi(CH_3)$ orbitals are responsible for the bond length difference: the $\pi(CH_2F)$ orbital (in FCH_2BH_2) is a considerably less effective hyperconjugative donor than $\pi(CH_3)$ (in CH_3BH_2). With the BH_2 group in its planar orientation (**49**), hyperconjugative interaction with the C-H bonds is restored via the $\pi(CH_2\bar{F})$ orbital (which, as noted in Section 2.2.4, is relatively unaffected by the substituent F), and now, the electrostatic bond-shortening effect of the F substituent leads to a C-B bond length shorter than in CH_3BH_2.

In a similar manner, the C-X and C-Y bond lengths of the remaining molecules listed in Table V can be rationalized in terms of a combination of hyperconjugative and electrostatic effects.

It is of interest to note that the interactions which give rise to the anomeric and exo-anomeric effects (*cf.* Section 3.6) also lead to changes in the individual C-O bond lengths in a manner analogous to that described for several of the molecules above. Calculated bond lengths [37c] for various conformations of the dimethoxymethane model system are shown in Figure 37.

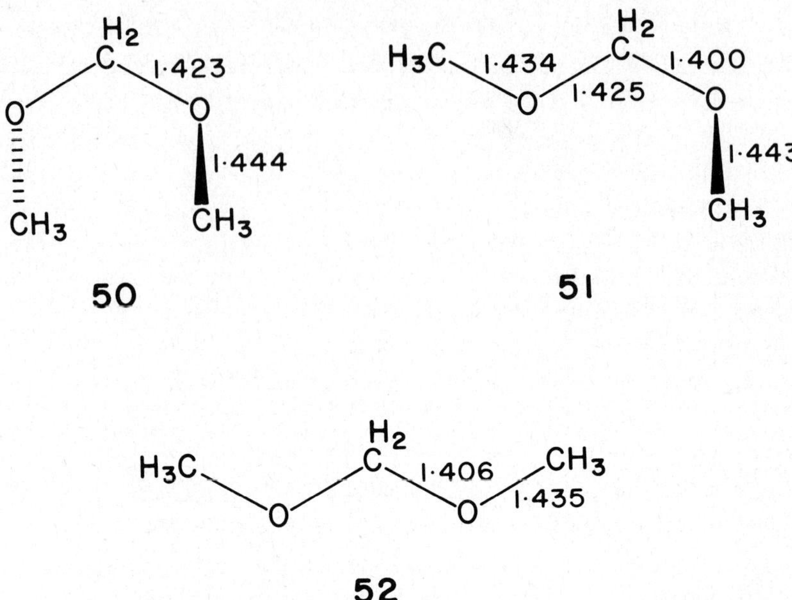

Fig. 37. Calculated bond lengths (4-31G, Ångstroms) for conformations of dimethoxymethane. The corresponding value for the reference system methanol is 1.437 Å.

There is a general shortening of the C-O bonds in the O-C-O segment which may be partly attributed to the electrostatic effect of electronegative substitution and partly to the favorable hyperconjugative interaction between a lone pair on oxygen and the low-lying $\pi^*(CH_2OCH_3)$ orbital. The effect is largest in the *gauche* bond of the asymmetric *gauche, trans* conformation 51, where there is less compensation from a similar interaction in the reverse direction, as shown in 53.

ACCEPTOR DONOR

CH_3—O — C — O — CH_3

53

There is also a lengthening of terminal O-C bonds when these are adjacent to a *gauche* bond as in **50** and **51**. This may be attributed to competition for the back-donation of the oxygen lone-pair electrons leading to reduced double-bond character in the terminal C-O bonds. The predicted structural changes are consistent with a large body of crystallographic data which are summarized comprehensively elsewhere [37d].

As noted in this section, the dependence of hyperconjugative interactions on the relative orientations of the interacting orbitals leads to a selective strengthening or weakening of particular bonds within a molecule. This property has been used by Lehn, Wipff and their associates [55,60] to provide a theoretical rationalization for experimentally observed [58] stereoelectronic control in the hydrolysis of carbonyl-containing compounds such as esters and amides.

4.3 Methyl Tilt Angles

A methyl group in an asymmetric environment is tilted, i.e. the approximate C_3 axis of the methyl group in the (non-C_{3v}) molecule CH_3X is tilted away from the C-X bond (*cf.* Figure 38).

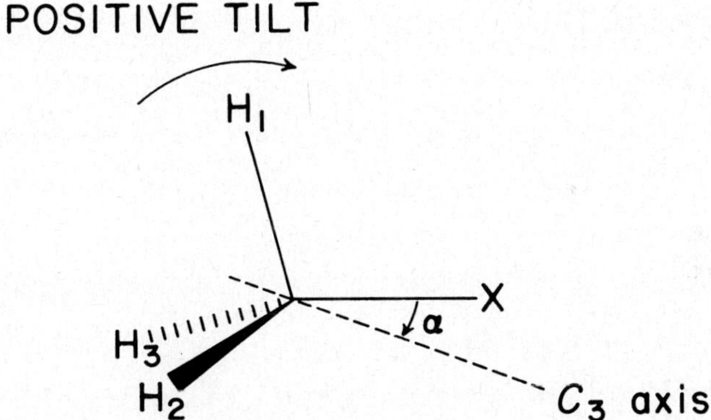

Fig. 38. The tilt of the approximate C_3 axis of a methyl group with respect to the C-X bond. The tilt angle, α, is defined as the angle between the C_3 axis and the C-X bond, and is positive as illustrated.

Although experimental and theoretical evidence for methyl tilts has been available for some time [61], it has only recently been demonstrated [62] that all the available data can be rationalized in terms of hyperconjugative interactions. An outline of the relevant arguments is included in this section.

Useful models for investigating methyl tilts are provided by perpendicular and eclipsed conformations of methylborane (CH_3BH_2) and methylamine (CH_3NH_2), again using constrained trigonal BH_2 and NH_2 groups. The calculated tilt angles (derived from STO-3G optimized structures) are shown in Figure 39 and may be rationalized straightforwardly in terms of hyperconjugative interactions.

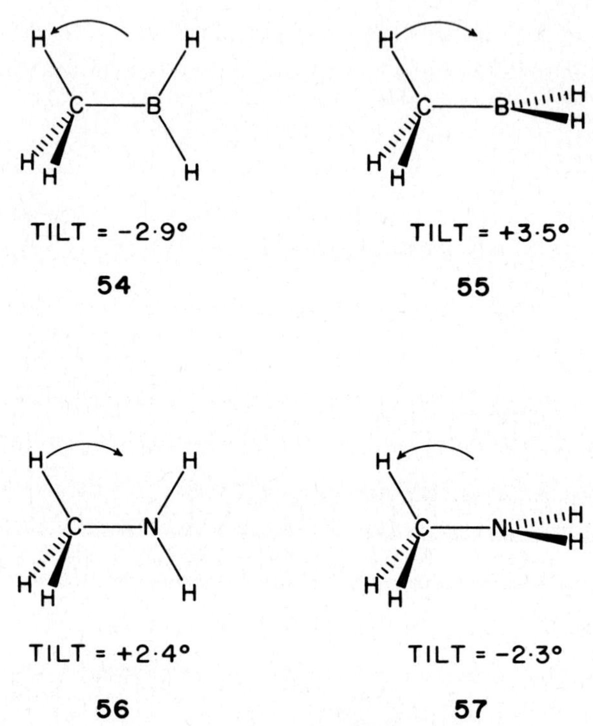

Fig. 39. Calculated tilt angles (derived from STO-3G optimized structures) for eclipsed and perpendicular conformations of methylborane and methylamine.

In eclipsed methylborane (**54**), the relevant interaction is that between the methyl $\pi_a(C\bar{H}_3)$ orbital and the formally vacant p orbital, 2p(B), on BH_2 (*cf.* Figure 40). Overlap of 2p(B) with $\pi_a(C\bar{H}_3)$ is clearly greater in the zone below the C-B bond than in the corresponding zone above the C-B bond owing to the contribution of orbitals on the out-of-plane methyl hydrogens to $\pi_a(C\bar{H}_3)$. There is thus a greater attractive interaction below the C-B bond than above it, and this leads immediately to a negative tilt as shown in **54**.

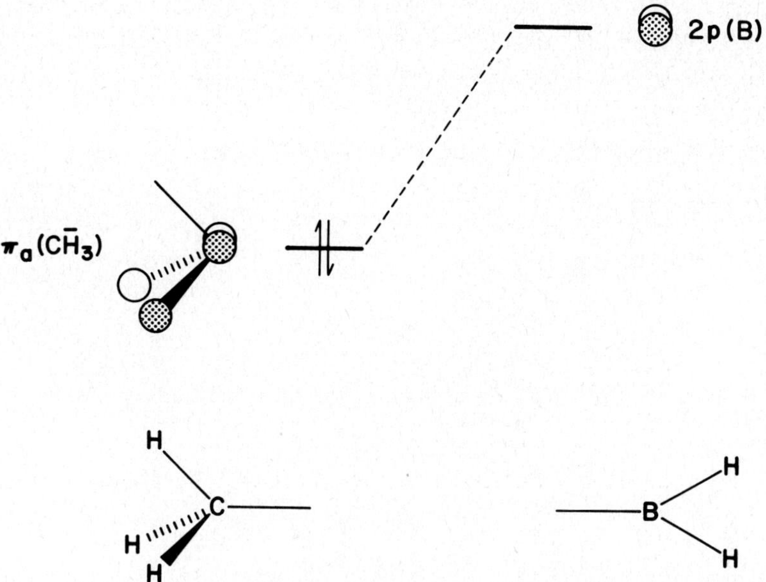

Fig. 40. Energy diagram showing the interaction of a methyl $\pi_a(C\bar{H}_3)$ orbital with a vacant p orbital, 2p(B), on B in eclipsed CH_3BH_2.

For perpendicular methylborane (**55**), the key interaction is that between $\pi_b(CH_3)$ and 2p(B) (*cf*. Figure 41).

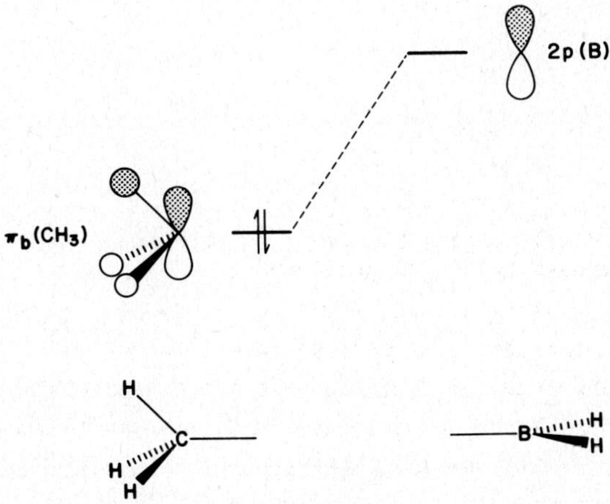

Fig. 41. Energy diagram showing the interaction of a methyl $\pi_b(CH_3)$ orbital with a vacant p orbital, 2p(B), on B in perpendicular CH_3BH_2.

In this case, owing to the greater proximity of the in-plane hydrogen, overlap is greater above the C-B bond than below it, leading to the observed positive tilt. The tilt of the methyl group in both conformations of methylborane may thus be attributed to attractive forces.

Turning now to the observed methyl tilts in perpendicular and eclipsed methylamine, we find that the direction of tilt is the opposite of that observed in the corresponding methylboranes. In the eclipsed conformation (**56**), the key interactions are those between the N lone pair, 2p(N), and the $\pi_a(C\bar{H}_3)$ orbital on the one hand and the $\pi_a{}^*(C\bar{H}_3)$ orbital on the other (Figure 42).

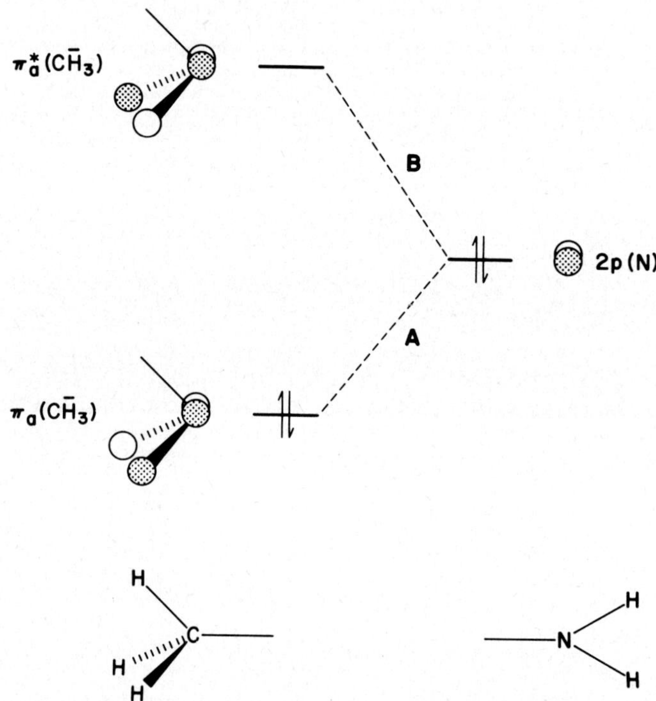

Fig. 42. Energy diagram showing the interaction of $\pi_a(C\bar{H}_3)$ and $\pi_a{}^*(C\bar{H}_3)$ orbitals with an occupied lone-pair orbital, 2p(N), on N in eclipsed CH_3NH_2.

Interaction **A** involves four electrons and is therefore repulsive; this generates a positive tilt. In the case of interaction **B**, overlap between the orbitals of the out-of-plane hydrogens and the 2p(N) orbital is *negative* so that, although interaction B is both stabilizing and attractive with respect to the methyl group as a whole, the opposite phases of the hydrogen orbitals and the 2p(N) orbital result in a repulsive force in the region *below* the C-N bond and a consequent positive tilt of the methyl group. Thus both interactions **A** and **B** contribute to the resultant positive tilt.

Similar arguments apply to the perpendicular conformation (**57**) of CH_3NH_2 (Figure 43).

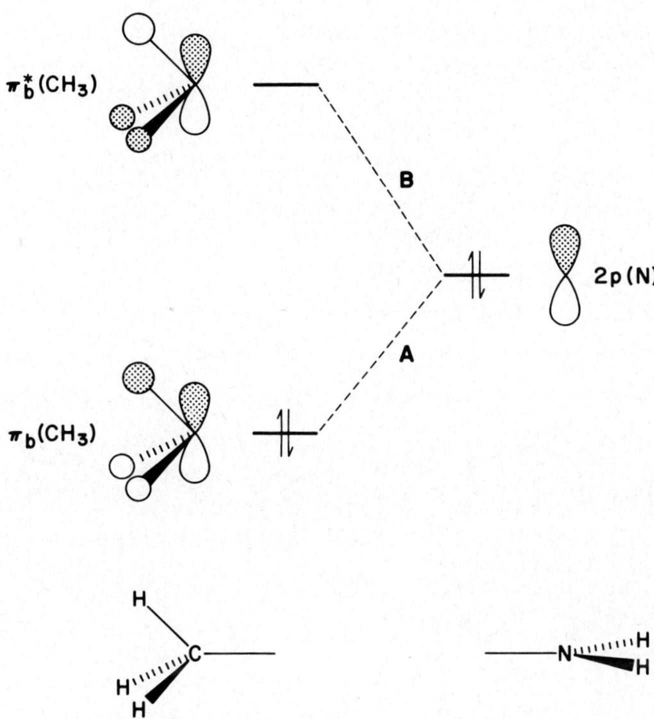

Fig. 43. Energy diagram showing the interaction of $\pi_b(CH_3)$ and $\pi_b^*(CH_3)$ orbitals with an occupied lone-pair orbital, 2p(N), on N in perpendicular CH_3NH_2.

Greater repulsion *above* the C-N bond via interaction **A**, and decreased attraction *above* the C-N bond (through the subtractive influence of the in-plane hydrogen orbital) both contribute to the observed negative tilt of the methyl group.

The above discussion suggests that both two-electron and four-electron interactions contribute to the methyl tilt in methylamine. Although this in fact may be the case, it should be pointed out that all the tilt angles discussed in this section are explicable solely in terms of two-electron effects.

Tilt angles calculated for a wide variety of systems are shown in Table VI [62-64].

TABLE VI
Calculated tilt angles[a]

Molecule	Conformation	Tilt Angle	Ref
CH_3BH_2	Perpendicular	+3.5°	62
CH_3BH_2	Eclipsed	-2.9°	62
CH_3NH_2	Perpendicular	-2.3°	62
CH_3NH_2	Eclipsed	+2.4°	62
$CH_3CH_2^+$	"Perpendicular"[b,c]	+5.4°	63
$CH_3CH_2^+$	Eclipsed[b]	-4.5°	63
$CH_3CH_2^-$	"Perpendicular"[b,c]	-5.7°	64
$CH_3CH_2^-$	Eclipsed[b]	+5.1°	64
$CH_3CH{=}CH_2$	H_1CCC *cis*	-0.3°	62
$CH_3CH{=}CH_2$	H_1CCC *trans*	+0.1°	62
$CH_3CH{=}NH$ (CCNH *anti*)	H_1CCN *cis*	+0.1°	62
$CH_3CH{=}NH$ (CCNH *anti*)	H_1CCN *trans*	-0.1°	62
$CH_3CH{=}NH$ (CCNH *syn*)	H_1CCN *cis*	-0.9°	62
$CH_3CH{=}NH$ (CCNH *syn*)	H_1CCN *trans*	-0.1°	62
$CH_3CH{=}O$	H_1CCO *cis*	-0.4°	62
$CH_3CH{=}O$	H_1CCO *trans*	-0.2°	62
$CH_3N{=}CH_2$	H_1CNC *cis*	-3.1°	62
$CH_3N{=}CH_2$	H_1CNC *trans*	+1.4°	62
$CH_3N{=}NH$ (CNNH *trans*)	H_1CNN *cis*	-2.6°	62
$CH_3N{=}NH$ (CNNH *trans*)	H_1CNN *trans*	+1.5°	62
$CH_3N{=}NH$ (CNNH *cis*)	H_1CNN *cis*	-2.6°	62
$CH_3N{=}NH$ (CNNH *cis*)	H_1CNN *trans*	+1.1°	62
$CH_3N{=}O$	HCNO *cis*	-2.3°	62
$CH_3N{=}O$	HCNO *trans*	+1.2°	62

[a]Defined using the sign convention of Figure 38. STO-3G values unless otherwise noted.
[b]4-31G values.
[c]These structures were optimized without a planarity constraint at C and therefore cannot strictly be called perpendicular; however, this notation is used to bring out more clearly the relationship with constrained conformations of CH_3BH_2 and CH_3NH_2.

These may all be rationalized [62] in terms of differential hyperconjugative interaction above and below the C-X bond; two examples are discussed in detail here.

The first concerns the conformational dependence of the tilt angle in perpendicular methylamine when the bonds at nitrogen are pyramidally distorted (Figure 44).

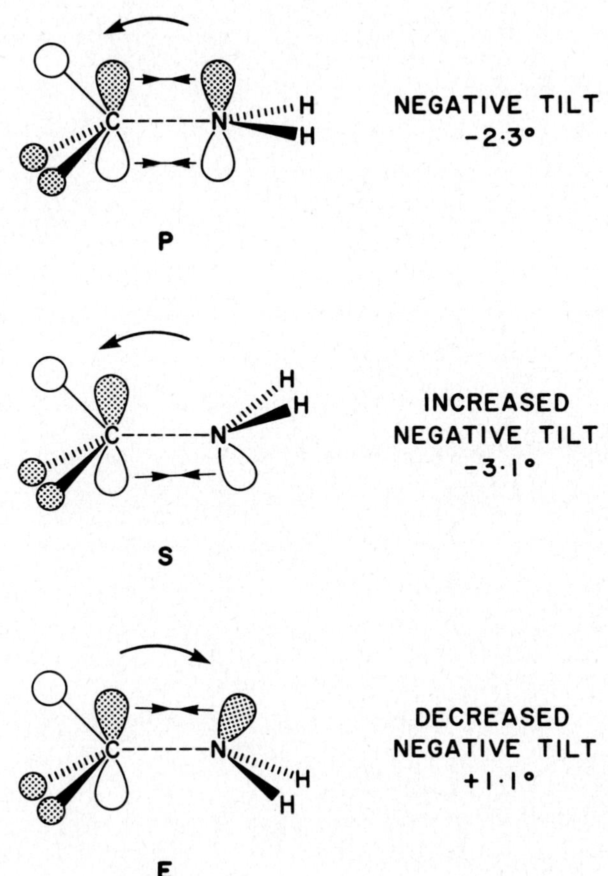

Fig. 44. Attractive two-electron interactions between the lone-pair orbital of N, 2p(N), and the methyl $\pi_b{}^*(CH_3)$ orbital, and corresponding tilts for perpendicular (P), pyramidal staggered (S) and pyramidal eclipsed (E) conformations of methylamine.

As noted above, in the perpendicular conformation (P) with a planar trigonal N a negative tilt (-2.3°) occurs owing to greater overlap below the C-N bond than above it. As we proceed to the pyramidal staggered conformation (S), this differential increases, which leads to an *enhanced* negative tilt (-3.1°). In the pyramidal eclipsed conformation (E), however, the differential decreases to the extent that the overlap is actually greater *above* the C-N bond than below it and, as a consequence, a positive tilt occurs.

A similar argument explains the conformational dependence of the tilt in methanol (Figure 45).

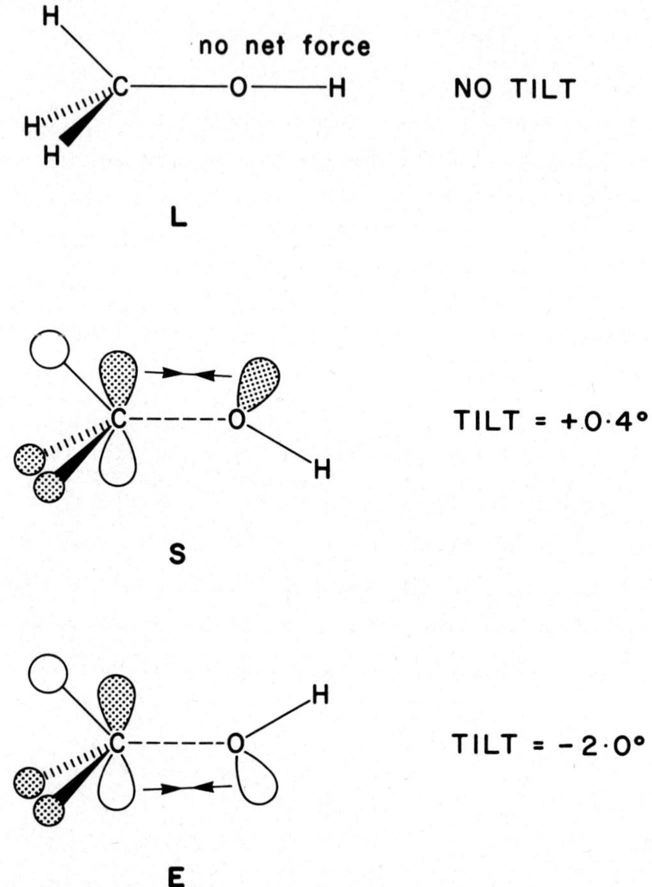

Fig. 45. Dominant attractive forces between an oxygen lone-pair orbital and the methyl π_b*(CH_3) orbital for linear (L), staggered (S) and eclipsed (E) conformations of methanol.

For methanol with a linear C-O-H linkage (L), the tilt is zero by symmetry: the force between one of the 2p(O) lone pairs and the π_a*($\bar{CH}_3$) orbital (similar to interaction B of Figure 42) must be exactly balanced by the opposing force between the second 2p(O) orbital and the methyl π_b*(CH_3) orbital (similar to interaction B in Figure 43). For nonlinear structures, the first of these interactions remains approximately the same but the second interaction changes, which leads to an imbalance of forces above and below the C-O bond and hence a tilt of the methyl group. In the staggered conformation (S) there is increased attraction above the C-O bond, which results in a positive tilt, while in the eclipsed conformation (E) there is increased attraction below the C-O bond, which leads to a negative tilt.

5. CONCLUDING REMARKS

The material presented in this review demonstrates the widespread structural manifestations of hyperconjugation. The effects are sometimes subtle (e.g. tilt angles of methyl groups) and sometimes large (e.g. conformational behavior of carbocations) but the picture that emerges is entirely self-consistent. The evidence of hyperconjugation, provided by the qualitative arguments of perturbation molecular orbital theory and supported by the quantitative *ab initio* calculations, is compelling.

6. ACKNOWLEDGEMENTS

I would like to acknowledge my debt to colleagues and students who contributed to the ideas and results presented in this article. My early thinking on the subject of hyperconjugation benefited from collaborative studies with Professors J.A. Pople, P.v.R. Schleyer, W.J. Hehre, R. Hoffmann and L. Salem. The more recent studies, carried out at the Australian National University, relied heavily on the contributions of Drs P.J. Stiles, M.A. Vincent and W.J. Bouma, Professor N.V. Riggs and, for the latest work particularly, Dr A. Pross.

7. REFERENCES

1 G.W. Wheland, J. Chem. Phys., 2 (1934) 474.
2 L. Pauling, H.D. Springall and K.J. Palmer, J. Am. Chem. Soc., 61 (1939) 927.
3 R.S. Mulliken, J. Chem. Phys., 7 (1939) 339.
4 R.S. Mulliken, C.A. Rieke and W.G. Brown, J. Am. Chem. Soc., 63 (1941) 41.
5 J.W. Baker, Hyperconjugation, Oxford University Press, 1952.
6 Conference on Hyperconjugation, Tetrahedron, 5 (1959) 105.
7 M.J.S. Dewar, Hyperconjugation, Ronald Press, New York, 1962.
8 D. Holtz, Prog. Phys. Org. Chem., 8 (1971) 1.
9 See, for example,
(a) H.F. Schaefer III, The Electronic Structure of Atoms and Molecules, Addison-Wesley, Reading, Mass. 1972.
(b) H.F. Schaefer III (Ed.), Methods of Electronic Structure Theory, Plenum, New York, 1977.
(c) H.F. Schaefer III (Ed.), Applications of Electronic Structure Theory, Plenum, New York, 1977.
(d) I.G. Csizmadia, Theory and Practice of MO Calculations on Organic Molecules, Elsevier, Amsterdam, 1976.
10 For reviews of the PMO method, see ref. 11 and
(a) I. Fleming, Frontier Orbitals and Organic Chemical Reactions, Wiley-Interscience, New York, 1976.

(b) N.D. Epiotis, W.R. Cherry, S. Shaik, R. Yates and F. Bernardi, Top. Curr. Chem., 70 (1977) 1.
(c) G. Klopman (Ed.), Chemical Reactivity and Reaction Paths, Wiley-Interscience, New York, 1974.
(d) R.F. Hudson, Angew. Chem. Int. Ed. Engl., 12 (1973) 36.
(e) M.J.S. Dewar and R.C. Doherty, The PMO Theory of Organic Chemistry, Plenum, New York, 1975.
(f) R. Hoffmann, Accounts Chem. Res., 4 (1971) 1.
(g) B.N. Gimarc, Molecular Structure and Bonding. The Qualitative Molecular Orbital Approach, Academic Press, New York, 1979.

11 A very readable outline is presented in W.L. Jorgensen and L. Salem, The Organic Chemist's Book of Orbitals, Academic Press, New York, 1973.

12 See, for example, F. Sherwood Taylor, Science Past and Present, Heinemann, London, 1949, p. 96.

13 C.C.J. Roothaan, Rev. Mod. Phys., 23 (1951) 69.

14 R. Ditchfield, W.J. Hehre and J.A. Pople, J. Chem. Phys., 54 (1971) 724.

15 W.J. Hehre and J.A. Pople, J. Chem. Phys., 56 (1972) 4233.

16 J.S. Binkley and J.A. Pople, J. Chem. Phys., 66 (1977) 879.

17 J.D. Dill and J.A. Pople, J. Chem. Phys., 62 (1975) 2921.

18 See, for example, J.A. Pople, in H.F. Schaefer III (Ed.), Applications of Electronic Structure Theory, Plenum, New York,1977.

19 J.A. Pople and M.S. Gordon, J. Am. Chem. Soc., 89 (1967) 4253.

20 For leading references, see O. Eisenstein, N.T. Anh, Y. Jean, A. Devaquet, J. Cantacuzene and L. Salem, Tetrahedron, 30 (1974) 1717.

21 L. Radom, W.J. Hehre and J.A. Pople, J. Am. Chem. Soc., 94 (1972) 2371.

22 (a) W.J. Hehre and J.A. Pople, J. Am. Chem. Soc., 97 (1975) 6941.
(b) A. Devaquet, R.E. Townshend and W.J. Hehre, J. Am. Chem. Soc., 98 (1976) 4068.

23 (a) J.P. Lowe, J. Am. Chem. Soc., 92 (1970) 3799.
(b) W.J. Hehre and L. Salem, J. Chem. Soc., Chem. Commun., (1973) 754.
(c) N.D. Epiotis, D. Bjorkquist, L. Bjorkquist and S. Sarkanen, J. Am. Chem. Soc., 95 (1973) 7558.

24 See for example,
(a) J.P. Lowe, J. Am. Chem. Soc., 96 (1974) 3759.
(b) B.M. Gimarc, J. Am. Chem. Soc., 95 (1973) 1417.

25 (a) L. Radom, J.A. Pople, V. Buss and P.v.R. Schleyer, J. Am. Chem. Soc., 92 (1970) 6380.
(b) L. Radom, J.A. Pople, V. Buss and P.v.R. Schleyer, J. Am. Chem. Soc., 92 (1970) 6987.
(c) L. Radom, J.A. Pople and P.v.R. Schleyer, J. Am. Chem. Soc., 94 (1972) 5935.
(d) Y. Apeloig, P.v.R. Schleyer and J.A. Pople, J. Am. Chem. Soc., 99 (1977) 5901

26 R. Hoffmann, L. Radom, J.A. Pople, P.v.R. Schleyer, W.J. Hehre and L. Salem, J. Am. Chem. Soc., 94 (1972) 6221.

27 A. Pross and L. Radom, Aust. J. Chem., 33 (1980) 241.

28 Y. Apeloig and Z. Rappoport, J. Am. Chem. Soc., 101 (1979) 5095.

29 (a) T. Kawamura, D.J. Edge and J.K. Kochi, J. Am. Chem. Soc., 94 (1972) 1752.
(b) D.J. Edge and J.K. Kochi, J. Am. Chem. Soc., 94 (1972) 6485, 7695.
(c) P.J. Krusik, P. Meakin and J.P. Jesson, J. Phys. Chem., 75 (1971) 3438.

30 L. Radom, J. Paviot, J.A. Pople and P.v.R. Schleyer, J. Chem. Soc., Chem. Commun., (1974) 58.

31 A. Pross and L. Radom, Tetrahedron, 36 (1980) 1999.

32 (a) J.K. Kochi, Adv. Free Radical Chem., 5 (1975) 189.
(b) P.J. Krusik and J.K. Kochi, J. Am. Chem. Soc., 93 (1971) 846.
(c) T. Kawamura and J.K. Kochi, J. Am. Chem. Soc., 94 (1972) 648.
(d) A.R. Lyons and M.C.R. Symons, J. Am. Chem. Soc., 93 (1971) 7330.
(e) A.R. Lyons, G.W. Nielson and M.C.R. Symons, J. Chem. Soc., Chem. Commun., (1972) 507.

33 J.D. Dill, P.v.R. Schleyer and J.A. Pople, J. Am. Chem. Soc., 98 (1976) 1663.

34 (a) S. Wolfe, A. Rauk, L.M. Tel and I.G. Csizmadia, J. Chem. Soc. B, (1971) 136.
(b) S. Wolfe, Accounts Chem. Res., 5 (1972) 102.

(c) S. Wolfe, L.M. Tel, W.J. Haines, M.A. Robb and I.G. Csizmadia, J. Am. Chem. Soc., 95 (1973) 4863.

35 For leading references, see
(a) S. Wolfe, M.H. Whangbo and D.J. Mitchell, Carbohyd. Res., 69 (1979) 1.
(b) W.A. Szarek and D. Horton (Eds.), Anomeric Effect. Origin and Consequences, A.C.S. Symposium Series No. 87, American Chemical Society, Washington, 1979.

36 (a) S. David, O. Eisenstein, W.J. Hehre, L. Salem and R. Hoffmann, J. Am. Chem. Soc., 95 (1973) 3806.
(b) S. David, in W.A. Szarek and D. Horton (Eds.), Anomeric Effect. Origin and Consequences, A.C.S. Symposium Series No. 87, American Chemical Society, Washington, 1979.

37 (a) G.A. Jeffrey, J.A. Pople and L. Radom, Carbohyd. Res., 25 (1972) 117.
(b) G.A. Jeffrey, J.A. Pople and L. Radom, Carbohyd. Res., 38 (1974) 81.
(c) G.A. Jeffrey, J.A. Pople, J.S. Binkley and S. Vishveshwara, J. Am. Chem. Soc., 100 (1978) 373.
(d) G.A. Jeffrey, in W.A. Szarek and D. Horton (Eds.), Anomeric Effect. Origin and Consequences, A.C.S. Symposium Series No. 87, American Chemical Society, Washington, 1979.
(e) S. Vishveshwara, Chem. Phys. Lett., 59 (1978) 30.
(f) G.A. Jeffrey and J.H. Yates, J. Am. Chem. Soc., 101 (1980) 820.

38 D.G. Gorenstein and D. Kar, J. Am. Chem. Soc., 99 (1977) 672.

39 E.E. Astrup, Acta Chem. Scand., 27 (1973) 3271.

40 C. Romers, C. Altona, H.R. Buys and E. Havinga, Top. Stereochem., 4 (1969) 39.

41 W.J. Bouma and L. Radom, Aust. J. Chem., 31 (1978) 1167.

42 (a) E. Saegebarth and L.C. Krisher, J. Chem. Phys., 52 (1970) 3555.
(b) R.G. Ford, J. Chem. Phys., 65 (1976) 354.

43 L. Radom, P.J. Stiles and M.A. Vincent, J. Mol. Struct., 48 (1978) 431.

44 L. Radom, P.J. Stiles and M.A. Vincent, Nouveau J. Chimie, 2 (1978) 115.

45 L. Radom, W.A. Lathan, W.J. Hehre and J.A. Pople, J. Am. Chem. Soc., 95 (1973) 693.

46 (a) S.S. Butcher, R.A. Cohen and T.C. Rounds, J. Chem. Phys., 54 (1971) 4123.
(b) P. Huber-Wälchli and H.H. Günthard, Chem. Phys. Lett., 30 (1975) 347.
(c) W.C. Harris, J.R. Holtzclaw and V.F. Kalasinsky, J. Chem. Phys., 67 (1977) 3330.
(d) D. Friesen and K. Hedberg, J. Am. Chem. Soc., 102 (1980) 3987.

47 (a) P.R. Bunker and H.C. Longuet-Higgins, Proc. Roy. Soc. Ser. A., 280 (1964) 340.
(b) R. Kopelman, J. Chem. Phys., 41 (1964) 1547.
(c) D. Papousek, Scr. Fac. Sci. Natur. Univ. Purkynianae Brun., 2 (1972) 159.

48 (a) L. Radom and J.A. Pople, J. Am. Chem. Soc., 92 (1970) 4786.
(b) P.A. Kollman, C.F. Bender and J. McKelvey, Chem. Phys. Lett., 28 (1974) 407.
(c) J.S. Binkley, J.A. Pople and W.J. Hehre, Chem. Phys. Lett., 36 (1975) 1.

49 L. Radom, P.J. Stiles and M.A. Vincent, J. Mol. Struct., 48 (1978) 259.

50 O.H. Ellestad and K. Kveseth, J. Mol. Struct., 25 (1975) 175.

51 A. Pross and L. Radom, J. Comput. Chem., 1 (1980) 295.

52 (a) H.A. Bent, Chem. Rev., 61 (1961) 275.
(b) E. Shustorovich, J. Am. Chem. Soc., 100 (1978) 7513.

53 D.J. DeFrees, M. Taagepera, B.A. Levi, S.K. Pollack, K.D. Summerhays, R.W. Taft, M. Wolfsberg and W.J. Hehre, J. Am. Chem. Soc., 101 (1979) 5532 and references therein.

54 L. Radom and P.J. Stiles, Tetrahedron Lett., (1975) 789.

55 (a) J.M. Lehn and G. Wipff, Helv. Chim. Acta., 61 (1978) 1274.
(b) G. Wipff, Tetrahedron Lett., (1978) 3269.
(c) J.M. Lehn and G. Wipff, J. Am. Chem. Soc., 102 (1980) 1347.
(d) J.M. Lehn and G. Wipff, J. Am. Chem. Soc., 96 (1974) 4048.
(e) J.M. Lehn, G. Wipff and H.B. Bürgi, Helv. Chim. Acta., 57 (1974) 493.

56 L.O. Brockway, J. Phys. Chem., 41 (1937) 185.

57 H.M. Berman, S.S.C. Chu and G.A. Jeffrey, Science, 157 (1967) 1576.

58 (a) P. Deslongchamps, Tetrahedron, 31 (1975) 2463.
(b) P. Deslongchamps, Heterocycles, 7 (1977) 1271.

59 (a) R. Livingston, J. Phys. Chem., 57 (1953) 496.
(b) E.A.C. Lucken, J. Chem. Soc., (1959) 2954.

60 H.B. Bürgi, J.D. Dunitz, J.M. Lehn and G. Wipff, Tetrahedron, 30 (1974) 1563 and references therein.
61 For leading references, see E. Flood, P. Pulay and J.E. Boggs, J. Am. Chem. Soc., 99 (1977) 5570.
62 A. Pross, L. Radom and N.V. Riggs, J. Am. Chem. Soc., 102 (1980) 2253.
63 P.C. Hariharan, W.A. Lathan and J.A. Pople, Chem. Phys. Lett., 14 (1972) 385.
64 A. Pross, D.J. DeFrees, B.A. Levi, S.K. Pollack, L. Radom and W.J. Hehre, J. Org. Chem., 46 (1981) 1693.

Molecular Structure and Conformation: Recent Advances,
I.G. Csizmadia (Ed.), *Progress in Theoretical Organic Chemistry*, Volume 3

QUANTITATIVE ORBITAL ANALYSIS OF THE CONFORMATIONAL PREFERENCES IN METHYL DERIVATIVES

Fernando Bernardi[1a] and Andrea Bottoni[1b]

1. INTRODUCTION

Recently various procedures have been described which allow to obtain estimates, in the framework of an ab-initio SCF-MO computation, of the energy effects associated with the orbital interactions occurring between the component fragments of a closed shell molecule [2-6]. In a first stage these procedures have been used only for the analyses of effects associated with π-type orbital interactions [2,3,7,8,9], and only very recently they have been modified to include in the analysis all kinds of orbital interactions [5,6]. At this level these procedures can be considered an important instrument for rationalizing various chemical problems and the preliminary results seem to be very promising [6,10].

In this paper we describe the application of such a procedure to the analysis of the conformational problems associated with the methyl group in molecules of the type CH_3-Y, with Y = XH_3 (X = C,Si), XH_2 (X = N,P), XH (X = O,S). In these analyses we have used different orbital representations based on fragment localized MO's and fragment canonical MO's. It is the purpose of this paper to examine the advantages and disadvantages of the various representations with respect to the interpretation which must be as clear as possible, and also with respect to the quantitative agreement with the total energy behaviour.

The different representations are simply inter-related and since they provide rationalizations in terms of different effects, their combined use allows to understand at a quantitative level also the relationship between effects which seem quite different such as steric and hyperconjugative effects.

In all cases the computations have been carried out at the STO-3G level [11]; in the case of second-row molecules they have been performed also at the STO-3G* level [12], i.e. using an STO-3G basis set implemented with a set of five 3d-type functions for the second row atom consisting of one second order Gaussian each. These computational levels provide a correct description of all the conformational problems investigated here, while in certain cases the magnitudes and trends of the rotational barriers are less satisfactory. The use of more sophisticated basis sets could certainly provide correct results also in these cases [13]. However the complexity of the quantitative analyses increases with the size of the basis set: at this preliminary stage we prefer to examine the use of these analyses at the simpler minimal basis level. Obviously, in those cases where this computational level does not provide a correct result, also the quantitative analysis can not provide a complete rationalization. However, it will be shown that also in these cases the quantitative analysis can provide useful information about the various factors which control the structural problem under examination.

The orbital interactions analyzed in this paper are in all cases "non-bonded" interactions. It is well accepted that these interactions play a very important role in molecular structure [14]. It has already been recognized that non-bonded interactions can be repulsive or attractive in nature. While repulsive interactions have been associated with steric effects, attractive interactions have been associated with Van der Waals interactions. However the modern theoretical organic chemistry has provided a more detailed understanding of non-bonded interactions [14], which is further significantly improved by the use of quantitative analysis, as shown in this paper.

2. QUANTITATIVE PMO ANALYSIS

Qualitative Perturbational MO (PMO) analysis have prooved to be a very useful instrument for analyzing a variety of chemical problems [15-25]. However a significant improvement over the qualitative understanding provided by these analyses can be obtained through the use of a "quantitative" PMO analysis [2-6], i.e. a PMO analysis where the various energy effects are computed using the results of ab-initio SCF-MO computations. The whole computational procedure follows as much as possible the line of the qualitative PMO approach and therefore involves a procedure for the computation of the MO's of the component fragments and a procedure for the computation of the energy effects associated with the orbital interactions under examination based on PMO expressions.

2.1 Computation of the Fragment MO's of a Closed Shell Molecule

The procedure used here for obtaining the fragment MO's is derived from that recently suggested by Wolfe et al. [2], which will be denoted hereafter as the WSW procedure. In the WSW procedure the MO's and related orbital energies of the various component fragments arising from the dissection of the molecule under consideration are obtained from the solution of the following eigenvalue problem

$$F^{\circ}C^{\circ} = S^{\circ}C^{\circ}\varepsilon^{\circ} \qquad (1)$$

where F° and S° are the Fock and overlap matrices for the molecule under consideration with all the non-diagonal matrix elements between atomic orbitals belonging to the different interacting fragments set equal to zero.

This procedure provides a set of fragment MO's which will be denoted here as fragment canonical MO's. When the interacting fragments are closed shell species, these MO's provide a correct basis for the quantitative orbital analysis. However when the dissection of the molecule under examination involves the breaking of bonds, these MO's are of limited usefulness for a quantitative analysis [4,5]. This problem is particularly important in the case of the σ fragment MO's and arises from their delocalized nature: in fact, in a fragment, the various doubly occupied and vacant σ MO's are usually mixed with the singly occupied MO arising from the breaking of a bond, with the consequence that the various σ MO's do not have anymore the correct orbital occupancies, i.e. 2 for a doubly occupied, 1 for a singly occupied and 0 for a vacant MO, required for the correct application of the PMO expressions. Therefore only the π canonical fragment MO's have correct orbital occupancies and, in fact, in the early applications of this procedure only this type of MO's has been used in the quantitative analysis of structural problems [2,3,7,8,9].

We have shown that this problem can be satisfactorily solved through the application of a localization procedure to the set of canonical fragment MO's [5]. The localization procedure is applied separately to the set of the occupied and to the set of the vacant fragment MO's. In our applications we have used the Boys' method of localization [26]. This new procedure, i.e. the WSW procedure followed by localization, provides for each fragment a new set of MO's that are still orthogonal and have now correct orbital occupancies. We have denoted this new set of MO's as fragment localized MO's.

The fragment localized MO's can now be used directly in the quantitative orbital analysis. This orbital representation allows to discuss the role of the various factors within a model which is very near to the qualitative description used by chemists. In fact, the various energy effects involving bond orbitals and lone pairs can be associated with steric repulsions, while those related to the interactions occurring between bond and antibond orbitals and between lone pairs and antibond orbitals to conjugative stabilizations.

2.2 Computation of the Energy Effects Associated with the Orbital Interactions

The interaction energy which obtains in the union of the component fragments is estimated on the basis of the following expression [5,27]:

$$\Delta E_T = \Sigma \Delta E^4_{ij} + \Sigma \Delta E^2_{ij} + E_c \qquad (2)$$

where

$$\Delta E^4_{ij} = 4 \frac{(\varepsilon_0 S^2_{ij} - H_{ij} S_{ij})}{(1 - S^2_{ij})} \qquad (3)$$

$$\Delta E_{ij}^{2} = 2\,\frac{(H_{ij} - S_{ij}\varepsilon_i)^2}{\varepsilon_i - \varepsilon_j} \tag{4}$$

Here ΔE_{ij}^{4} represents the destabilization energy associated with the interaction between two doubly occupied MO's, ϕ_i and ϕ_j, ΔE_{ij}^{2} the stabilization energy associated with the interaction between a doubly occupied MO, ϕ_i, and a vacant MO, ϕ_j, and E_c the Coulomb Energy associated with the interaction of the component fragments. In these expressions ε_i and ε_j denote the energies of the two unperturbed MO's, ϕ_i and ϕ_j, S_{ij} their overlap integral, H_{ij} their matrix element and ε_o the mean of the energies ε_i and ε_j. Eq. (3) is obtained by application of the variational method to the case of a two-orbital interaction problem [14], while eq. (4) is a well known second order perturbation expression [15,24].

The various terms appearing in these expressions are then computed using the results of the ab-initio SCF-MO computations [5].
In particular:
(i) the term E_c is simply computed according to the following expression:

$$E_c = \sum_{r<r'} \frac{q_r q_{r'}}{R_{rr'}} \tag{5}$$

which represents the Coulomb Energy associated with the interaction between the net atomic charges of the interacting fragments, with the net charges taken from the results of the Mulliken population analysis;
(ii) the matrix elements H_{ij} and the overlap integrals S_{ij} between the interacting fragments MO's are computed according to the following relations:

$$H = (C^{\circ})^{t}\, FC^{\circ} \tag{6}$$

$$S = (C^{\circ})^{t}\, \bar{S}C^{\circ} \tag{7}$$

where C° denotes the coefficient matrix of the fragment MO's, which can be localized or canonical, and F and $\bar{S}$ the Fock and the overlap matrices for the composite system over the atomic orbital basis;
(iii) the values of the energies of the fragment MO's, ε_i, are chosen to be, for a given representation, the diagonal elements of the H matrix in that representation.

Since this PMO treatment includes explicitly, besides the term E_c whose effect in the type of problems investigated here is usually small, only second order effects, it can provide reliable results only in those cases where the first order effects are not important. In structural problems the largest variations of the first order effects are caused by geometrical changes: therefore it can be expected that these effects are minimal when the analysis is performed for a same molecular species in the framework of a rigid model, where the geometrical changes are also minimal. Preliminary results seem to support this expectation [5,6] and suggest that the analysis of a structural problem can be performed with satisfactory results using a two-step procedure, which involves first, a quantitative PMO analysis in the rigid model, choosing as a starting point the less crowded optimized geometry, followed by a quantitative comparative analysis of the effects of geometry relaxation in the more crowded geometry. In this approach, the first order change which accompanies the geometry relaxation, is usually destabilizing. Even though, at the present stage, the computational procedure used here does not provide a quantitative estimate of this effect, information about its trend are provided by the following term:

$$IS = \sum_{i}^{occ} \eta_i \varepsilon_i \tag{8}$$

where ε_i denotes the energy of the various fragment localized MO's, η_i the corresponding occupation number (1 or 2) and the sum is taken over all the occupied fragment MO's.

This problem is more complicated when we compare different molecular species: in these cases, in fact, at the present stage we can discuss the first order changes only at a qualitative level.

2.3 Localized and Canonical Representations

The use of fragment localized MO's has the positive feature of providing a clear interpretation of the various energy effects, but has also the negative feature of neglecting in the quantitative analysis the intra-fragment non diagonal matrix elements, which have non zero values and whose effect becomes particularly significant when the fragment localized MO's are degenerate and centered on the same atom.

Another positive feature of the fragment localized MO's is their use as a starting point for obtaining a set of fragment canonical-type MO's which have correct orbital occupancies and intra-fragment non-diagonal matrix elements with zero value. These MO's can be obtained, in fact, from the diagonalization of the submatrices of the Fock matrix in the fragment localized representation, associated only with the doubly occupied and vacant localized MO's of each fragment. We refer to the latter orbital representation as to the canonical representation.

Obviously for each fragment these two sets of MO's are simply inter-related.

For illustrative purposes, let we consider the construction of the MO's of the H_3C- fragment arising from the dissection of ethane, in the localized and canonical representations. We compute first the fragment localized MO's with the procedure previously described. The orbital energies and orbital occupancies of the localized MO's of the H_3C- fragment in various conformations of ethane, computed at the STO-3G level, are listed in Table 1.

The valence localized MO's, which are shown in Fig. 1, involve, in order of increasing energy:

(i) three degenerate doubly occupied σ MO's localized along the three C-H bonds and bonding between C and H (the bond MO's σ_{CH});

(ii) a singly occupied σ MO localized along the C-C axis and pointing toward the adjacent fragment (σ_C);

(iii) three degenerate vacant σ MO's localized along the three C-H bonds and antibonding between C and H (the antibond MO's σ^*_{CH}).

We consider now the submatrix H° of H associated with the 1s carbon atomic orbital, the three σ_{CH} and the three σ^*_{CH} MO's, where H is the Fock matrix for ethane expressed over the fragment localized MO's: from the diagonalization of this submatrix we obtain canonical-type MO's for the H_3C- fragment with correct orbital occupancies, as shown in Table 1. Here, in fact, we have also listed the orbital energies and occupancies of these MO's, which are also illustrated in Fig. 1. When expressed in terms of atomic orbitals, they correspond to the already familiar orbitals of a methyl group, whose qualitative construction is clearly described in ref. [28]. However an equivalent description of these MO's can be given in terms of the fragment localized basis: this description is also illustrated in Fig. 1 and shows the type of relationship occurring between the two representations.

3. CONFORMATIONAL PREFERENCES IN CH_3-Y MOLECULES

In the following sections we discuss the factors which control the conformational preferences and the magnitude of the rotational barriers in the simple methyl derivatives CH_3-XH_3 (X = C, Si), CH_3-XH_2 (X = N, P) and CH_3-XH (X = O,S). For all molecular species, the computations have been performed for the eclipsed and staggered conformations (see Figure 2) at the STO-3G level. For the species involving a second-row atom (X = Si, P, S), the computations have also been performed at the STO-3G* level. The STO-3G computations have been carried out at the STO-3G fully optimized geometries and those at the STO-3G* level at STO-3G* partially optimized geometries: the values of the various geometrical parameters used in these computations are listed in Tables 2, 3 and 4. In each case, we have also performed a computation at the eclipsed geometry obtained through a rigid rotation around the C-X bond of the staggered optimized geometry

TABLE 1 - Energies (ε_i, a.u.) and Gross Populations (Q_i) of the Orbitals of the H_3C- Fragment in the Staggered and Eclipsed Conformations of Ethane.

		STAGGERED		ECLIPSED			
				Rigid Model		Optimized Model	
		ε_i	Q_i	ε_i	Q_i	ε_i	Q_i
Localized Representation	$1s_C$	-10.95258	1.99823	-10.95085	1.99823	-10.95173	1.99818
	σ_{CH}	-0.64492	2.00151	-0.62445	2.00231	-0.62401	2.00204
	σ_C	-0.51735	0.98582	-0.51657	0.98584	-0.50861	0.98693
	σ^*_{CH}	0.72141	0.00380	0.72260	0.00299	0.72291	0.00292
Canonical Representation	σ_{1s}	-10.97089	1.99838	-10.96915	1.99838	-10.97010	1.99829
	σ_{CH_3}	-0.81298	2.01560	-0.81223	2.01556	-0.81308	2.01457
	π_{CH_3}	-0.52175	1.99440	-0.52142	1.99562	-0.52029	1.99572
	$\pi^*_{CH_3}$	0.70508	0.00560	0.70650	0.00438	0.70554	0.00428
	$\sigma^*_{CH_3}$	0.75407	0.00020	0.75481	0.00022	0.75765	0.00021

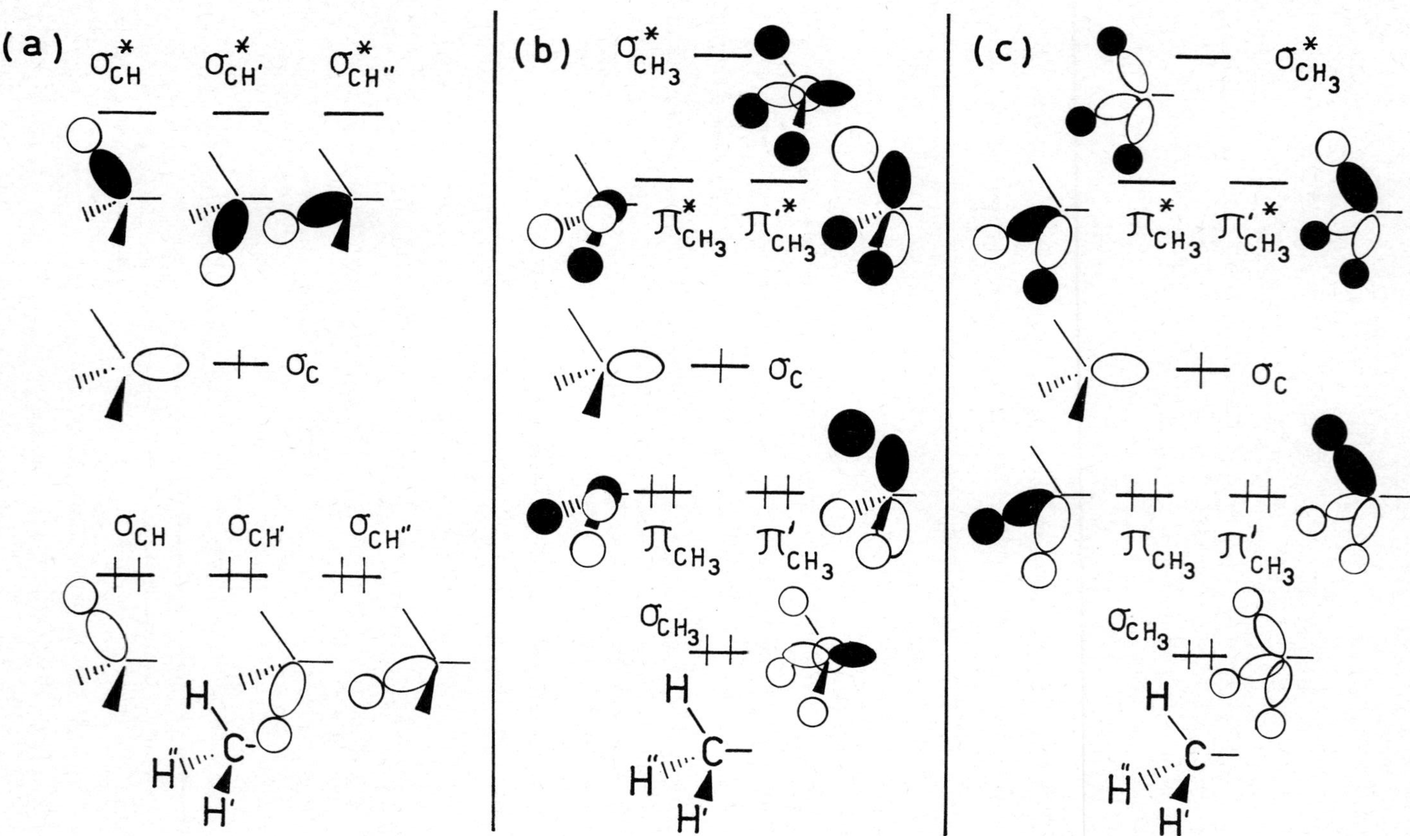

Fig. 1. Valence MO's of the H_3C- fragment: (a) localized representation; (b) canonical representation in terms of the atomic basis; (c) canonical representation in terms of the localized basis.

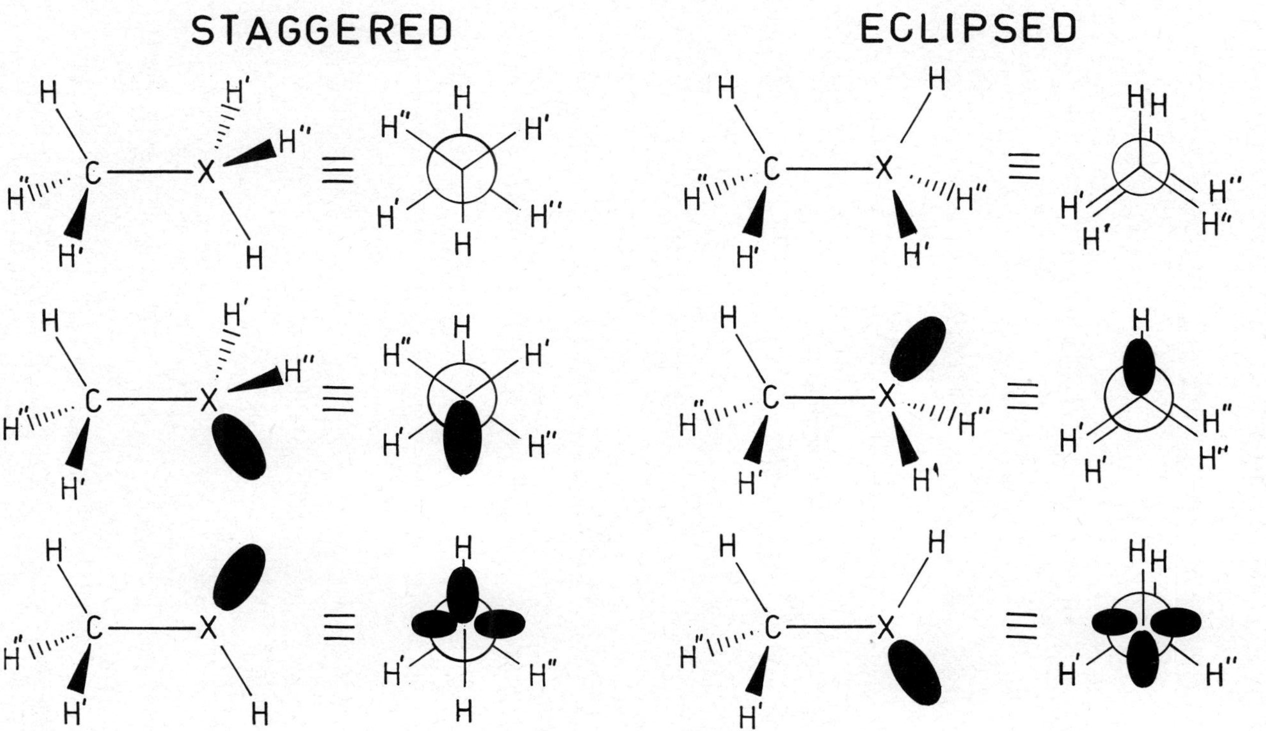

Fig. 2. Indices of the hydrogen atoms in the staggered and eclipsed conformations of the methyl derivatives $H_3C\text{-}XH_3$, $H_3C\text{-}XH_2$ and $H_3C\text{-}XH$.

TABLE 2 - Geometric Parameters[a] of the Staggered and Eclipsed Conformers of Ethane and Methylsilane, Computed with the STO-3G and STO-3G* Basis Sets.

	ETHANE[b]		METHYLSILANE			
	STO-3G Level		STO-3G Level[c]		STO-3G* Level	
	Staggered	Eclipsed	Staggered	Eclipsed	Staggered[d]	Eclipsed[e]
r(C-X)	1.538	1.548	1.861	1.868	1.855	1.868
r(C-H)	1.086	1.086	1.082	1.082	1.085	(1.082)
r(X-H)	1.086	1.086	1.423	1.423	1.425	(1.423)
∠ HCH'	108.2	107.8	107.6	107.4	107.8	107.5
∠ HXH'	108.2	107.8	108.8	108.5	110.1	108.7
∠ HCX	110.7	111.1	111.3	111.5	111.1	111.4
∠ HXC	110.7	111.1	110.2	110.4	108.9	110.2

a. Bond lengths in Å and bond angles in degrees. The indices of the hydrogen atoms are specified in Fig. 2.

b. From ref. 13.

c. From ref. 29.

d. The values of the parameters r(C-Si), r(Si-H), HSiH' and HSiC are taken from ref. 12; the remaining parameters have been optimized with the axial iteration tecnique implemented in Gaussian 76.

e. Parameters optimized with the axial iteration tecnique (values in brakets kept fixed at the values optimized at the STO-3G level).

TABLE 3 - Geometric Parameters[a] of the Staggered and Eclipsed Conformers of Methylamine and Methylphosphine, Computed with the STO-3G and STO-3G* Basis Sets.

	METHYLAMINE[b]		METHYLPHOSPHINE			
	STO-3G Level		STO-3G Level		STO-3G* Level	
	Staggered	Eclipsed	Staggered[c]	Eclipsed[d]	Staggered[c]	Eclipsed[e]
r(C-X)	1.486	1.493	1.841	1.850	1.830	1.841
r(C-H)	1.093	1.091	1.084	1.083	1.085	(1.083)
r(C-H')	1.089	1.089	1.083	1.083	1.086	(1.083)
r(X-H')	1.033	1.031	1.381	1.381	1.379	(1.381)
∠ HCH'	108.3	107.8	107.9	107.4	108.0	(107.4)
∠ H'CH"	108.2	108.0	107.4	107.7	107.3	(107.7)
∠ H'XH"	104.4	104.5	93.7	93.3	92.8	93.4
∠ HCX	113.7	110.0	113.1	110.9	113.4	(110.9)
∠ H'CX	109.1	111.6	110.2	111.7	110.0	(111.7)
∠ H'XC	107.3	108.2	96.1	96.7	95.7	96.4

a. Bond lengths in Å and bond angles in degrees. The indices of the hydrogen atoms are specified in Fig. 2.

b. From ref. 13.

c. From ref. 32.

d. From ref. 29.

e. Parameters optimized with the axial iteration tecnique (values in brakets kept fixed at the values optimized at the STO-3G level).

TABLE 4 - Geometrical Parameters[a] of the Staggered and Eclipsed Conformers of Methanol and Methanethiol, Computed with the STO-3G and STO-3G* Basis Sets.

	METHANOL[b]		METHANETHIOL			
	STO-3G Level		STO-3G Level		STO-3G* Level	
	Staggered	Eclipsed	Staggered[c]	Eclipsed[d]	Staggered[c]	Eclipsed[e]
r(C-X)	1.433	1.439	1.798	1.804	1.782	1.791
r(C-H)	1.092	1.092	1.085	1.084	1.088	(1.084)
r(C-H')	1.095	1.094	1.087	1.086	1.089	(1.086)
r(X-H)	0.991	0.989	1.331	1.329	1.320	1.319
∠ HCH'	108.1	107.9	107.8	107.9	107.6	107.9
∠ H'CH"	108.1	107.9	108.2	107.6	108.3	(107.6)
∠ HCX	107.7	113.0	108.6	111.3	108.2	111.4
∠ H'CX	112.4	110.0	112.2	111.0	112.5	(111.0)
∠ HXC	103.8	104.6	95.4	95.9	95.1	95.7

a. Bond lengths in Å and bond angles in degrees. The indices of the hydrogen atoms are specified in Fig. 2.

b. From ref. 13.

c. From ref. 32.

d. From ref. 29.

e. Parameters optimized with the axial iteration tecnique (values in brakets kept fixed at the values optimized at the STO-3G level).

TABLE 5

Total Energies (a.u.) of the Staggered and Eclipsed Conformations of Various Methyl Derivatives Computed at the STO-3G and STO-3G* Levels.

	STAGGERED	ECLIPSED	
		Rigid Geometry	Optimized Geometry
STO-3G Level			
CH_3-CH_3	-78.30618	-78.30142	-78.30160
CH_3-NH_2	-94.03286	-94.02730	-94.02844
CH_3-OH	-113.54919	-113.54430	-113.54598
CH_3-SiH_3	-326.51180	-326.50969	-326.50973
CH_3-PH_2	-377.22385	-377.22037	-377.22081
CH_3-SH	-432.89607	-432.89338	-432.89375
STO-3G* Level			
CH_3-SiH_3	-326.57125	-326.56851	-326.56888
CH_3-PH_2	-377.27661	-377.27243	-377.27294
CH_3-SH	-432.92985	-432.92669	-432.92716

TABLE 6

Computed and Experimental Rotational Barriers (kcal/mol) of Various Methyl Derivatives.

	COMPUTED		EXPERIMENTAL
	Rigid Model	Optimized Model	
STO-3G Level			
CH_3-CH_3	2.99	2.87	2.928[a]
CH_3-NH_2	3.49	2.77	1.98[b]
CH_3-OH	3.07	2.01	1.07[c]
CH_3-SiH_3	1.32	1.30	1.665[d]
CH_3-PH_2	2.18	1.91	1.959[e]
CH_3-SH	1.69	1.45	1.27[f]
STO-3G* Level			
CH_3-SiH_3	1.72	1.49	1.665
CH_3-PH_2	2.62	2.30	1.959
CH_3-SH	1.98	1.69	1.27

a. See ref. 33; b. see ref. 34; c. see ref. 35; d. see ref. 36; e. see ref. 37; f. see ref. 38.

(eclipsed rigid geometry).

The total energy values obtained in this way are listed in Table 5. These results show that in all cases the preferred conformation is the staggered in agreement with the experimental findings. In Table 6 we have listed the various computed rotational barriers which in these molecules are the differences between the total energy values of the eclipsed and staggered conformations. For each molecular species we have listed two values for the barrier, one associated with the eclipsed rigid geometry (rigid model) and the other with the eclipsed optimized geometry (optimized model). The computed barriers agree moderately well with the experimental values also listed in Table 6. In all cases the SCF results have been obtained with the GAUSSIAN 76 series of programs [39].

In order to perform the quantitative analyses, these molecules have been dissected into the fragments H_3C- and $-XH_3$, H_3C- and $-XH_2$, H_3C- and -XH respectively. The relevant MO's of the H_3C- fragment in ethane have already been described (see Figure 1). Similar MO's has also the H_3C- fragment in methylsilane, while in the other molecules the MO's of this fragment show a reduced degeneracy: in particular, in the localized representation, only two of the three σ_{CH} MO's and only two of the three σ^*_{CH} MO's are degenerate, while in the canonical representation the two π and the two π^* MO's are not degenerate anymore. The MO's of the $-SiH_3$ fragment in methylsilane are similar to those of the H_3C- fragment in the same molecule, while those of the $-XH_2$ (X = N, P) and of the -XH (X = O, S) fragments are illustrated in Figures 3 and 4. In these Figures, the positioning of the singly occupied MO arising from the breaking of the C-X bond (σ_X) seems to be anomalous and is probably an artifact of the procedure used for the construction of the fragment MO's. However this problem should not affect the results presented here, since we do not take into examination the orbital interactions involving the singly occupied MO. In order to obtain the localized MO's of the $-XH_2$ fragment the localization procedure has been applied to all the fragment MO's obtained from the WSW procedure, while in the case of the -XH fragment the localization procedure has been applied only to the σ-type fragment MO's. In these analysis we focus our attention just upon the non-bonded interactions, i.e. upon the interactions between the valence doubly occupied MO's of the two fragments and between the valence doubly occupied MO's of one fragment and the vacant MO's of the other fragment.

As previously pointed out, the analyses presented here are based on the use of eq. (2), which represents the interaction energy which obtains in the union of the two interacting fragments. Therefore, according to this equation, in addition to the energy effects associated with the various orbital non-bonded interactions, we have estimated also the Coulomb Energy E_C between the component fragments. The analysis of the E_C values, which are listed in Tables 7, 9, 11,

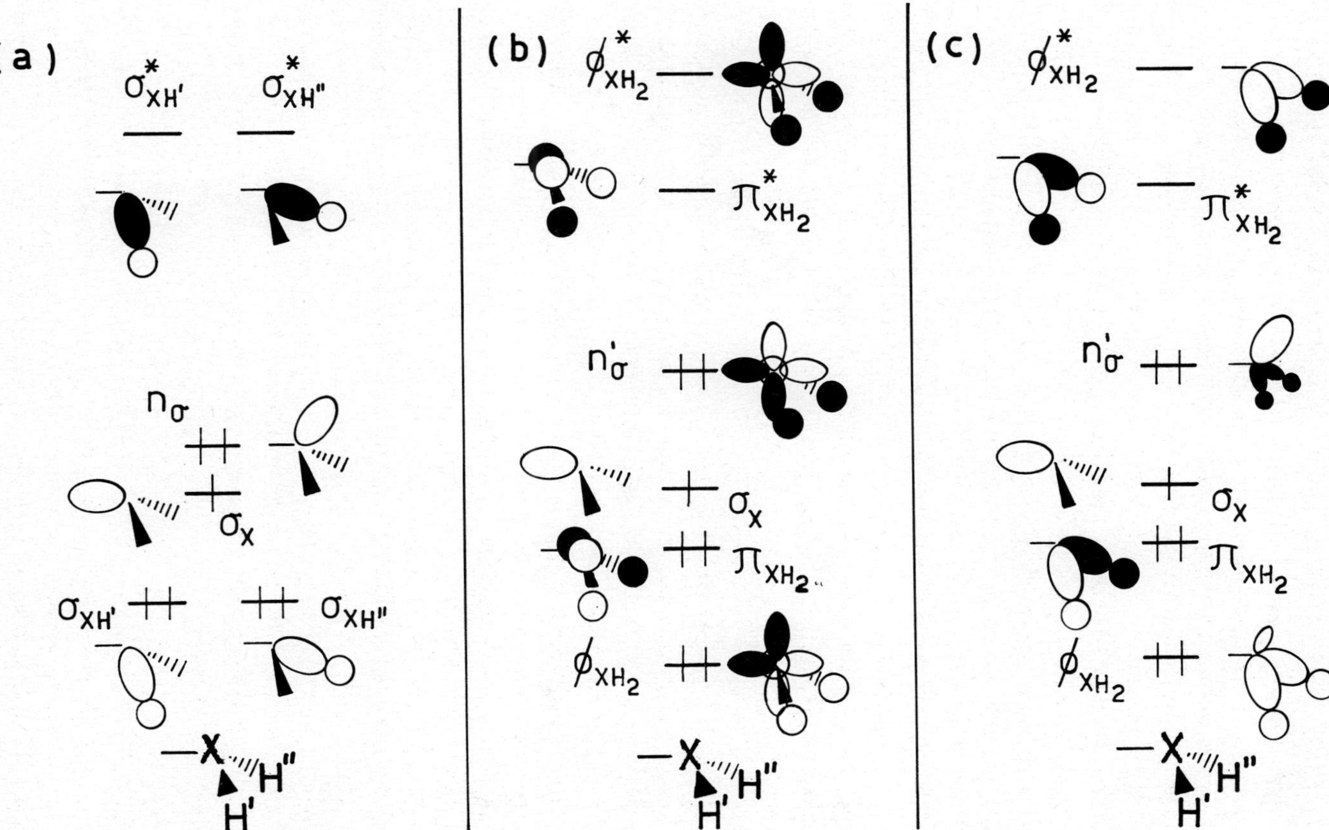

Fig. 3. Valence MO's of the H_2X- fragment: (a) localized representation; (b) canonical representation in terms of the atomic basis; (c) canonical representation in terms of the localized basis.

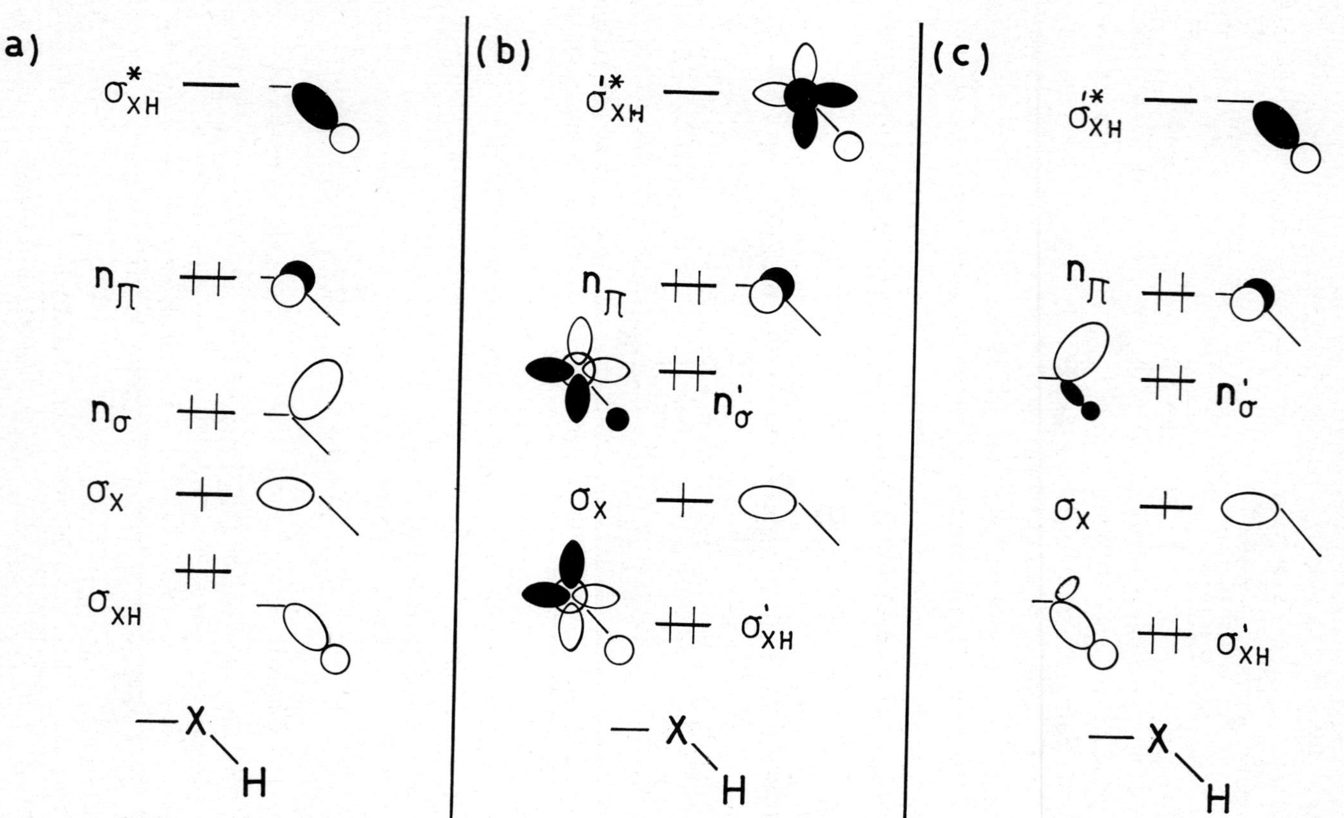

Fig. 4. Valence MO's of the HX- fragment: (a) localized representation; (b) canonical representation in terms of the atomic basis; (c) canonical representation in terms of the localized basis.

14, 17 and 20, reveals that:
(i) In the first-row molecules, the term E_c is small and almost identical in the two conformations. Its value is destabilizing in ethane and stabilizing in methylamine and methanol, with the stabilization increasing with the increase of the electronegativity of the atom X.
(ii) In the second-row molecules the term E_c is stabilizing and the stabilization decreases with the increase of the electronegativity of the atom X. In all cases the absolute value of this term is significantly smaller at the STO-3G* than at the STO-3G level. In the rigid model the E_c values are almost identical in the two conformations, while its effect becomes more significant in the optimized model. The larger differential contribution is found for methylsilane, where it favors the staggered conformation by ~0.2 kcal/mol.

Therefore, the term E_c does not play a significant role in the structural problems under examination. Consequently, in the following sections we limit our discussion to the analysis of the effects of the non-bonded orbital interactions.

3.1 Conformational Isomerism in CH_3-XH_3 Molecules

3.1.1 Ethane.

At the STO-3G level, the staggered conformer with optimized geometry has been found to be more stable by 2.99 kcal/mol than the eclipsed with rigid geometry and more stable by 2.87 kcal/mol than the eclipsed with optimized geometry. Both these energy differences agree well with the value of 2.93 kcal/mol experimentally found for the rotational barrier [33].

We examine now the energy effects associated with the non-bonded interactions occurring between the valence MO's of the two H_3C- fragments in the localized and canonical representations. These interactions are depicted in Figure 5, while the values of the energy effects are listed in Tables 7 and 8.

We discuss first the results obtained with the localized representation (see Table 7), in the framework of the rigid model. The following points are of interest:
(i) The overall energy effect associated with the non-bonded interactions ($\Sigma\Delta E$) is destabilizing and less destabilizing in the staggered geometry. Therefore the trend of the overall energy effect parallels the total energy behaviour. Furthermore the difference between the $\Sigma\Delta E$ values computed for the eclipsed and staggered conformations, 2.31 kcal/mol, agrees well with the corresponding total energy difference of 2.99 kcal/mol.
(ii) Both the overall destabilizing ($\Sigma\Delta E^4$) and stabilizing ($\Sigma\Delta E^2$) effects favor the staggered conformation. However the preferential stabilization is mainly determined by the destabilizing effect as indicated by the contributions of

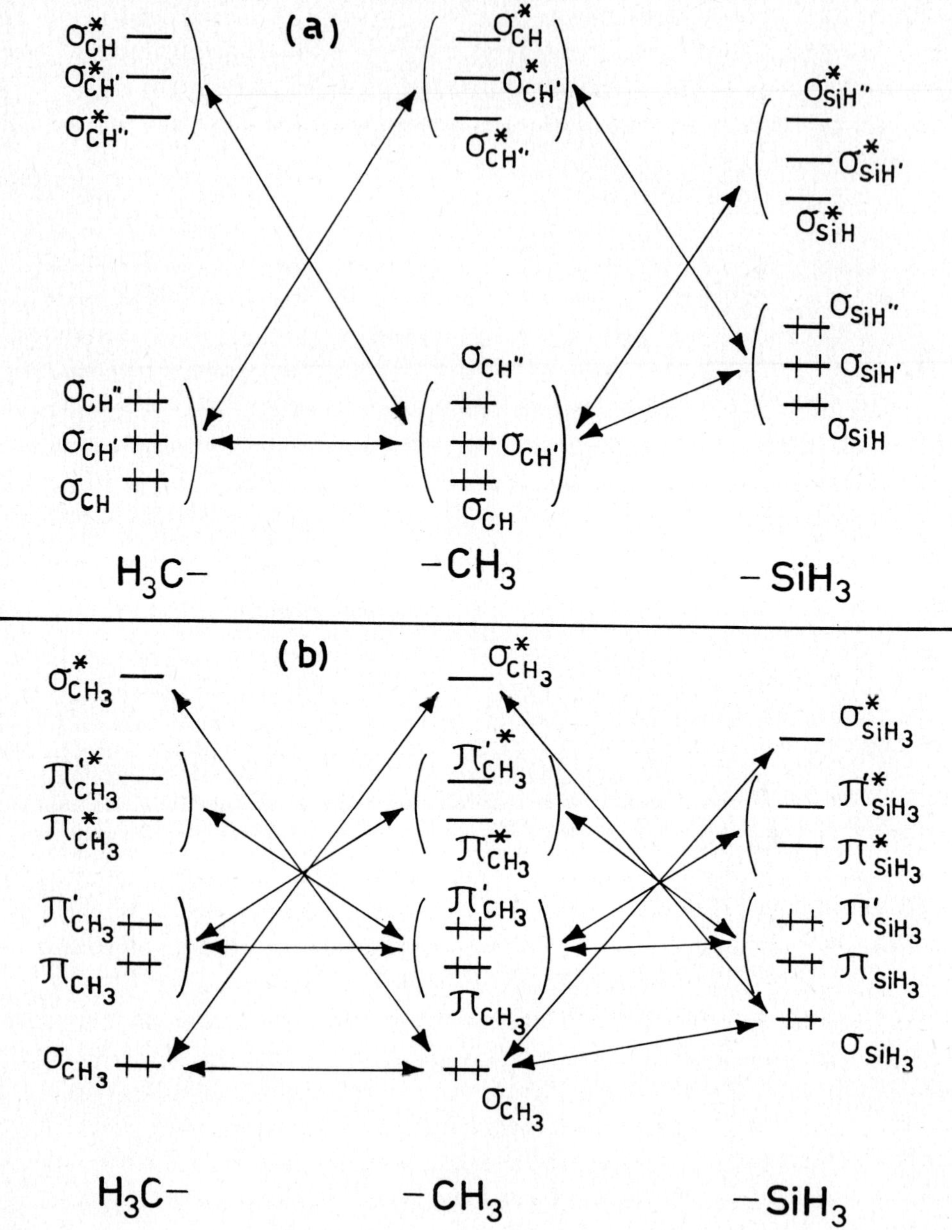

Fig. 5. Interaction diagram describing the non-bonded interactions between the localized MO's (a) and the canonical MO's (b) of the H_3C- and $-XH_3$ fragments.

TABLE 7

Stabilization (ΔE^2_{ij}) and Destabilization (ΔE^4_{ij}) Energies (kcal/mol) Associated with the Non-Bonded Interactions Occurring in the Staggered and Eclipsed Conformations of Ethane, Computed in the Localized Representation.

	STAGGERED	ECLIPSED	
		Rigid Geometry	Optimized Geometry
$\Delta E^4_{\sigma_{CH},\sigma_{CH}}$	3.61	8.50	7.98
$\Delta E^4_{\sigma_{CH},\sigma_{CH'}}$	2.64	0.53	0.52
$\Delta E^2_{\sigma_{CH},\sigma^*_{CH}}$	-0.29	-0.14	-0.13
$\Delta E^2_{\sigma_{CH},\sigma^*_{CH'}}$	-0.03	-0.08	-0.08
$\Sigma\Delta E^4$	26.67	28.68	27.06
$\Sigma\Delta E^2$	-2.10	-1.80	-1.74
$\Sigma\Delta E$	24.57	26.88	25.32
E_C [a]	1.05	1.09	1.07

a. Coulomb Energy (kcal/mol) computed according to eq. 5.

$\Sigma\Delta E^4$ and $\Sigma\Delta E^2$ to the $\Sigma\Delta E$ difference, which are 2.01 kcal/mol for $\Sigma\Delta E^4$ and 0.30 kcal/mol for $\Sigma\Delta E^2$.

(iii) Since the repulsions between bond MO's can be associated with the steric repulsions and the stabilizing effects arising from the interactions between bond and antibond MO's with the conjugative stabilizations, the magnitude and trend of $\Sigma\Delta E^4$ and $\Sigma\Delta E^2$ provide an estimate of the relative importance of the steric and conjugative effects. Therefore it follows that in each conformation the steric repulsions are significantly larger than the conjugative stabilizations, so that the overall energy effect associated with the non-bonded interactions is destabilizing. The steric effects play also a major role in determining the conformational preference and the magnitude of the rotational barrier.

(iv) The trends of all the various orbital interactions are overlap controlled. Consequently in plane bond MO's repulsions and conjugative stabilizations are larger than those associated with out of plane MO's.

The comparison of these results with those obtained at the eclipsed optimized geometry shows that in the eclipsed conformation the geometry tends to change in order to reduce the large repulsions between mutually eclipsing bond MO's. This result is obtained mainly through an increase of the C-C bond length (see

TABLE 8

Stabilization (ΔE^2_{ij}) and Destabilization (ΔE^4_{ij}) Energies (kcal/mol) Associated with the Non-Bonded Interactions Occurring in the Staggered and Eclipsed Conformations of Ethane, Computed in the Canonical Representation.

	STAGGERED	ECLIPSED	
		Rigid Geometry	Optimized Geometry
$\Delta E^4_{\sigma,\sigma}$	1.23	1.27	1.14
$\Delta E^2_{\sigma,\sigma^*}$	0.00	0.00	0.00
$\Sigma\Delta E(\sigma)$	1.23	1.27	1.14
$\Delta E^4_{\pi,\pi}$	16.16	17.46	16.56
$\Delta E^2_{\pi,\pi^*}$	-1.00	-0.82	-0.80
$\Sigma\Delta E^4(\pi)$	32.32	34.92	33.12
$\Sigma\Delta E^2(\pi)$	-4.00	-3.28	-3.20
$\Sigma\Delta E\ (\pi)$	28.32	31.64	29.92
$\Sigma\Delta E^4$	33.55	36.19	34.26
$\Sigma\Delta E^2$	-4.00	-3.28	-3.20
$\Sigma\Delta E$	29.55	32.91	31.06

Table 2). Consequently at the eclipsed optimized geometry the overall energy effect is less destabilizing. However this reduction has been obtained at the expenses of a destabilization of the various fragment MO's and, in particular, of the singly occupied σ_C MO, which accompanies the geometry relaxation, as indicated by the values of the index IS, which are -26.1649 a.u. at the eclipsed rigid geometry and -26.1561 a.u. at the eclipsed optimized geometry. The result is that the overall gain in total energy obtained in the geometry relaxation is very small (~0.1 kcal/mol). Therefore, in the optimized model, while the contribution of the stabilizing effects remains almost unaltered, that of the destabilizing effects is significantly reduced, and in this model the conformational preference is determined, in addition to the conjugative and steric effects which have now similar weight, also by a significant first order effect associated with the destabilization of the fragment MO's which accompanies the geometry relaxation in the eclipsed conformation.

We consider now the results obtained in the canonical representation (see Table 8), which provide, in particular, information about the energy effects associated with hyperconjugation. The $\Sigma\Delta E$ values computed with this orbital representation show the same trend of those computed with the localized representation; however the $\Sigma\Delta E$ values are now larger and the $\Sigma\Delta E$ difference is in better agreement with the total energy difference.

In this representation the $\Sigma\Delta E$ value can be decomposed in the contributions $\Sigma\Delta E(\sigma)$ and $\Sigma\Delta E(\pi)$ associated with the σ-type and π-type orbital interactions. It is found that the difference between the $\Sigma\Delta E(\sigma)$ values for the eclipsed and staggered conformations is negligible. Therefore the conformational preference and the magnitude of the rotational barrier are completely determined by the energy effects related to the π orbital interactions, i.e. by hyperconjugative effects. The energy effect associated with the hyperconjugative interactions $\Sigma\Delta E(\pi)$, can be decomposed in a destabilizing contribution, $\Sigma\Delta E^4(\pi)$, arising from the interactions between the doubly occupied π MO's, and in a stabilizing contribution, $\Sigma\Delta E^2(\pi)$, arising from the π-π^* interactions. The destabilizing contribution, which is just a steric effect since this term is associated with π MO's that are linear combinations of bond MO's, is significantly larger than the stabilizing contribution, so that the overall hyperconjugative energy effect is destabilizing. In the rigid model the contributions of $\Sigma\Delta E^4(\pi)$ and $\Sigma\Delta E^2(\pi)$ to the difference between the $\Sigma\Delta E(\pi)$ values computed for the eclipsed and staggered conformations are 2.60 and 0.72 kcal/mol respectively. These values agree well with the contributions of the steric and conjugative effects obtained with the localized representation and show a clear correspondence between the two representations, confirming that in a rigid model the conformational preference and the rotational barrier are determined by conjugative and steric contributions, with the latter largely dominant.

The values of the various energy effects obtained with the canonical representation support completely the conclusions reached on the basis of a qualitative analysis [14] and agree also with the results of preliminary quantitative analysis based only on π orbital interactions [2,9].

3.1.2 Methylsilane.

The staggered conformer with optimized geometry has been found to be more stable than the eclipsed with rigid geometry by 1.32 kcal/mol at the STO-3G level and by 1.72 kcal/mol at the STO-3G* level, and more stable than the eclipsed with optimized geometry by 1.30 kcal/mol at the STO-3G level and by 1.49 kcal/mol at the STO-3G* level. All these energy differences agree well with the value of 1.665 kcal/mol experimentally found for the rotational barrier [36].

The energy effects associated with the non-bonded interactions in the localized representation have been computed at the STO-3G and STO-3G* levels and are listed in Table 9. We proceed to discuss the results in the following order:

TABLE 9

Stabilization (ΔE^2_{ij}) and Destabilization (ΔE^4_{ij}) Energies (kcal/mol) Associated with the Non-Bonded Interactions Occurring in the Staggered and Eclipsed Conformations of Methylsilane, Computed in the Localized Representation.

	STO-3G LEVEL			STO-3G* LEVEL		
	STAGGERED	ECLIPSED		STAGGERED	ECLIPSED	
		Rigid Geometry	Optimized Geometry		Rigid Geometry	Optimized Geometry
$\Delta E^4_{\sigma_{CH},\sigma_{SiH}}$	1.78	2.60	2.50	1.81	2.73	2.44
$\Delta E^4_{\sigma_{CH},\sigma_{SiH'}}$	0.67	0.41	0.40	0.70	0.41	0.40
$\Delta E^2_{\sigma_{CH},\sigma^*_{SiH}}$	-0.27	-0.12	-0.12	-0.27	-0.13	-0.11
$\Delta E^2_{\sigma_{CH},\sigma^*_{SiH'}}$	-0.02	-0.08	-0.08	-0.03	-0.08	-0.09
$\Delta E^2_{\sigma_{SiH},\sigma^*_{CH}}$	-0.11	-0.02	-0.02	-0.10	-0.01	0.00
$\Delta E^2_{\sigma_{SiH},\sigma^*_{CH'}}$	0.00	-0.03	-0.03	0.00	-0.03	-0.03
$\Sigma\Delta E^4$	9.36	10.26	9.90	9.63	10.65	9.72
$\Sigma\Delta E^2$	-1.26	-1.08	-1.08	-1.29	-1.08	-1.05
$\Sigma\Delta E$	8.10	9.18	8.82	8.34	9.57	8.67
$\Sigma\Delta E^2_{3d}$	-	-	-	-10.99	-10.98	-10.35
$\Sigma\Delta E_T$	-	-	-	-2.65	-1.41	-1.68
E^a_C	-17.73	-17.79	-17.51	-11.92	-11.99	-11.65

a. Coulomb Energy (kcal/mol) computed according to eq. 5.

(a) Rigid Model. At the STO-3G level, in each conformation the steric repulsions ($\Sigma\Delta E^4$) are significantly larger than the conjugative stabilizations ($\Sigma\Delta E^2$) and both effects favor a preferential stabilization of the staggered conformer. Consequently the overall energy effect $\Sigma\Delta E$ is destabilizing and less destabilizing in the staggered geometry in agreement with the total energy behaviour. The difference between the $\Sigma\Delta E$ values computed for the eclipsed and staggered conformations is 1.08 kcal/mol and agrees well with the corresponding total energy difference of 1.32 kcal/mol. The contributions of $\Sigma\Delta E^4$ and $\Sigma\Delta E^2$ to the $\Sigma\Delta E$ difference are 0.90 and 0.18 kcal/mol respectively. Therefore also in this case the steric effects play the major role in determining the conformational preference and the magnitude of the rotational barrier.

At the STO-3G* level, the overall energy effect (denoted at this level $\Sigma\Delta E_T$) becomes stabilizing; however the difference between the values of $\Sigma\Delta E_T$ computed for the eclipsed and staggered conformations is 1.24 kcal/mol and therefore almost identical with the value of 1.08 kcal/mol computed at the STO-3G level. Therefore, the inclusion of the silicon 3d orbitals has just a significant stabilizing effect which is of the same order of magnitude in the two conformations (see the $\Sigma\Delta E^2_{3d}$ values in Table 9) and does not modify the magnitude and trend of the energy effects associated with the interactions of the other orbitals, as shown by the values of $\Sigma\Delta E^4$, $\Sigma\Delta E^2$ and $\Sigma\Delta E$ computed at the STO-3G* level, which are almost identical to the values computed at the STO-3G level.

(b) Optimized Model. At the STO-3G level, the effects associated with the geometry relaxation are very similar to those already pointed out in the case of ethane. The geometry of the eclipsed conformer, in fact, tends to change in order to reduce the repulsions between the mutually eclipsing bond MO's. This result is obtained mainly through a lenthening of the C-Si bond (see Table 2). Consequently the overall energy effect is less destabilizing at the eclipsed optimized geometry than at the eclipsed rigid geometry. However this reduction is almost completely counterbalanced by the destabilization of the various fragment MO's which accompanies the geometry relaxation, as indicated by the overall gain in total energy which is just 0.03 kcal/mol. Therefore also in methylsilane, the relaxation process of the eclipsed geometry is associated with a decrease in the contribution of the steric effects and a related increase in the contribution of the first order effects, while the contribution of the conjugative effects remains unaltered.

At the STO-3G* level, in addition to the previous effects we have to consider also that of the 3d orbitals, whose stabilizing contribution decreases with the geometry relaxation as a consequence of the increase in the C-Si bond length. This energy effects operates in the same direction of the first order effects and both tend to counterbalance the decrease of the steric effects which accompanies the geometry relaxation, so that the gain in total energy is again very small (~0.2 kcal/mol).

Therefore, while in the rigid model the conformational preference can be satisfactorily discussed only in terms of conjugative and steric effects, in the optimized model we have to consider also the first order effects and, at the STO-3G* level, the effect of the silicon 3d orbitals. All these effects favor the staggered geometry.

We proceed now to discuss the results obtained in the canonical representation, which are listed in Table 10. Since the purpose of this type of analysis here is to establish the relative importance of the σ and π contributions and since the silicon 3d orbitals have only a minor effect upon the conformational preference, the analysis has been performed only at the STO-3G level.

The $\Sigma\Delta E$ values agree very well with those computed with the localized representation; the only difference is that the values computed with the canonical representation are slightly larger, particularly those for the eclipsed geometries, so that also the $\Sigma\Delta E$ differences are slightly larger. It is also found that in methylsilane the energy effects associated with the σ- type orbital interactions are negligible (see the $\Sigma\Delta E(\sigma)$ values in Table 10) and the conformational preference is controlled only by the hyperconjugative interactions. The contributions of the destabilizing $\Sigma\Delta E^4(\pi)$ and stabilizing $\Sigma\Delta E^2(\pi)$ components to the difference between the $\Sigma\Delta E(\pi)$ values computed for the eclipsed and staggered conformations are 1.10 and 0.42 kcal/mol respectively. These values agree well with the contributions of the steric and conjugative effects obtained in the localized representation and confirm that in a rigid model the conformational preference and the rotational barrier are determined by conjugative and steric contributions, with the latter largely dominant.

3.1.3 Effects of the Replacement of Silicon for Carbon.

The experimental results show that the replacement in a H_3C-XH_3 molecule of a first row atom (X = C) with a second row atom (X = Si) has the effect of reducing significantly the rotational barrier (from 2.928 to 1.665 kcal/mol respectively). This trend is well reproduced at the computational levels used here. Since we have already pointed out that the inclusion of the silicon 3d orbitals has only a minor effect upon the magnitude of the rotational barrier, the comparative analysis can be performed at the STO-3G level.

The previous discussion has shown that in the rigid model, in both molecular species, the magnitude of the rotational barrier is controlled by conjugative and steric effects, with the latter largely dominant. In this model, also the decrease of the rotational barrier in going from ethane to methylsilane is mainly caused by the steric contribution which varies from 2.01 kcal/mol in ethane

TABLE 10

Stabilization (ΔE^2_{ij}) and Destabilization (ΔE^4_{ij}) Energies (kcal/mol) Associated with the Non-Bonded Interactions Occurring in the Staggered and Eclipsed Conformations of Methylsilane, Computed in the Canonical Representation.

	STAGGERED	ECLIPSED	
		Rigid Geometry	Optimized Geometry
$\Delta E^4_{\sigma_{CH_3},\sigma_{SiH_3}}$	0.07	0.08	0.07
$\Delta E^2_{\sigma_{CH_3},\sigma^*_{SiH_3}}$	-0.01	-0.01	-0.01
$\Delta E^2_{\sigma_{SiH_3},\sigma^*_{CH_3}}$	-0.02	-0.02	-0.02
$\Sigma\Delta E(\sigma)$	0.04	0.05	0.04
$\Delta E^4_{\pi_{CH_3},\pi_{SiH_3}}$	5.72	6.27	6.06
$\Delta E^2_{\pi_{CH_3},\pi^*_{SiH_3}}$	-1.20	-1.07	-1.05
$\Delta E^2_{\pi_{SiH_3}\pi^*_{CH_3}}$	-0.23	-0.15	-0.15
$\Sigma\Delta E^4(\pi)$	11.44	12.54	12.12
$\Sigma\Delta E^2(\pi)$	-2.86	-2.44	-2.40
$\Sigma\Delta E\ (\pi)$	8.58	10.01	9.72
$\Sigma\Delta E^4$	11.51	12.62	12.19
$\Sigma\Delta E^2$	-2.89	-2.47	-2.43
$\Sigma\Delta E$	8.62	10.15	9.76

to 0.90 kcal/mol in methylsilane. This decrease is a consequence of the lengthening of the C-X bond which accompanies the replacement of a first row with a second row atom and which has the effect of increasing the distance between the bond MO's and of decreasing the repulsion between these MO's and also the difference between the repulsions computed at the eclipsed and staggered conformations. While in fact the repulsions between two mutually staggering and between two mutually eclipsing bond MO's in ethane are 3.61 and 8.50 kcal/mol, respectively, the corresponding values in methylsilane become 1.78 and 2.60 kcal/mol.

In the optimized model, however, the contributions of the steric effects to the two barriers become of similar order of magnitude, and the main cause of the barrier decrease seems to be a smaller first order effect which accompanies the geometry relaxation of the eclipsed conformer in methylsilane.

3.2 Conformational Isomerism in CH_3-XH_2 Molecules

3.2.1 Methylamine

At the STO-3G level the staggered conformer with optimized geometry has been found to be more stable by 3.49 kcal/mol than the eclipsed with rigid geometry and more stable by 2.77 kcal/mol than the eclipsed with optimized geometry. These values have to be compared with the value of 1.98 kcal/mol experimentally found for the rotational barrier [34]: therefore, in this molecule, while the optimized model provides a satisfactory description of the rotational barrier, the rigid model is much less accurate.

The energy effects associated with the non-bonded interactions occurring between the valence MO's of the H_3C- and -NH_2 fragments in the localized representation are listed in Table 11, while the related interaction diagram is shown in Figure 6. We discuss first the results obtained in the framework of the rigid model. The following points are of interest:

(i) In each conformation, the steric repulsions ($\Sigma\Delta E^4$) are larger than the conjugative stabilizations ($\Sigma\Delta E^2$), so that the overall energy effect associated with the non-bonded interactions ($\Sigma\Delta E$) is destabilizing. Furthermore, both $\Sigma\Delta E^4$ and $\Sigma\Delta E^2$ favor the staggered geometry, with the consequence that $\Sigma\Delta E$ is less destabilizing in the staggered than in the eclipsed geometry. Therefore the trend of $\Sigma\Delta E$ parallels also in this case the total energy behaviour.

(ii) The difference between the $\Sigma\Delta E$ values computed for the eclipsed and staggered geometries is 2.88 kcal/mol and agrees well with the value of 3.49 kcal/mol computed in terms of the total energies. The contributions of $\Sigma\Delta E^4$ and $\Sigma\Delta E^2$ to the $\Sigma\Delta E$ difference are 2.40 and 0.48 kcal/mol respectively. Therefore the steric repulsions play also in this case the major role in determining the conformational preference and the magnitude of the rotational barrier.

(iii) The results of the quantitative analysis suggest that the dominant factor is the repulsion between the σ_{CH} and the σ_{NH} bond MO's: these interactions become, in fact, very destabilizing when these bonds are mutually eclipsing.

The comparison between the results obtained at the rigid and optimized eclipsed geometries shows that in the eclipsed conformation the geometry tends to change in order to reduce the large steric repulsions. In fact the energy effect associated with the interaction between mutually eclipsing σ_{CH} and σ_{NH} bond MO's varies from 11.24 kcal/mol in the rigid geometry to 9.29 kcal/mol in the optimized geometry. This result is obtained mainly through a lengthening of the C-N bond and changes of the bond angles (see Table 3), which reduce the overlap between the mutually eclipsing bond MO's. Consequently the overall energy effect associated with the non-bonded interactions shows a significant decrease at the optimized geometry (from 31.29 to 29.03 kcal/mol) arising almost completely from a decrease of the steric component $\Sigma\Delta E^4$. However this energy

TABLE 11

Stabilization (ΔE^2_{ij}) and Destabilization (ΔE^4_{ij}) Energies (kcal/mol) Associated with the Non-Bonded Interactions Occurring in the Staggered and Eclipsed Conformations of Methylamine, Computed in the Localized Representation.

	STAGGERED	ECLIPSED	
		Rigid Geometry	Optimized Geometry
$\Delta E^4_{\sigma_{NH'},\sigma_{CH}}$	3.06	0.85	0.61
$\Delta E^4_{\sigma_{NH'},\sigma_{CH'}}$	3.73	11.24	9.29
$\Delta E^4_{\sigma_{NH'},\sigma_{CH''}}$	3.15	0.21	0.41
$\Delta E^4_{n_\sigma,\sigma_{CH}}$	7.06	7.96	9.94
$\Delta E^4_{n_\sigma,\sigma_{CH'}}$	3.02	1.41	1.30
$\Delta E^2_{\sigma_{NH'},\sigma^*_{CH}}$	-0.04	-0.18	-0.11
$\Delta E^2_{\sigma_{NH'},\sigma^*_{CH'}}$	-0.31	-0.23	-0.19
$\Delta E^2_{\sigma_{NH'},\sigma^*_{CH''}}$	-0.04	-0.05	-0.08
$\Delta E^2_{n_\sigma,\sigma^*_{CH}}$	-1.97	-1.21	-1.29
$\Delta E^2_{n_\sigma,\sigma^*_{CH'}}$	-0.30	-0.43	-0.50
$\Delta E^2_{\sigma_{CH},\sigma^*_{NH'}}$	0.00	-0.25	-0.25
$\Delta E^2_{\sigma_{CH'},\sigma^*_{NH'}}$	-0.61	-0.07	-0.06
$\Delta E^2_{\sigma_{CH'},\sigma^*_{NH''}}$	0.00	-0.23	-0.23
$\Sigma\Delta E^4$	32.98	35.38	33.16
$\Sigma\Delta E^2$	-4.57	-4.09	-4.13
$\Sigma\Delta E$	28.41	31.29	29.03
E_c^a	-0.27	-0.21	-0.21

a. Coulomb Energy (kcal/mol) computed according to eq. 5.

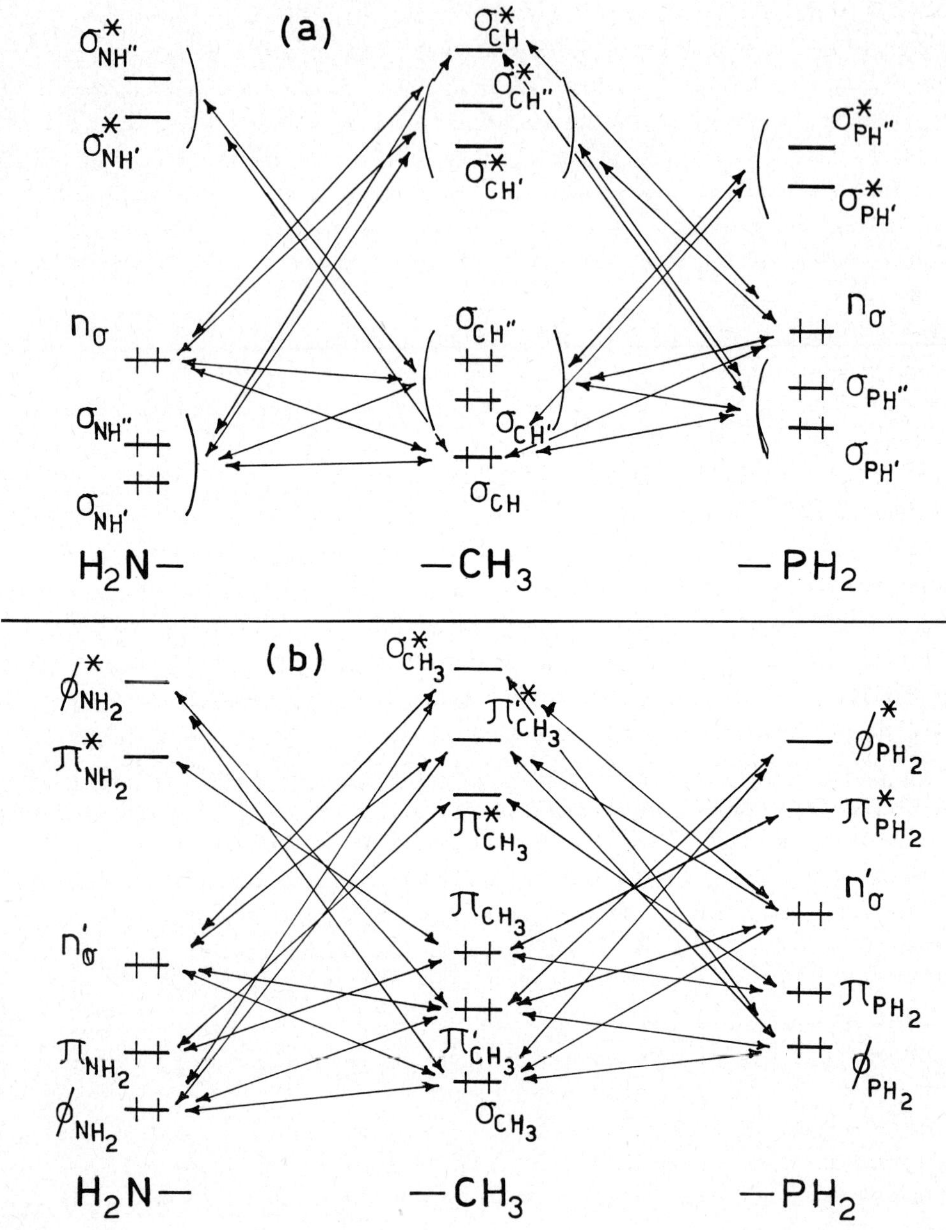

Fig. 6. Interaction diagram describing the non-bonded interactions between the localized MO's (a) and the canonical MO's (b) of the H_3C- and $-XH_2$ fragments.

variation is largely counterbalanced by the destabilization of the various fragment MO's which accompanies the geometry relaxation, as indicated by the index IS which varies from the value of -61.1192 a.u. in the rigid geometry to the value of -61.0971 a.u. in the optimized geometry. The final gain in total energy is just 0.71 kcal/mol and has the effect of reducing the rotational barrier from the value of 3.49 kcal/mol in the rigid model to the value of 2.77 kcal/mol in the optimized model. The various energy changes which accompany the geometry relaxation have also significant consequences upon the contributions of the various effects to the rotational barrier. While, in fact, in the rigid model the rotational barrier is almost completely controlled by the non-bonded interactions, with a dominant role of the steric repulsions, in the optimized model the contribution of the non-bonded interactions to the rotational barrier becomes very small (just 0.62 kcal/mol) and the dominant role is played by the first order effects, i.e. by the destabilization of the fragment MO's, associated with the geometry relaxation.

The results obtained in the canonical representation are listed in Table 12, while the related interaction diagram is shown again in Figure 6. The information provided by the values of $\Sigma\Delta E$ are similar to those obtained in the framework of the localized representation. In fact the overall energy effect is destabilizing and less destabilizing in the staggered conformation. The only difference is that in the canonical representation the $\Sigma\Delta E$ values are slightly larger and also the differences between the $\Sigma\Delta E$ values computed for the eclipsed and staggered conformations are slightly larger.

As previously pointed out this representation allows to compute the contributions associated with the σ-type and the π-type orbital interactions, denoted as $\Sigma\Delta E(\sigma)$ and $\Sigma\Delta E(\pi)$. The values listed in Table 12 show that the magnitude of these two terms is very similar and that the contribution of both terms to the $\Sigma\Delta E$ difference is significant in both the rigid and optimized models. In fact the contributions of $\Sigma\Delta E(\sigma)$ and $\Sigma\Delta E(\pi)$ are in the rigid model 1.48 and 2.60 kcal/mol respectively and in the optimized model 0.85 and 0.72 kcal/mol respectively.

Therefore, in methylamine the conformational preference and the magnitude of the rotational barrier can not be discussed in terms of hyperconjugative interactions only, since also the σ-type interactions play a significant role. However the interpretation of the σ-type interactions is less immediate than that of the interactions in the localized representation, since the various σ MO's are significantly delocalized. For instance, the HOMO of the $-NH_2$ fragment in the canonical representation, n'_σ , has the following expression in terms of the localized basis:

TABLE 12
Stabilization (ΔE^2_{ij}) and Destabilization (ΔE^4_{ij}) Energies (kcal/mol) Associated with the Non-Bonded Interactions Occurring in the Staggered and Eclipsed Conformations of Methylamine, Computed in the Canonical Representation.

	STAGGERED	ECLIPSED	
		Rigid Geometry	Optimized Geometry
$\Delta E^4_{\sigma_{CH_3},\phi_{NH_2}}$	1.30	1.62	1.27
$\Delta E^4_{\sigma_{CH_3},n'_\sigma}$	0.00	0.10	0.00
$\Delta E^4_{\pi'_{CH_3},\phi_{NH_2}}$	0.91	2.20	1.21
$\Delta E^4_{\pi'_{CH_3},n'_\sigma}$	20.85	20.38	20.96
$\Delta E^4_{\pi_{CH_3},\pi_{NH_2}}$	18.44	20.47	18.65
$\Delta E^2_{\sigma_{CH_3},\phi^*_{NH_2}}$	-0.05	-0.04	-0.05
$\Delta E^2_{\pi'_{CH_3},\phi^*_{NH_2}}$	-0.31	-0.27	-0.27
$\Delta E^2_{\phi_{NH_2},\pi'^*_{CH_3}}$	0.00	-0.04	-0.01
$\Delta E^2_{\phi_{NH_2},\sigma^*_{CH_3}}$	0.00	0.00	0.00
$\Delta E^2_{n'_\sigma,\pi'^*_{CH_3}}$	-4.82	-4.60	-4.38
$\Delta E^2_{n'_\sigma,\sigma^*_{CH_3}}$	-0.03	-0.02	-0.03
$\Delta E^2_{\pi_{CH_3},\pi^*_{NH_2}}$	-0.94	-0.70	-0.68
$\Delta E^2_{\pi_{NH_2},\pi^*_{CH_3}}$	-1.44	-1.11	-1.19
$\Sigma\Delta E^4(\sigma)$	23.06	24.30	23.44
$\Sigma\Delta E^2(\sigma)$	-5.21	-4.97	-4.74
$\Sigma\Delta E\ (\sigma)$	17.85	19.33	18.70
$\Sigma\Delta E^4(\pi)$	18.44	20.47	18.65
$\Sigma\Delta E^2(\pi)$	-2.38	-1.81	-1.87
$\Sigma\Delta E\ (\pi)$	16.06	18.66	16.78
$\Sigma\Delta E^4$	41.50	44.77	42.09
$\Sigma\Delta E^2$	-7.59	-6.78	-6.61
$\Sigma\Delta E$	33.91	37.99	35.48

$$n'_\sigma = 0.95\ n_\sigma - 0.23\ (\sigma_{NH'} + \sigma_{NH''}) \tag{9}$$

where n_σ is a lone pair and the σ_{NH}'s bond MO's (see Figure 3). A similar delocalized expression has also the other σ MO, ϕ_{NH_2}. Therefore, the interactions involving these orbitals can not be easily interpreted in terms of simple models. Furthermore, since the various coefficients change in different molecules, these MO's do not seem to be suitable for comparative analyses.

3.2.2 Methylphosphine

The staggered conformer with optimized geometry has been found to be more stable than the eclipsed with rigid geometry by 2.18 kcal/mol at the STO-3G level and by 2.62 kcal/mol at the STO-3G* level, and more stable than the eclipsed with optimized geometry by 1.91 kcal/mol at the STO-3G level and by 2.30 kcal/mol at the STO-3G* level. All these values agree well with the experimental rotational barrier [37], which is found to be 1.959 kcal/mol.

The energy effects associated with the non-bonded interactions occurring between the valence MO's of the two interacting fragments, H_3C- and -PH_2, computed in the localized representation at the STO-3G and STO-3G* level, are listed in Tables 13 and 14 while the related interaction diagram is shown in Figure 6. We proceed to discuss these results in the following order:

(a) Rigid Model. At the STO-3G level the overall energy effect associated with the non-bonded interactions $\Sigma\Delta E$ is destabilizing and less destabilizing in the staggered conformation in agreement with the trend of the total energy values. The contribution associated with the conjugative interactions is much smaller than that associated with the steric repulsions (compare the values of $\Sigma\Delta E^2$ and $\Sigma\Delta E^4$). Both these terms favor a preferential stabilization of the staggered geometry. The difference between the $\Sigma\Delta E$ values computed for the eclipsed and staggered conformations is 3.47 kcal/mol and is somewhat larger than the corresponding total energy difference of 2.18 kcal/mol. The contributions of $\Sigma\Delta E^2$ and $\Sigma\Delta E^4$ to the $\Sigma\Delta E$ difference are 0.27 and 3.20 kcal/mol respectively: these values show that the conformational preference and the magnitude of the rotational barrier is mainly controlled by the steric repulsions. The results of the quantitative analysis show that the dominant factor is the repulsion between the σ_{CH} and the σ_{PH} bond MO's, while the repulsions involving the phosphorus lone pair are less important.

The results obtained at the STO-3G* level provide the information that the stabilization energy associated with the interactions occurring between the MO's of the H_3C- fragment and the 3d orbitals of phosphorus is conformationally independent (compare the $\Sigma\Delta E^2_{3d}$ values in Table 14). However the inclusion of the phosphorus 3d orbitals has some indirect effects upon the various terms associ-

TABLE 13

Stabilization (ΔE^2_{ij}) and Destabilization (ΔE^4_{ij}) Energies (kcal/mol) Associated with the Non-Bonded Interactions Occurring in the Staggered and Eclipsed Conformations of Methylphosphine, Computed in the Localized Representation.

	STO-3G LEVEL			STO-3G* LEVEL		
	STAGGERED	ECLIPSED		STAGGERED	ECLIPSED	
		Rigid Geometry	Optimized Geometry		Rigid Geometry	Optimized Geometry
$\Delta E^4_{\sigma_{PH'},\sigma_{CH}}$	3.04	0.40	0.35	3.23	0.41	0.34
$\Delta E^4_{\sigma_{PH'},\sigma_{CH'}}$	1.58	7.26	6.32	1.64	7.89	6.73
$\Delta E^4_{\sigma_{PH'},\sigma_{CH''}}$	1.55	0.00	0.00	1.68	0.00	0.00
$\Delta E^4_{n_\sigma,\sigma_{CH}}$	5.27	1.13	1.30	5.41	1.01	1.22
$\Delta E^4_{n_\sigma,\sigma_{CH'}}$	0.06	2.24	2.02	0.03	2.37	2.11
$\Delta E^2_{\sigma_{PH'},\sigma^*_{CH}}$	-0.03	-0.13	-0.10	-0.03	-0.13	-0.11
$\Delta E^2_{\sigma_{PH'},\sigma^*_{CH'}}$	-0.23	-0.06	-0.06	-0.24	-0.05	-0.05
$\Delta E^2_{\sigma_{PH'},\sigma^*_{CH''}}$	-0.01	-0.03	-0.03	-0.01	-0.03	-0.03
$\Delta E^2_{n_\sigma,\sigma^*_{CH}}$	-0.62	-0.27	-0.26	-0.58	-0.23	-0.22
$\Delta E^2_{n_\sigma,\sigma^*_{CH'}}$	-0.04	-0.18	-0.19	-0.03	-0.16	-0.18
$\Delta E^2_{\sigma_{CH},\sigma^*_{PH'}}$	-0.13	-0.01	-0.02	-0.13	-0.01	-0.01
$\Delta E^2_{\sigma_{CH'},\sigma^*_{PH'}}$	-0.10	-0.26	-0.24	-0.10	-0.28	-0.25
$\Delta E^2_{\sigma_{CH'},\sigma^*_{PH''}}$	-0.09	0.00	0.00	-0.11	0.00	0.00

TABLE 14

Total Stabilization ($\Sigma\Delta E^2$) and Destabilization ($\Sigma\Delta E^4$) Energies (kcal/mol) Associated with the Non-Bonded Interactions Occurring in the Staggered and Eclipsed Conformations of Methylphosphine, Computed in the Localized Representation.

	STAGGERED	ECLIPSED	
		Rigid Geometry	Optimized Geometry
STO-3G Level			
$\Sigma\Delta E^4$	17.73	20.93	18.68
$\Sigma\Delta E^2$	-1.88	-1.61	-1.54
$\Sigma\Delta E$	15.85	19.32	17.14
E_c^a	-8.68	-8.68	-8.55
STO-3G* Level			
$\Sigma\Delta E^4$	18.57	22.35	19.58
$\Sigma\Delta E^2$	-1.88	-1.55	-1.48
$\Sigma\Delta E$	16.69	20.80	18.10
$\Sigma\Delta E^2_{3d}$	-12.44	-12.45	-11.80
$\Sigma\Delta E_T$	4.25	8.35	6.30
E_c^a	-4.40	-4.41	-4.45

a. Coulomb Energy (kcal/mol) computed according to eq. 5.

ated with the non-bonded interactions. In particular the term which is mostly affected is $\Sigma\Delta E^4$, which becomes more destabilizing at the STO-3G* level. The stabilizing nature of the interaction between the MO's of the H_3C- fragment and the phosphorus 3d orbitals causes, in fact, a decrease of the C-P bond length (see Table 3): but a decrease in this bond length has the effect of increasing the steric repulsions between the bond MO's, with the consequence that the eclipsed conformer becomes relatively more destabilized . However this effect is not large (the $\Sigma\Delta E$ difference varies from 3.47 kcal/mol at the STO-3G level to 4.11 kcal/mol at the STO-3G* level) and therefore the inclusion of the phosphorus 3d orbitals has mainly the effect of making the overall energy effect less destabilizing, without altering its trend.

(b) Optimized Model. The comparison of these results with those obtained at the optimized eclipsed geometry shows a trend similar to that already observed in methylamine. Also in this case the geometry of the eclipsed conformer tends to change in order to reduce the large steric effects (see the values of $\Sigma\Delta E^4$ in Table 14). Again this energy variation is largely counterbalanced by the destabilization of the various fragment MO's which accompanies the geometry relaxation. The values of the index IS, in fact, are, at the STO-3G level, -232.0951 a.u. at the rigid eclipsed geometry and -232.0822 a.u. at the optimized eclipsed geometry. The final gain in total energy associated with the geometry relaxation is just 0.28 kcal/mol. Therefore also in methylphosphine, while in the rigid model the conformational preference can be satisfactorily discussed only in terms of conjugative and steric effects, in the optimized model we have to consider also the first order effects associated with the geometry relaxation.

The results obtained in the canonical representation are listed in Table 15, while the related interaction diagram is shown in Figure 6. Again the trend and magnitude of the $\Sigma\Delta E$ values are similar to those obtained with the localized representation, the only difference being that in the canonical representation the $\Sigma\Delta E$ values are slightly larger.

The decomposition into the σ and π contributions shows that also in methylphosphine the conformational preference and the magnitude of the rotational barrier can not be discussed in terms of hyperconjugative interactions only, since also the σ-type interactions play a significant role.

3.2.3 Effects of the Replacement of Phosphorus for Nitrogen

The experimental results show that the replacement in a $H_3C\text{-}XH_2$ molecule of a first row atom (X = N) with a second row atom (X = P) has no effect upon the conformational preference and only a very small effect upon the magnitude of the rotational barrier which varies from 1.98 kcal/mol in methylamine to 1.96 kcal/mol in methylphosphine. The computations reproduce the observed conformational preference and the trend of the rotational barriers (see Tables 5 and 6), even though the rotational barrier of methylamine is relatively overestimated. Since we have found that the inclusion of the phosphorus 3d orbitals has only minor effects upon the magnitude of the rotational barrier in methylphosphine, we perform the comparative analysis in terms of the results obtained at the STO-3G level. Furthermore we use the results obtained in terms of the fragment localized MO's, which are more suitable for such a comparative analysis.

We have already pointed out that in the rigid model the conformational preference and the rotational barrier are determined in both molecular species

TABLE 15

Stabilization (ΔE^{2}_{ij}) and Destabilization (ΔE^{4}_{ij}) Energies (kcal/mol) Associated with the Non-Bonded Interactions Occurring in the Staggered and Eclipsed Conformations of Methylphosphine, Computed in the Canonical Representation.

	STAGGERED	ECLIPSED	
		Rigid Geometry	Optimized Geometry
$\Delta E^{4}_{\sigma_{CH_3},\phi_{PH_2}}$	0.79	1.02	0.80
$\Delta E^{4}_{\sigma_{CH_3},n'_{\sigma}}$	1.26	1.70	1.30
$\Delta E^{4}_{\pi'_{CH_3},\phi_{PH_2}}$	1.98	2.66	2.13
$\Delta E^{4}_{\pi'_{CH_3},n'_{\sigma}}$	8.35	8.74	8.41
$\Delta E^{4}_{\pi_{CH_3},\pi_{PH_2}}$	8.20	9.92	9.09
$\Delta E^{2}_{\sigma_{CH_3},\phi^{*}_{PH_2}}$	-0.05	-0.05	-0.05
$\Delta E^{2}_{\pi'_{CH_3},\phi^{*}_{PH_2}}$	-0.36	-0.35	-0.35
$\Delta E^{2}_{\phi_{PH_2},\pi'^{*}_{CH_3}}$	-0.03	-0.05	-0.03
$\Delta E^{2}_{\phi_{PH_2},\sigma^{*}_{CH_3}}$	-0.02	-0.02	-0.02
$\Delta E^{2}_{n'_{\sigma},\pi'^{*}_{CH_3}}$	-1.17	-1.12	-1.08
$\Delta E^{2}_{n'_{\sigma},\sigma^{*}_{CH_3}}$	-0.01	0.00	0.00
$\Delta E^{2}_{\pi_{CH_3},\pi^{*}_{PH_2}}$	-0.87	-0.63	-0.63
$\Delta E^{2}_{\pi_{PH_2},\pi^{*}_{CH_3}}$	-0.55	-0.28	-0.31
$\Sigma\Delta E^{4}(\sigma)$	12.38	14.12	12.64
$\Sigma\Delta E^{2}(\sigma)$	-1.64	-1.59	-1.53
$\Sigma\Delta E\ (\sigma)$	10.74	12.53	11.11
$\Sigma\Delta E^{4}(\pi)$	8.20	9.92	9.09
$\Sigma\Delta E^{2}(\pi)$	-1.42	-0.91	-0.94
$\Sigma\Delta E\ (\pi)$	6.78	9.01	8.15
$\Sigma\Delta E^{4}$	20.58	24.04	21.73
$\Sigma\Delta E^{2}$	-3.06	-2.50	-2.47
$\Sigma\Delta E$	17.52	21.54	19.26

mainly by the steric repulsions. We attempt now to understand why these two molecular species have very similar rotational barriers and therefore display a behaviour different from that of the H_3C-XH_3 molecules, where the replacement of a first-row with a second-row atom has the effect of reducing significantly the barrier. To this purpose we analize the various energy effects in the rigid model.

To obtain a better understanding of the role played by the various terms, it is convenient to decompose the steric contribution to the rotational barrier into a term associated with the repulsions between a X-H bond MO and the three C-H bond MO's of the H_3C- fragment (denoted here as ΔE_{XH}) and a term associated with the repulsions between the lone pair centered at X and the three C-H bond MO's of the H_3C- fragment (denoted here as ΔE_{n_X}). Therefore the steric contribution to the barrier (denoted here as SC) can be expressed in the following form

$$SC = 2\, D\Delta E_{XH} + D\Delta E_{n_X} \tag{10}$$

where $D\Delta E_{XH}$ denotes the difference between the ΔE_{XH} values computed at the eclipsed and staggered geometries and $D\Delta E_{n_X}$ the similar difference between the ΔE_{n_X} values. The values of the two quantities ΔE_{XH} and ΔE_{n_X} are given in Table 16. These results show that the ΔE_{XH} values display the same trend in both molecular species and $D\Delta E_{XH}$ is larger in methylamine than in methylphosphine (2.36 kcal/mol versus 1.49 kcal/mol). Therefore this term, which represents the contribution to the rotational barrier of the steric effect arising from the repulsions between bond MO's, favors a larger barrier in methylamine than in methylphosphine and shows a behaviour similar to that found in the H_3C-XH_3 molecular species. On the other hand the ΔE_{n_X} values show opposite trends in the two molecules: in particular, while in methylphosphine this term favors the staggered conformation and $D\Delta E_{n_P}$ has the same sign of $D\Delta E_{PH}$, in methylamine this term favors significantly the eclipsed conformation and $D\Delta E_{n_N}$ has now sign opposite to that of $D\Delta E_{NH}$, with the result of reducing significantly the rotational barrier.

The different trends of ΔE_{n_X} are a consequence of the different orientations of the nitrogen and phosphorus σ lone pairs. Indicative values for these orientations are shown below in terms of the angles β between the C-X axis and the axis of the lone pair, the latter determined using the computed centroids of charge:

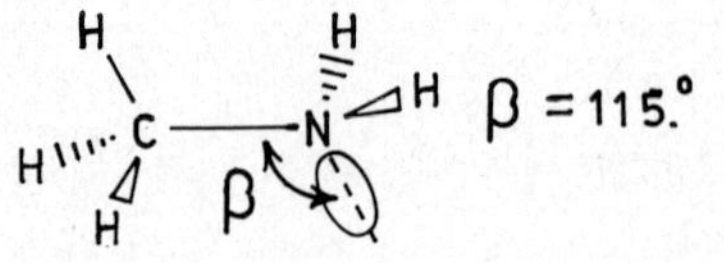

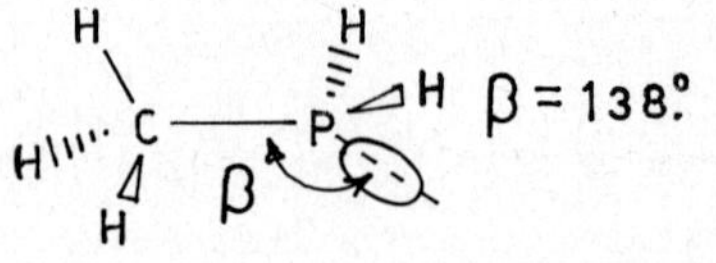

$\beta = 115°$ $\beta = 138°$

TABLE 16

The Repulsion Energies[a] ΔE_{XH} and ΔE_{n_X} Associated with the Interactions of the Three C-H Bond MO's of the H_3C-Fragment with the X-H Bond MO and the Lone Pair of the $-XH_2$ Fragment (X = N,P).

	METHYLAMINE		METHYLPHOSPHINE	
	Staggered	Eclipsed	Staggered	Eclipsed
ΔE_{XH}	9.94	12.30	6.17	7.66
ΔE_{n_X}	13.10	10.78	5.39	5.61

a. Values in kcal/mol

These different orientations cause completely different trends of the repulsive effects associated with the interactions between these σ lone pairs and the C-H bond MO's, as indicated by the $\Delta E^4_{n_\sigma \sigma_{CH}}$ values listed in Tables 11 and 13.

In the optimized model we have to consider also the contributions to the rotational barriers associated with the first order effects which accompany the geometry relaxation of the eclipsed conformers. These effects seem to be larger in the first-row molecules and therefore should favor a larger barrier in methylamine. On the other hand, the contributions of the conjugative effects remain of the order of magnitude found in the rigid model, while the contributions of the steric effects become particularly small and smaller in methylamine (0.18 kcal/mol in methylamine and 0.95 kcal/mol in methylphosphine). Therefore, also in this model, the main cause of the relatively smaller barrier in methylamine is the smaller steric contribution due to the large cancellation between the component terms of the steric effect, as previously pointed out.

3.3 Conformational Isomerism in CH_3-XH Molecules.

3.3.1 Methanol

At the STO-3G level the staggered conformer with optimized geometry has been found to be more stable by 3.07 kcal/mol than the eclipsed with rigid geometry and more stable by 2.01 kcal/mol than the eclipsed with optimized geometry. These values have to be compared with the value of 1.07 kcal/mol experimentally found for the rotational barrier [35]: therefore the accuracy in this case is much less than in the other first-row and second-row molecules previously investigated.

The energy effects associated with the non-bonded interactions occurring between the valence MO's of the H_3C- and -OH fragments in the localized representation are listed in Table 17, while the related interaction diagram is shown in Figure 7. Following the procedure applied in this paper, we discuss first the results obtained in the rigid model. Again the overall energy effect associated with the non-bonded interactions, $\Sigma\Delta E$, is destabilizing and less destabilizing in the staggered conformation: therefore the trend of $\Sigma\Delta E$ parallels also in this case the total energy behaviour. Also the difference between the $\Sigma\Delta E$ values computed at the eclipsed and staggered geometries, 4.46 kcal/mol, compares well with the corresponding total energy difference of 3.07 kcal/mol. Both the steric and conjugative effects favor the staggered conformation (see the $\Sigma\Delta E^4$ and $\Sigma\Delta E^2$ values in Table 17); however the conformational preference is determined almost exclusively by the steric effects. The dominant term is the repulsion between the in plane bond MO's of the two fragments, σ_{CH} and σ_{OH}, which favors by a large amount an anti over a syn alignement.

The comparison between the results obtained at the rigid and optimized eclipsed geometries shows that in the eclipsed conformation the geometry tends to change in order to reduce the large steric repulsion between σ_{CH} and σ_{OH}. In fact the energy effect associated with this interaction, $\Delta E^4_{\sigma_{OH}\sigma_{CH}}$, varies from a value of 15.93 kcal/mol at the rigid geometry to a value of 11.33 kcal/mol at the optimized geometry. This result is obtained mainly through an increase of the HCO angle (see Table 4), which reduces the overlap between the two interacting MO's. Consequently the overall energy effect associated with the non-bonded interactions, $\Sigma\Delta E$, shows a significant decrease at the optimized geometry (from 39.92 to 35.92 kcal/mol). However, as previously observed in the other molecules, this energy variation is counterbalanced by a destabilization of various fragment MO's which accompanies the geometry relaxation: in fact, the index IS varies from the value of -71.1689 a.u. in the rigid geometry to the value of -71.1495 a.u. in the optimized geometry. The resulting gain in total energy is again small, 1.05 kcal/mol, even though larger than in the other molecules previously investigated. Therefore, in the optimized model, there is a significant decrease in the contribution of the steric effect to the rotational barrier and a related increase in the contribution of the first order effects. Furthermore there is also an increase in the contribution of the conjugative effects.

The results obtained in the canonical representation are listed in Table 18, while the related interaction diagram is shown in Figure 7. Also in this case, the $\Sigma\Delta E$ values are only slightly larger than those obtained in the localized representation. Also the difference between the $\Sigma\Delta E$ values computed at the eclipsed and staggered geometries is larger than the corresponding value computed in the localized representation (5.70 and 4.46 kcal/mol respectively) and also significantly larger than the corresponding value computed in terms of

TABLE 17

Stabilization (ΔE^2_{ij}) and Destabilization (ΔE^4_{ij}) Energies (kcal/mol) Associated with the Non-Bonded Interactions Occurring in the Staggered and Eclipsed Conformations of Methanol, Computed in the Localized Representation.

	STAGGERED	ECLIPSED	
		Rigid Geometry	Optimized Geometry
$\Delta E^4_{\sigma_{OH},\sigma_{CH}}$	2.85	15.93	11.33
$\Delta E^4_{\sigma_{OH},\sigma_{CH'}}$	4.17	0.24	0.17
$\Delta E^4_{n_\sigma,\sigma_{CH}}$	2.61	3.65	3.95
$\Delta E^4_{n_\sigma,\sigma_{CH'}}$	1.29	0.24	0.44
$\Delta E^4_{n_\pi,\sigma_{CH'}}$	9.62	9.69	9.71
$\Delta E^2_{\sigma_{OH},\sigma^*_{CH}}$	-0.33	-0.41	-0.29
$\Delta E^2_{\sigma_{OH},\sigma^*_{CH'}}$	-0.06	-0.14	-0.10
$\Delta E^2_{n_\sigma,\sigma^*_{CH}}$	-0.45	-0.48	-0.83
$\Delta E^2_{n_\sigma,\sigma^*_{CH'}}$	-0.28	-0.03	-0.07
$\Delta E^2_{n_\pi,\sigma^*_{CH'}}$	-3.97	-4.07	-3.63
$\Delta E^2_{\sigma_{CH},\sigma^*_{OH}}$	-1.12	-0.03	-0.01
$\Delta E^2_{\sigma_{CH'},\sigma^*_{OH}}$	-0.02	-0.50	-0.51
$\Sigma\Delta E^4$	35.62	39.92	35.92
$\Sigma\Delta E^2$	-10.56	-10.40	-9.75
$\Sigma\Delta E$	25.06	29.52	26.17
E_c^a	-1.31	-1.27	-1.23

a. Coulomb Energy (kcal/mol) computed according to eq. 5.

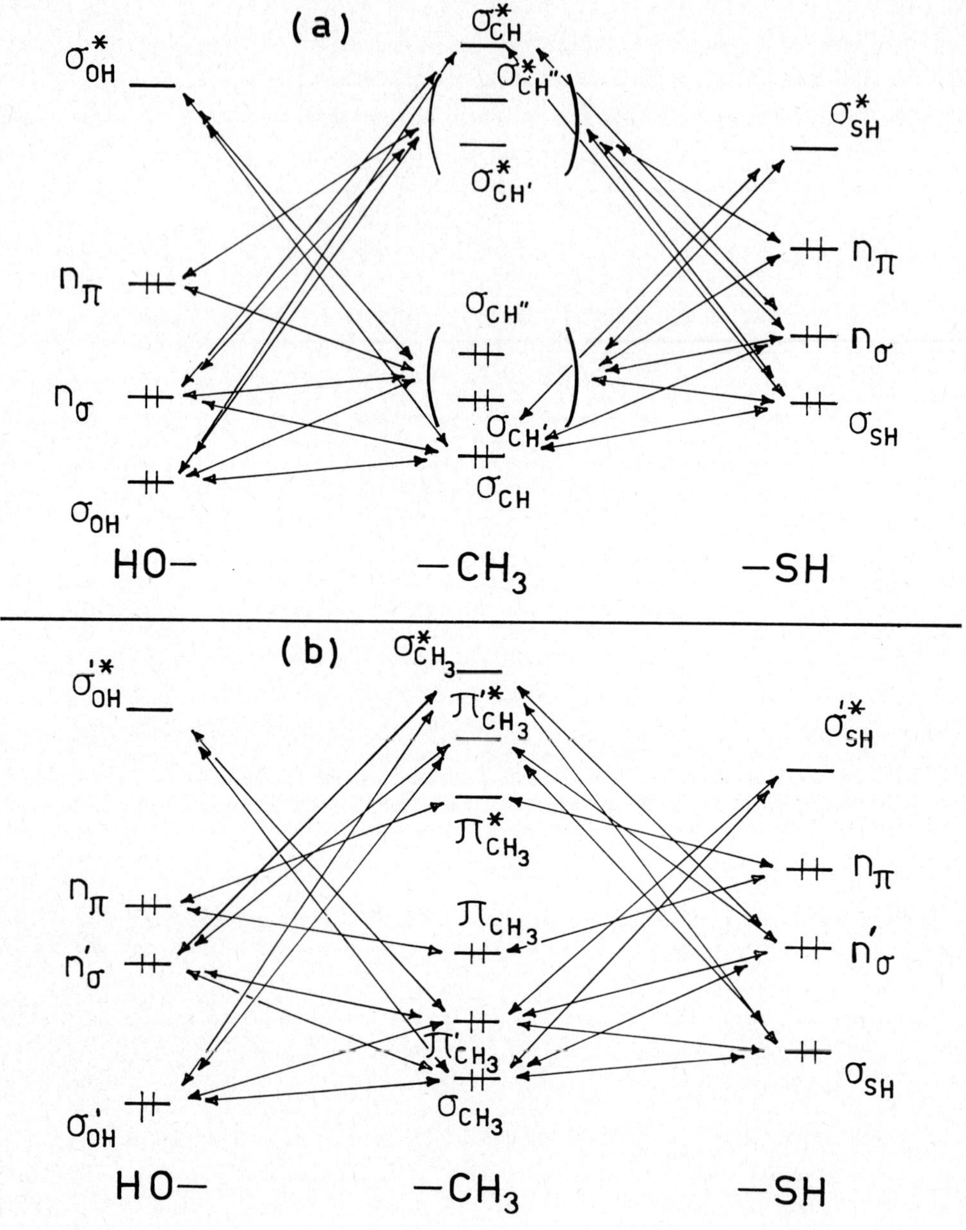

Fig. 7. Interaction diagram describing the non-bonded interactions between the localized MO's (a) and the canonical MO's (b) of the H_3C- and -XH fragments.

TABLE 18

Stabilization (ΔE^2_{ij}) and Destabilization (ΔE^4_{ij}) Energies (kcal/mol) Associated with the Non-Bonded Interactions Occurring in the Staggered and Eclipsed Conformations of Methanol, Computed in the Canonical Representation.

	STAGGERED	ECLIPSED	
		Rigid Geometry	Optimized Geometry
$\Delta E^4_{\sigma_{CH_3},\sigma'_{OH}}$	0.81	1.19	0.83
$\Delta E^4_{\sigma_{CH_3},n'_\sigma}$	0.86	1.59	0.94
$\Delta E^4_{\pi'_{CH_3},\sigma'_{OH}}$	2.70	5.07	3.01
$\Delta E^4_{\pi'_{CH_3},n'_\sigma}$	16.35	17.77	16.21
$\Delta E^4_{\pi_{CH_3},n_\pi}$	20.68	20.82	20.87
$\Delta E^2_{\sigma_{CH_3},\sigma'^*_{OH}}$	-0.11	-0.08	-0.11
$\Delta E^2_{\pi'_{CH_3},\sigma'^*_{OH}}$	-0.66	-0.48	-0.47
$\Delta E^2_{\sigma'_{OH},\pi'^*_{CH_3}}$	-0.02	-0.13	-0.04
$\Delta E^2_{\sigma'_{OH},\sigma^*_{CH_3}}$	0.00	0.00	0.00
$\Delta E^2_{n'_\sigma,\pi'^*_{CH_3}}$	-3.84	-3.04	-3.63
$\Delta E^2_{n'_\sigma,\sigma^*_{CH_3}}$	0.00	0.00	0.00
$\Delta E^2_{n_\pi,\pi^*_{CH_3}}$	-8.06	-8.30	-7.40
$\Sigma\Delta E^4(\sigma)$	20.72	25.62	20.99
$\Sigma\Delta E^2(\sigma)$	-4.63	-3.73	-4.25
$\Sigma\Delta E\ (\sigma)$	16.09	21.89	16.74
$\Sigma\Delta E^4(\pi)$	20.68	20.82	20.87
$\Sigma\Delta E^2(\pi)$	-8.06	-8.30	-7.40
$\Sigma\Delta E\ (\pi)$	12.62	12.52	13.47
$\Sigma\Delta E^4$	41.40	46.44	41.86
$\Sigma\Delta E^2$	-12.69	-12.03	-11.65
$\Sigma\Delta E$	28.71	34.41	30.21

total energies (5.70 and 3.07 kcal/mol respectively). This latter comparison suggests that in this case the first order effects can have a certain importance already in the rigid model.

The decomposition into the σ and π contributions shows that in the rigid model the conformational problem is controlled only by the σ-type interactions, while the π interactions, i.e. the hyperconjugative interactions, have no conformational effect. However in the optimized model both types of interactions show a similar effect favoring the staggered conformation.

3.3.2 Methanethiol

The staggered conformer with optimized geometry has been found to be more stable than the eclipsed with rigid geometry by 1.69 kcal/mol at the STO-3G level and by 1.98 kcal/mol at the STO-3G* level, and more stable than the eclipsed with optimized geometry by 1.45 kcal/mol at the STO-3G level and by 1.69 kcal/mol at the STO-3G* level. Therefore the values computed in both the rigid and optimized models agree well with the experimental rotational barrier [38], which is found to be 1.27 kcal/mol. The accuracy is greater for the values computed in the optimized model and in both models for the values computed at the STO-3G level.

The energy effects associated with the non-bonded interactions occurring between the valence MO's of the two interacting fragments, H_3C- and -SH, computed in the localized representation at the STO-3G and STO-3G* levels, are listed in Tables 19 and 20, while the relative interaction diagram is shown in Figure 7. We proceed to discuss these results in the following order:

(a) Rigid Model. As in the previous cases, the overall steric effect ($\Sigma\Delta E^4$) is larger than the overall conjugative effect ($\Sigma\Delta E^2$), so that the overall energy effect associated with the non-bonded interactions, $\Sigma\Delta E$, is destabilizing. Both the steric and conjugative effects favor the staggered conformation, with the consequence that $\Sigma\Delta E$ is less destabilizing in the staggered than in the eclipsed geometry. Therefore the trend of $\Sigma\Delta E$ parallels also in this case that of the total energy. Also the difference between the $\Sigma\Delta E$ values computed at the eclipsed and staggered geometries, 3.06 kcal/mol, compares well with the corresponding total energy difference of 1.69 kcal/mol. The conformational preference is determined almost exclusively by the steric effects, as indicated by the contributions of $\Sigma\Delta E^4$ and $\Sigma\Delta E^2$ to the $\Sigma\Delta E$ difference (2.93 and 0.13 kcal/mol respectively). The dominant term is the repulsion between the in plane bond MO's of the two fragments, which is found to be the dominant term also in methanol. This term favors, in fact, by a large amount an anti over a syn alignement of σ_{CH} and σ_{SH}, as indicated by the values of $\Delta E^4_{\sigma_{SH}\sigma_{CH}}$ listed in Table 19.

The analysis of the results obtained at the STO-3G* level shows that the

TABLE 19

Stabilization (ΔE^2_{ij}) and Destabilization (ΔE^4_{ij}) Energies (kcal/mol) Associated with the Non-Bonded Interactions Occurring in the Staggered and Eclipsed Conformations of Methanethiol, Computed in the Localized Representation.

	STO-3G LEVEL			STO-3G* LEVEL		
	STAGGERED	ECLIPSED		STAGGERED	ECLIPSED	
		Rigid Geometry	Optimized Geometry		Rigid Geometry	Optimized Geometry
$\Delta E^4_{\sigma_{SH},\sigma_{CH}}$	1.53	9.06	7.51	1.59	10.10	8.03
$\Delta E^4_{\sigma_{SH},\sigma_{CH'}}$	2.50	0.06	0.06	2.69	0.05	0.05
$\Delta E^4_{n_\sigma,\sigma_{CH}}$	0.35	3.23	3.03	0.37	3.48	3.21
$\Delta E^4_{n_\sigma,\sigma_{CH'}}$	1.29	0.00	0.00	1.38	0.00	0.00
$\Delta E^4_{n_\pi,\sigma_{CH'}}$	6.97	6.96	6.92	7.35	7.32	7.23
$\Delta E^2_{\sigma_{SH},\sigma^*_{CH}}$	-0.22	-0.08	-0.08	-0.23	-0.08	-0.07
$\Delta E^2_{\sigma_{SH},\sigma^*_{CH'}}$	-0.02	-0.07	-0.07	-0.02	-0.07	-0.06
$\Delta E^2_{n_\sigma,\sigma^*_{CH}}$	-0.08	-0.22	-0.27	-0.08	-0.20	-0.25
$\Delta E^2_{n_\sigma,\sigma^*_{CH'}}$	-0.13	0.00	0.00	-0.13	0.00	0.00
$\Delta E^2_{n_\pi,\sigma^*_{CH'}}$	-1.60	-1.64	-1.51	-1.70	-1.76	-1.58
$\Delta E^2_{\sigma_{CH},\sigma^*_{SH}}$	-0.19	-0.20	-0.18	-0.19	-0.23	-0.20
$\Delta E^2_{\sigma_{CH'},\sigma^*_{SH}}$	-0.06	-0.03	-0.04	-0.08	-0.02	-0.03

TABLE 20
Total Stabilization ($\Sigma\Delta E^2$) and Destabilization ($\Sigma\Delta E^4$) Energies (kcal/mol) Associated with the Non-Bonded Interactions Occurring in the Staggered and Eclipsed Conformations of Methanethiol, Computed in the Localized Representation.

	STAGGERED	ECLIPSED	
		Rigid Geometry	Optimized Geometry
STO-3G Level			
$\Sigma\Delta E^4$	23.40	26.33	24.50
$\Sigma\Delta E^2$	-4.11	-3.98	-3.77
$\Sigma\Delta E$	19.29	22.35	20.73
E_c^a	-2.07	-2.04	-2.00
STO-3G* Level			
$\Sigma\Delta E^4$	24.80	28.32	25.80
$\Sigma\Delta E^2$	-4.36	-4.21	-3.86
$\Sigma\Delta E$	20.44	24.11	21.94
$\Sigma\Delta E^2_{3d}$	-14.59	-14.58	-14.06
$\Sigma\Delta E_T$	5.85	9.53	7.88
E_c^a	-0.20	-0.17	-0.18

a. Coulomb Energy (kcal/mol) computed according to eq. 5.

inclusion of the sulphur 3d orbitals has a significant stabilizing effect which, however, is the same in the two conformations. Therefore also in this case the inclusion of the 3d orbitals has mainly the effect of making the overall energy effect less destabilizing without altering its trend.

(b) Optimized Model. The comparison between the results obtained in the rigid and optimized eclipsed geometries shows a trend similar to that observed in the previous molecules and in particular in methanol. Also in this case the geometry of the eclipsed conformer tends to change in order to reduce the large steric repulsions, in particular that between the two in-plane fragment MO's σ_{CH} and σ_{SH}. In fact $\Delta E^4_{\sigma_{SH}\sigma_{CH}}$ varies from 9.06 kcal/mol in the rigid geometry to 7.51 kcal/mol in the optimized geometry at the STO-3G level and from

10.10 kcal/mol in the rigid geometry to 8.03 kcal/mol in the optimized geometry at the STO-3G* level. This result is obtained mainly through an increase of the HCS angle (see Table 4), which reduces the overlap between the two interacting MO's σ_{SH} and σ_{CH}. Again this energy variation is largely counterbalanced by a destabilization of various fragment MO's: the values of the index IS at the STO-3G level are, in fact, -265.8238 a.u. in the rigid eclipsed geometry and -265.8115 a.u. in the optimized eclipsed geometry. The final gain in total energy at the STO-3G level is just 0.23 kcal/mol. The STO-3G* results show that in the geometry relaxation of the eclipsed conformer, also the stabilization energy associated with the 3d orbitals slightly decreases (from -14.58 kcal/mol in the rigid geometry to -14.06 kcal/mol in the optimized geometry): therefore, at this level, also this effect concurs with the destabilization of fragment MO's to counterbalance the decrease of steric repulsions and also at this level the final gain in total energy is very small, just 0.29 kcal/mol.

The results obtained in the canonical representation are listed in Table 21, while the interaction diagram is again shown in Figure 7. The analysis has been also in this case performed only at the STO-3G level, since we have previously shown that the 3d orbitals have no significant effect upon this structural problem. The information provided by these results are similar to those already pointed out in methanol. In particular the relevant points are the following:

(i) The trend and magnitude of the $\Sigma\Delta E$ values are very similar to those obtained with the localized representation. Again the only difference is that the $\Sigma\Delta E$ values obtained in the canonical representation are slightly larger.

(ii) The decomposition into the σ and π contributions shows that in the rigid model the conformational problem is controlled only by the σ-type interactions, while in the optimized model the dominant factors are the σ-type interactions again and the first order effects which accompany the geometry relaxation.

(iii) In both the rigid and optimized models, the π interactions have no conformational effects.

3.3.3 Effects of the Replacement of Sulphur for Oxygen

Experimentally it is found that in a H_3C-XH molecule, the replacement of a first-row atom (X = O) with a second-row atom (X = S) does not alter the conformational preference, since in both species the stable conformation is found to be the staggered, but has the effect of increasing the rotational barrier which varies from a value of 1.07 kcal/mol in methanol to a value of 1.27 kcal/mol in methanethiol. This trend is different from that observed in the H_3C-XH_3 molecules, where the replacement of carbon with silicon has the effect of reducing significantly the barrier, and more similar to that observed in the H_3C-XH_2 molecules, where the replacement of nitrogen with phosphorus has only a small

TABLE 21
Stabilization (ΔE^2_{ij}) and Destabilization (ΔE^4_{ij}) Energies (kcal/mol) Associated with the Non-Bonded Interactions Occurring in the Staggered and Eclipsed Conformations of Methanethiol, Computed in the Canonical Representation.

	STAGGERED	ECLIPSED	
		Rigid Geometry	Optimized Geometry
$\Delta E^4_{\sigma_{CH_3},\sigma'_{SH}}$	0.29	0.51	0.37
$\Delta E^4_{\sigma_{CH_3},n'_\sigma}$	1.24	1.77	1.42
$\Delta E^4_{\pi'_{CH_3},\sigma'_{SH}}$	2.79	4.22	3.43
$\Delta E^4_{\pi'_{CH_3},n'_\sigma}$	6.69	7.73	7.09
$\Delta E^4_{\pi_{CH_3},n_\pi}$	15.30	15.28	15.21
$\Delta E^2_{\sigma_{CH_3},\sigma'^*_{SH}}$	-0.01	-0.01	-0.01
$\Delta E^2_{\pi'_{CH_3},\sigma'^*_{SH}}$	-0.58	-0.44	-0.43
$\Delta E^2_{\sigma'_{SH},\pi'^*_{CH_3}}$	-0.06	-0.08	-0.06
$\Delta E^2_{\sigma'_{SH},\sigma^*_{CH_3}}$	-0.02	-0.02	-0.02
$\Delta E^2_{n'_\sigma,\pi'^*_{CH_3}}$	-0.92	-0.61	-0.70
$\Delta E^2_{n'_\sigma,\sigma^*_{CH_3}}$	-0.01	0.00	-0.01
$\Delta E^2_{n_\pi,\pi^*_{CH_3}}$	-3.25	-3.34	-3.09
$\Sigma\Delta E^4(\sigma)$	11.01	14.23	12.31
$\Sigma\Delta E^2(\sigma)$	-1.60	-1.16	-1.23
$\Sigma\Delta E\ (\sigma)$	9.41	13.07	11.08
$\Sigma\Delta E^4(\pi)$	15.30	15.28	15.21
$\Sigma\Delta E^2(\pi)$	-3.25	-3.34	-3.09
$\Sigma\Delta E\ (\pi)$	12.05	11.94	12.12
$\Sigma\Delta E^4$	26.31	29.51	27.52
$\Sigma\Delta E^2$	-4.85	-4.50	-4.32
$\Sigma\Delta E$	21.46	25.01	23.20

effect upon the rotational barrier.

The present computations, as shown in Tables 5 and 6, reproduce the observed conformational preferences, but not the trend of the rotational barriers. However, despite of these computational results, we attempt to understand why the rotational barrier is larger in methanethiol than in methanol. In fact, also in these cases where the computations do not provide a correct result, the analysis of the energy effects obtained in the quantitative PMO treatment can provide useful information about the role played by the various factors. The analysis is performed using the results obtained at the STO-3G level in the localized representation.

We consider first the steric contribution to the rotational barrier, which can be expressed in the following form

$$SC = D\Delta E_{XH} + D\Delta E_{n_X} + D\Delta E_{\pi_X} \tag{11}$$

Here ΔE_{XH}, ΔE_{n_X} and ΔE_{π_X} denote the repulsion energies associated with the interactions of the three C-H bond MO's of the H_3C- fragment with the X-H bond MO, the σ lone pair and the π lone pair of the -XH fragment, respectively, while $D\Delta E_{XH}$, $D\Delta E_{n_X}$ and $D\Delta E_{\pi_X}$ denote the differences between the corresponding ΔE values computed at the eclipsed and staggered geometries. The values of ΔE_{XH}, ΔE_{n_X} and ΔE_{π_X} are listed in Table 22.

These results show that in both molecular species the ΔE_{XH} values display the same trend, while the ΔE_{n_X} values show opposite trends: in particular, while $D\Delta E_{n_S}$ has the same sign of $D\Delta E_{SH}$, $D\Delta E_{n_O}$ has sign opposite to that of $D\Delta E_{OH}$. Also the ΔE_{π_X} values show opposite trends in the two molecular species; however their contribution seems to be less significant.

In the rigid model, where the $D\Delta E_{XH}$ term is largely dominant, the different trend of ΔE_{n_X} has only a minor effect: in this case $D\Delta E_{XH}$ and consequently the steric contribution is larger in methanol than in methanethiol. Also the overall contribution of the non-bonded interactions shows the same trend, since the conjugative contribution is negligible, and this is also the trend shown by the computed rotational barriers, since in this model the non-bonded interactions represent the dominant factor.

However, in the optimized model, there is a significant decrease of the steric repulsions between bond MO's in the eclipsed conformations and consequently a significant decrease of the values of $D\Delta E_{XH}$, which are now 0.48 kcal/mol in methanol and 1.10 kcal/mol in methanethiol. In methanol, this value is further reduced by the addition of $D\Delta E_{n_O}$, since $D\Delta E_{OH}$ and $D\Delta E_{n_O}$ have opposite signs, while in methanethiol, where $D\Delta E_{n_S}$ has the same sign of $D\Delta E_{SH}$, the previous value is further increased. Consequently the steric contribution and also the overall

TABLE 22
The Repulsion Energies[a] ΔE_{XH}, ΔE_{n_X} and ΔE_{π_X} Associated with the Interactions of the Three C-H Bond MO's of the H_3C- Fragment with the MO's of the -XH Fragment (X = O,S).

	METHANOL		METHANETHIOL	
	Staggered	Eclipsed[b]	Staggered	Eclipsed[b]
ΔE_{XH}	11.19	16.41(11.67)	6.53	9.18(7.63)
ΔE_{n_X}	5.19	4.13(4.83)	2.93	3.23(3.03)
ΔE_{π_X}	19.24	19.38(19.42)	13.94	13.92(13.84)

a. Values in kcal/mol.
b. The values in parentheses refer to the optimized model and the other to the rigid model.

contribution of the non-bonded interactions are smaller in methanol than in methanethiol, although the conjugative contribution shows opposite trend. Obviously, in this model we have to consider also the contribution of the first order effects, that is expected to be larger in methanol and to be responsible for the final trend of the rotational barriers.

Therefore, even though the STO-3G computations do not reproduce the experimental trend of the rotational barriers in methanol and methanethiol, the quantitative PMO analysis of these results has shown some interesting trends and, in particular, a difference in the signs of the component terms of the steric contribution, which can make this contribution significantly smaller in methanol than in methanethiol and which can be therefore responsible for the experimentally observed trend of the rotational barriers. On the other hand, the computational trend may well be caused by an overestimate of the other terms, for example the contribution of the first order effects.

3.4 Effects of the Replacement Along a Row

In this section we attempt to rationalize the trends of the rotational barriers in the two series: ethane, methylamine, methanol and methylsilane, methylphosphine, methanethiol. In other words, we attempt to analyze the effects caused by the replacement of a bond with a lone pair in this type of molecules.

The experimental results listed in Table 6 show that in the first-row molecules either the replacement of a bond with a σ-type lone pair (replacement of $-NH_2$ for $-CH_3$) or the additional replacement of another bond with a π-type lone pair (replacement of -OH for $-NH_2$) has the effect of decreasing the barrier.

On the other hand, in the second-row molecules, the replacement of a bond with a σ-type lone pair (replacement of $-PH_2$ for $-SiH_3$) causes a barrier increase while the additional replacement of another bond with a π-type lone pair (replacement of -SH for $-PH_2$) has again the effect of decreasing the barrier.

The present comparative analysis will be based upon the results obtained in the framework of the optimized model, since only the rotational barriers computed at this level can be related to the experimental data. Furthermore, we use here only the results obtained with the localized representation, since only these values allow to discuss quantitatively the energy effects associated with the replacement of a bond with a lone pair. In this analysis we attempt to decompose each rotational barrier into the contributions arising from the conjugative, steric and first order effects and we analyze the magnitude and trend of the various contributions in the two series of molecules. In these discussions we will make use also of the results obtained in the rigid model, particularly for understanding the relationship between steric and first-order effects.

3.4.1 First-Row Molecules

The STO-3G computations reproduce correctly the experimental trend of the rotational barriers in the series ethane, methylamine, methanol, even though the degree of accuracy of the computed barriers decreases significantly along the series (see Table 6). As previously pointed out, in the optimized model we can distinguish three different types of contributions to the rotational barrier arising from the conjugative, steric and first order effects.

The contributions to the rotational barriers of the conjugative effects are as follows:

Ethane: 0.36 kcal/mol; Methylamine: 0.44 kcal/mol; Methanol: 0.81 kcal/mol

Therefore the conjugative contributions show a clear trend, which is opposite to that experimentally and computationally found for the barriers. However the various component terms do not show equally clear trends: these quantities are, in fact, quite small and their values can be significantly altered by geometrical modifications. In particular it is found that the contribution arising from the nitrogen lone pair is larger than that associated with the MO's of a C-H bond, while that associated with the oxygen σ-type lone pair is almost negligible, so that in methanol the conjugative contribution arises almost completely from the oxygen π-type lone pair.

The corresponding contributions of the steric effects are as follows:

Ethane: 0.39 kcal/mol; Methylamine: 0.18 kcal/mol; Methanol: 0.30 kcal/mol

These contributions do not show a clear trend. However, in an optimized model, the steric effects are inter-related with the first order effects, whose contribution can not be estimated quantitatively with the PMO approach. As previously pointed out, to understand how these two effects operate, it is convenient to start the discussion from the values of the steric repulsions obtained in the rigid model where it is expected that the contribution of the first order effects is much smaller. At this level, the steric contributions to the rotational barriers are as follows:

Ethane: 2.01 kcal/mol; Methylamine: 2.40 kcal/mol; Methanol: 4.30 kcal/mol.

Therefore in the rigid model the steric contributions are significantly larger than those computed in the optimized model and display a clear trend. In order to understand how this trend originates, we analyze the behaviour of the various component terms. These, in the case of methylamine and methanol have already been defined in the equations (10) and (11), while the relevant values are listed in Tables 16 and 22. For ethane the corresponding expression of the steric contribution is as follows:

$$SC = 3\ D\Delta E_{CH} \tag{12}$$

and the values of ΔE_{CH} at the staggered and rigid eclipsed geometries are 8.09 and 9.56 kcal/mol respectively. The comparison of these values shows that:
(i) The ΔE_{XH} term, which represents, as previously pointed out, the energy effect associated with the interactions between a X-H bond MO and the C-H bond MO's of the H_3C- fragment, favors in all cases a staggered rather than an eclipsed alignement of the X-H bond with the methyl C-H bonds. Along the series ethane, methylamine, methanol, with the decrease of the distance between the two interacting fragments, either ΔE_{XH} or the related contribution to the barrier $D\Delta E_{XH}$ increase.
(ii) The ΔE_{n_X} term, which represents the energy effect associated with the interactions between a σ lone pair and the C-H bond MO's of the H_3C- fragment, favors, in both methylamine and methanol, an eclipsed rather than a staggered alignement of the σ lone pair with the methyl C-H bonds. Therefore the related contribution to the barrier has a sign opposite to that of the corresponding term $D\Delta E_{XH}$.
(iii) The ΔE_{π_O} term, which represents the corresponding energy effect when a π-type lone pair is involved and occurs only in methanol, has almost no effect upon the barrier.

Therefore the increase of the steric contribution along the series ethane, methylamine, methanol, is mainly determined by $D\Delta E_{XH}$, the term related to the repulsion between the bond MO's, whose value increases significantly with the decrease of the distance between the two interacting fragments along the series. In fact the contribution associated with the three C-H bonds (2.01 kcal/mol) is smaller than that associated with two N-H bonds (4.72 kcal/mol) and the latter is smaller than that associated with a O-H bond (5.22 kcal/mol). The addition of the term ΔE_{n_X} does not modify this trend.

When we relax the geometry of the eclipsed conformer, there are two main effects acting in opposite directions, which are a reduction of the large steric repulsions associated with the interactions of mutually eclipsing bond MO's and a destabilization of the fragment MO's of the eclipsed conformer. The former effect favors a decrease of the barrier; this trend becomes more pronounced with the increase of the number of lone pairs in the fragment, since the presence of lone pairs allows to attain more significant geometrical changes. On the other hand the destabilization of the fragment MO's tends to increase the barrier and is more pronounced for a bond MO than for a lone pair. Therefore both these effects favor a larger decrease of the barrier in methanol than in methylamine, and in the latter than in ethane. In addition, in methylamine and methanol the contributions to the barrier associated with the bond MO's are even further reduced by the contributions associated with the σ lone pairs, which are of opposite sign. All these factors concur to determine the trend of the rotational barriers in these molecules and dominate the conjugative effects, which just contribute to define the actual values of these barriers.

This rationalization agrees well with the results recently obtained in terms of a total energy approach [40] which shows that the contribution of the repulsive effects to the rotational barrier, which includes here the contributions of both the steric and first order effects, decreases with the decrease of the number of bond-bond repulsions, while the contribution of the conjugative effects increases with the increase of the number of heteroatom lone pairs. However the latter increase is not large enough to compensate for the decrease associated with the lowering of the number of bond-bond repulsions, and consequently the rotational barrier decreases along the sequence ethane, methylamine, methanol. Therefore, these results provide additional quantitative support to the proposed rationalization. They show also that the two treatments, total energy approach and PMO, can complete each other and their combined use can provide a better understanding of the problem under examination.

3.4.2 Second-Row Molecules

In the series methylsilane, methylphosphine, methanethiol, both the STO-3G and STO-3G* computations show the same trends of the rotational barriers (see Table 6), which agree only partially with that experimentally observed. However both types of computations reproduce the main features of the experimental trend, i.e. the barrier increase which accompanies the replacement of the $-SiH_3$ with the $-PH_2$ group and the subsequent barrier decrease which accompanies the replacement of the $-PH_2$ with the -SH group. Therefore we attempt here to understand the origin of these trends and we discuss only the STO-3G results, since we have already pointed out that the effect of the second-row 3d orbitals upon the barrier is small.

The contribution to the rotational barrier of the conjugative effects is as follows:

Methylsilane: 0.18 kcal/mol; Methylphosphine: 0.34 kcal/mol;
Methanethiol: 0.34 kcal/mol.

Therefore, the conjugative effects contribute to determine the barrier increase which accompanies the replacement of $-SiH_3$ with $-PH_2$, but do not contribute to determine the barrier decrease which accompanies the replacement of the $-PH_2$ with the -SH group.

The contribution of the steric effects is as follows:

Methylsilane: 0.54 kcal/mol; Methylphosphine: 0.95 kcal/mol;
Methanethiol: 1.10 kcal/mol.

As previously pointed out in an optimized model these effects are somewhat related to the first order effects and to understand better the role of both these effects it is convenient to start the discussion from the values of the steric repulsions obtained in the rigid model where the contribution of the first order effects should be much smaller. At this level the contribution of the steric effects to the barrier is:

Methylsilane: 0.90 kcal/mol; Methylphosphine: 3.20 kcal/mol;
Methanethiol: 2.93 kcal/mol.

Therefore in the rigid model the trend of the steric contributions parallel that of the rotational barriers. This trend is mainly determined by the contribution associated with the repulsions between the bond MO's of the H_3C- group and the bond MO's of the adjacent moiety. In particular it is found that in all cases the interaction of an X-H bond MO (X = Si,P,S) with the bond MO's of a H_3C-fragment favors a staggered rather than an eclipsed alignement of the X-H bond with the methyl C-H bonds. The preference for the staggered conformation is 0.3, 1.49 and 2.65 kcal/mol for an Si-H, P-H and S-H bond MO respectively and therefore increases significantly with the overlap increase which accompanies the decrease of the distance between the two interacting fragments along the series methylsilane, methylphosphine, methanethiol. Also the repulsive effects associated with the σ lone pairs favor the staggered conformer, even though the energy preference is, in this case, very small (~0.3 kcal/mol), while the π lone pair does not show any conformational preference. Consequently the barrier increase which accompanies the replacement of the $-SiH_3$ with the $-PH_2$ group is caused by the fact that the contribution associated with two P-H bond MO's is larger than that associated with three Si-H bond MO's (2.98 versus 0.9 kcal/mol), while the subsequent decrease which accompanies the replacement of the PH_2 with the -SH group is caused by the fact that the contribution associated with one S-H bond MO is smaller than that associated with two P-H bond MO's (2.65 versus 2.98 kcal/mol).

When we relax the geometry of the eclipsed conformer, we have to consider again the two main effects already pointed out in the discussion concerning the first-row molecules, i.e. the reduction of the large steric repulsions associated with the interactions of mutually eclipsing bond MO's and the related destabilization of the fragment MO's of the eclipsed conformer. The net result of these two effects which act in opposite directions is a reduction of the steric contribution computed in the rigid model which should be larger in methanethiol than in methylphosphine and in the latter than in methylsilane. However in the case of second row molecules both these effects are less pronounced, because of the larger distance between the two interacting fragments, than in the corresponding series of first row molecules. Consequently, the trend of the rotational barriers remains that determined by the steric effects in the rigid model and the other effects just concur to determine the actual values.

4. CONCLUSIONS.

In this paper we have described the results obtained in an application of a quantitative PMO analysis to the study of conformational problems in methylderivatives. In addition to these structural problems, we have also investigated some basic problems connected with the quantitative PMO procedure: particular attention has, in fact, been devoted to a comparative analysis of the advantages and disadvantages associated with the use of fragment localized MO's and fragment canonical MO's. As previously pointed out, the use of fragment localized MO's has the advantage of providing a clear interpretation of the various nonbonded interactions in terms of steric and conjugative interactions. However, in this representation, the intra-fragment non diagonal matrix elements have non zero values and are neglected in the quantitative analysis. Therefore, one of the purposes of the present paper has been to assess the effect, upon the results of the quantitative PMO analysis, of neglecting these matrix elements.

As reference values we have taken those obtained with a canonical representation based on canonical-type fragment MO's. These latter MO's have been obtained starting from the fragment localized MO's and have correct orbital occupancies. The comparison has shown that in all cases the magnitude and trend of the energy effects computed with the localized representation are very similar to those obtained with the canonical representation. The only difference is that the overall energy effects computed in the canonical representation are slightly larger.

The present results show also that the interpretative model based on the canonical representation is, in general, quite complicated. In fact such a model is quite simple when only the π interactions are taken into consideration, but it becomes complicated in all cases where also the σ interactions play a significant role, because of the delocalized nature of the σ fragment MO's. Therefore the present results suggest that it is preferable to use the interpretative model based on the localized representation, where all problems can be discussed in terms of steric and conjugative effects. This model is particularly suited also for comparative analysis on related molecules, as shown in the previous sections.

The various conformational problems have been investigated here using the two-step procedure previously suggested, which involves, first a quantitative PMO analysis in the rigid model, choosing as a starting point the less crowded optimized geometry which is in this case the staggered geometry, followed by a quantitative analysis of the effects of the geometry relaxation in the eclipsed conformer. We have found that in the rigid model all the conformational problems are satisfactorily rationalized only in terms of steric and conjugative effects: both these effects favor a preferential stabilization of the staggered conformation, with the steric effects largely dominant in all cases.

The results support the expectation that in the rigid model the first order effects are not significant. However these effects become important in the optimized model: in fact, the geometry relaxation of the eclipsed conformer is accompanied by a decrease in the steric effects, which is largely counterbalanced by a destabilization of fragment MO's, while the contribution of the conjugative effects remain almost unaltered. Consequently, the gain in total energy which accompanies the geometry relaxation is usually small: in particular, it is very small when only bond MO's are involved and tends to increase increasing the number of lone pairs, since the latter MO's can undergo geometrical modifications at a minor energetical cost than the bond MO's. Therefore, in an optimized model, in addition to the steric and conjugative effects, we have to consider also the first order effects, given by the destabilization of fragment MO's.

However, also in this case, the various structural problems can be rationalized only in terms of conjugative and steric effects, simply through an appropriate extension of the definition of steric effect. To this purpose the term steric effect has to include either the repulsions involving bond orbitals and lone pairs or the destabilization of the fragment MO's.

On this basis we have attempted to rationalize also the trends of the rotational barriers in the various classes of molecules and also in the two series of first-row and second-row molecules.

The present results provide a detailed understanding of the factors which control these conformational problems and show that the quantitative PMO analysis illustrated here, even though it is yet at a preliminary stage, is already an important instrument for the study of structural problems.

5. REFERENCES

1 (a) Istituto Chimico G.Ciamician, Via Selmi 2, Bologna (Italy).
(b) Istituto di Chimica Organica, Viale Risorgimento 4, Bologna (Italy).
2 M.H. Whangbo, H.B. Schlegel and S. Wolfe, J. Am. Chem. Soc., 99 (1977) 1296.
3 F. Bernardi, A. Bottoni, N.D. Epiotis and M. Guerra, J. Am. Chem. Soc., 100 (1978) 6018.
4 D. Kost, H.B. Schlegel, D.J. Mitchell and S. Wolfe, Can. J. Chem. 57 (1979) 729.
5 F. Bernardi and A. Bottoni, Theoret. Chim. Acta, 58 (1981) 245.
6 F. Bernardi and A. Bottoni, in R. Daudel and I.G. Csizmadia (Eds.), Computational Theoretical Organic Chemistry, Reidel, Holland, 1981.
7 (a) M.H. Whangbo and S. Wolfe, Can. J. Chem., 55 (1977) 2778.
(b) S. Wolfe, D.J. Mitchell and M.H. Whangbo, J. Am. Chem. Soc., 100 (1978) 1936.
(c) S. Wolfe, D.J. Mitchell and M.H. Whangbo, J. Am. Chem. Soc., 100 (1978) 3698.
8 F. Bernardi, A. Bottoni and N.D. Epiotis, J. Am. Chem. Soc., 100 (1978) 7205.
9 F. Bernardi, A. Bottoni and G. Tonachini, Theoret. Chim. Acta, 52 (1979) 37.
10 F. Bernardi, A. Bottoni, A. Mangini and G. Tonachini, Theochem. in press.
11 W.J. Hehre, R.F. Stewart and J.A. Pople, J. Chem. Phys., 51 (1969) 2657.
12 J.B. Collins, P.v.R. Schleyer, J.S. Binkley and J.A. Pople, J. Chem. Phys., 64 (1976) 5142.
13 W.A. Lathan, L.A. Curtiss, W.J. Hehre, J.B. Lisle and J.A. Pople, Progr. Phys. Org. Chem., 11 (1974) 175.
14 N.D. Epiotis, W.R. Cherry, S. Shaik, R. Yates and F. Bernardi, Top. Curr. Chem., 70 (1977) 1.
15 R. Hoffmann, Acc. Chem. Res., 4 (1971) 1.
16 N.D. Epiotis, J. Am. Chem. Soc., 95 (1973) 3087.
17 J.P. Lowe, Science, 179 (1973) 527.
18 R. Hoffmann, C.C. Levin and R.A. Moss, J. Am. Chem. Soc., 95 (1973) 629.
19 W.J. Hehre and L. Salem, Chem. Comm., (1973) 754.
20 O. Eisenstein, N.T. Anh, Y. Jean, A. Devaquet, J. Cantacuzene and L. Salem, Tetrahedron, 30 (1974) 1717.
21 I. Fleming, Frontier Orbitals and Organic Chemical Reactions, Wiley-Interscience, New York, 1976.
22 G. Klopman (Ed.), Chemical Reactivity and Reaction Paths, Wiley-Interscience, New York, 1974.
23 R.F. Hudson, Angew. Chem. Int. Ed. Engl., 12 (1973) 36.
24 M.J.S. Dewar and R.C. Doherty, The PMO Theory of Organic Chemistry, Plenum, New York, 1975.
25 B.N. Imarc, Molecular Structure and Bonding. The Qualitative Molecular Orbital

Approach, Academic Press, New York, 1973.
26 S.F. Boys, Rev. Mod. Phys., 32 (1960) 296; S.F. Boys and J.M. Foster, Rev. Mod. Phys., 32 (1960) 300.
27 A. Devaquet, Mol. Phys., 18 (1970) 233.
28 W.L. Jorgensen and L. Salem, The Organic Chemist's Book of Orbitals, Academic Press, New York, 1973.
29 G. Tonachini: unpublished results [30].
30 Geometries computed with the recently suggested computational procedure based on the ab-initio analytical computation of the forces [31]. At the computed geometries the forces are all below 0.005 mdyne, giving bond lengths accurate to 0.001 Å and bond angles accurate to 0.1°.
31 F. Bernardi, H.B. Schlegel and G. Tonachini, J. Mol. Structure, 48 (1978) 243.
32 D.J. Mitchell, S. Wolfe and H.B. Schlegel, submitted for publication.
33 S. Weiss and G.E. Leroi, J. Chem. Phys., 48 (1968) 962.
34 T. Nishikawa, T. Itoh and K. Shimoda, J. Chem. Phys., 23 (1955) 1735; D.R. Lide, Jr., J. Chem. Phys., 27 (1957) 343.
35 K.T. Hecht and D.M. Dennison, J. Chem. Phys., 26 (1957) 48; T. Nishikawa, J. Phys. Soc. Japan, 11 (1956) 781; E.V. Ivash and D.M. Dennison, J. Chem. Phys., 21 (1953) 1804.
36 D.R. Herschbach, J. Chem. Phys., 31 (1959) 91.
37 T. Kojima, E.L. Breig and C.C. Lin, J. Chem. Phys., 35 (1961) 2139.
38 B. Kirtman, J. Chem. Phys., 37 (1962) 2516; T. Kojima and T. Nishikawa, J. Phys. Soc. Japan, 12 (1957) 680; N. Solimene and B.P. Dailey, J. Chem. Phys., 23 (1955) 124; J.L. Binder, J. Chem. Phys., 17 (1949) 499.
39 J.S. Binkley, R.A. Whitehead, P.C. Hariharan, R. Seeger and J.A. Pople, Program No. 368, Quantum Chemistry Program Exchange, Indiana University, Bloomington, Ind.
40 F. Bernardi and A. Bottoni, J. Mol. Structure, 70 (1981) 105.

Molecular Structure and Conformation: Recent Advances,
I.G. Csizmadia (Ed.), *Progress in Theoretical Organic Chemistry*, Volume 3

THE NONCLASSICAL POLYHEDRAL ORGANIC MOLECULES AND IONS

VLADIMIR I.MINKIN and RUSLAN M.MINYAEV
Institute of Physical Organic Chemistry,Rostov on Don University, Rostov on Don,344711,USSR.

1. INTRODUCTION

The frameworks of the innumerable variety of organic molecules are constructed from carbon chains and cycles,whereas three-dimensional cage and polyhedral structures are most typical for inorganic and organometallic compounds,e.g.boron hydrides,carbo and heteroboranes,transition metal carbonyl clusters,etc.Compounds 1-6 represent some simple polyhedral structural forms: tetrahedron,square,pentagonal pyramid and cube.

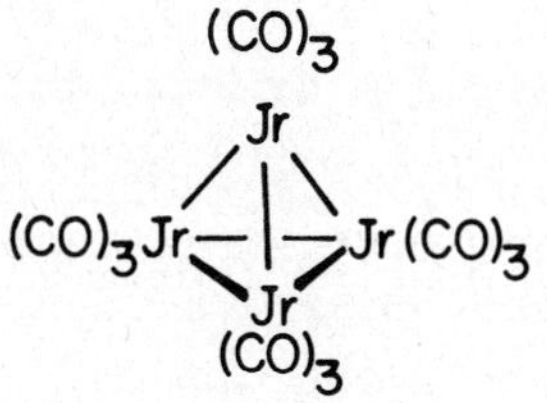

1 $Jr_4(CO)_{12}$

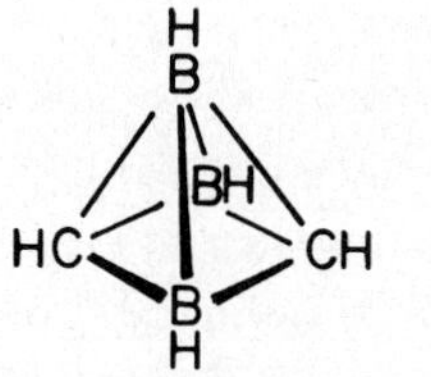

2 $C_2B_3H_5^{2-}$

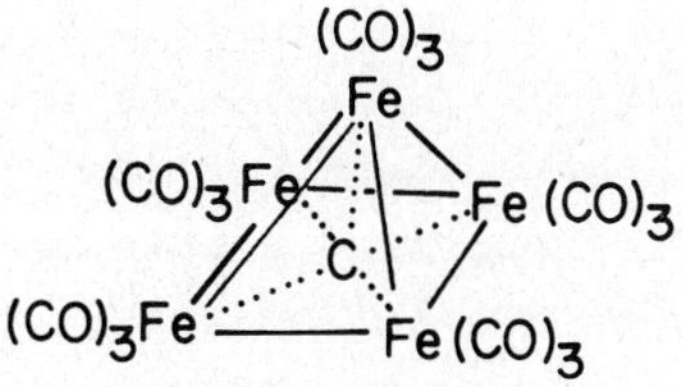

3 $CFe_5(CO)_{15}$

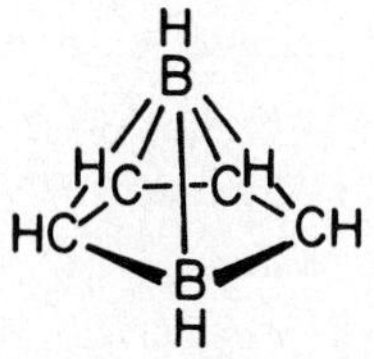

4 $C_4B_2H_6$

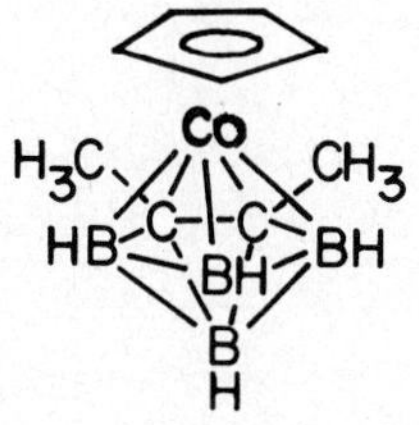

5

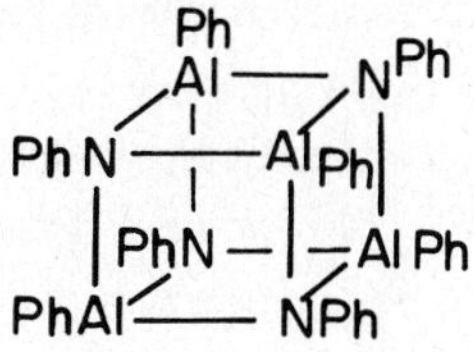

6

The origin of such a structural dichotomy is obviously related to peculiarities of an electronic structure of organic and organometallic compounds.While each separate bond in the carbon chains or cycles of organic molecules may be ascribed to the two-center two-electron bonds,the above shown compounds 1-6 serve as examples of the so-called electron deficient (orbital redundant) structures. The number of valence electrons in these structures is not sufficient for a formation of "normal" two-center two-electron bonds between all the neighbouring atoms.However,molecular geometries provide a full occupation of all bonding multicenter orbitals (ref.1-5).

Therefore the three-dimensionality of the molecular framework of electron deficient compounds,as well as the significant deviations from the standard bond angles (carbon angles of compounds 2-5) and the unusually high coordination numbers (from 5 for the carbon in carbonyl carbide cluster 3 and in a series of closo-carboranes (ref. 5,6) till 8 in analogous Os and Co clusters (ref.7)) should be considered as their compelled adaptation to the main requirement of the molecular structure stability - maximal electron population of all bonding (and only bonding) molecular orbitals.This condition may be fulfilled within geometries providing the attraction between the group orbitals of molecular fragments (ref.8-12).Not infrequently these geometries are in a striking contrast to the standard valency restrictions and normal bond angles. That is why the classical structure theory often loses its predictable power in case of the electron deficient molecules.

Two important groups of organic compounds are to be considered in this connection.These include nonclassical carbonium ions (ref.13) and strained structures (ref.14,15).Tetrahedrane 7a and cubane 11 serve as examples of the latter type of compounds and may be considered as the hydrocarbon analogues of tetrahedral cluster 1 and cluster 6 respectively.The square pyramidal nonclassical cations 8 and 9 (ref.16,17) possess the same type of polyhedral structure as isoelectronic nido-carboranes 2 and 4.The pyramidal structure 10 is also predicted by calculations (ref.18) for a metastable isomer of the neutral hydrocarbon C_5H_4 for which an electron deficiency is due to a distraction of the apical carbon electron pair from the framework bond electron cloud.

The structural analogies in the two groups of compounds considered above (1-6 and 7-11) are obviously not casual.These are determined by a likeness in their electronic configurations.Meanwhile communications about synthesis of the hydrocarbon clusters 7-9,11 were perceived by the organic chemists as sensations.

At present much more information, both of experimental and theoretical character,on the nonclassical polyhedral structures is availab[le] and the similarity of their electronic nature to that of structurall[y] analogous organometallic and inorganic compounds becomes more eviden[t]

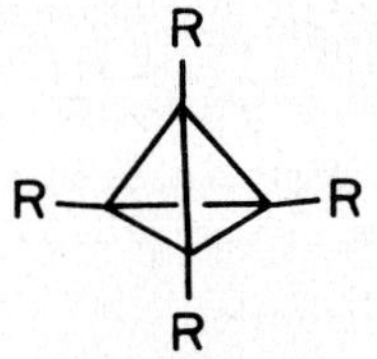

7a, (R=H)
7b, (R=t-Bu)
8 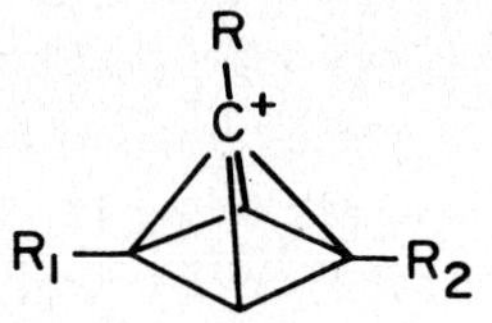
9 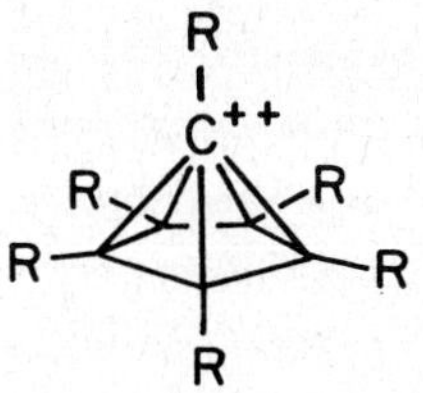
10 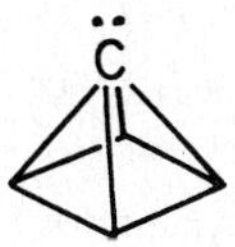
11 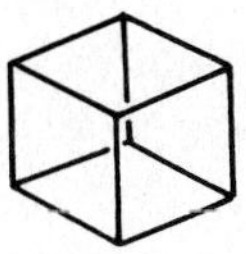

This condensed account aims 1)to survey briefly studies of nonclassical polyhedral organic structures with the accent on theoretical results; 2) to present a simple qualitative approach to understanding and predicting the stability of such structures; 3) to explain on the basis of such an approach the principal reasons for the structural similarities between organic and organometallic polyhedral compounds.

2. PYRAMIDAL STRUCTURES

2.1. Pyramidal hydrocarbons and heterocyclic compounds

Tetrahedrane 7a represents the simplest pyramidal structure. The problem of synthesizing 7a had been formulated about half a century ago and stimulated an extensive investigation of the chemistry of its possible precursors (ref.14,19). Nevertheless, at present tetra tret-butyl tetrahedrane 7b remains the only reliably characterized compound with a tetrahedral carbon framework 7 (ref.20,21). According to ab intio STO-3G calculations (ref.22), the strain energy of the tetrahedrane molecule is equal to 148.8 kcal/mol, i.e. about 25 kcal/mol for each CC bond. In a more rigorous ab initio PMC SCF calculation employing an extended basis set of double zeta quality and taking into account the electron correlation contributions the energy of the CC bond rupture in 7a has been estimated to be equal to only 20 kcal/mol (ref.23)(this value should be compared with 88 kcal/mol for the cc bond rupture in ethane ,ref.24).

The next member of the series of pyramidal hydrocarbons $(CH)_n$ is a square pyramidal cation $(CH)_5^+$ 8 ($R=R_1=R_2=H$) which relates not only to the highly strained, 8 but also to the nonclassical structure with a pentacoordinated carbon atom at an apex. Such a structure has been first proposed by Williams (ref.25), who has pointed out that 8 is isoelectronic with the pyramidal pentaborane 12. The origin of the stability of the nonclassical structure 11 has been elucidated by Stohrer and Hoffmann (ref.26) on the basis of the EHT calculations and the orbital interaction diagrams. At the same time Masamune (ref. 27) reported the experimental observation by NMR spectroscopy of the dimethyl derivative 8 ($R=R_1=CH_3, R_2=H$)

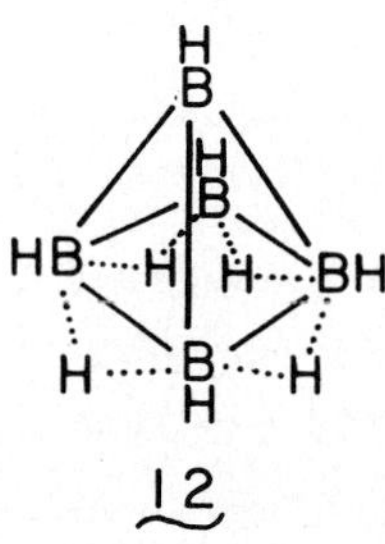

The cation 8 has the homotetrahedranyl precursor 13 which may be derived as an elusive species in several various ways (ref.16,17, 28). Two distinct reaction path conserving the C_s (1) and C_{2v}(2) symmetry may be suggested for the 13 → 8 transformation.

The MINDO/3 calculated potential energy surface (PES) of the above rearrangement (ref.29) (Figure 1) reveals that 13 corresponds to a saddle point for the 8a → 8b interconversion. As seen from Figure 1, the C_{2v} path (2) requires surmounting an additional energy barrier for the 13 → 8c rearrangement, which was evaluated to be 17 kcal/mol.

Theoretical predictions are in full accordance with experimantal findings. Solvolysis of 5-hydroxy-1,3-dimethylhomotetrahedrane in a mixture of SO_2ClF and FSO_3H at -80° provides the 1,3-dimethyl derivative of the cation 8a, but not 8c. Similarly, 1,3-dimethyl-5-CD_2H-homotetrahedranyl cation 13 which is formed by protonation of 5-deuteromethylydene-1,3-dimethylhomotetrahedrane (ref.28,29) undergoes rearragements to the cation 8a or 8b with the CD_2H group

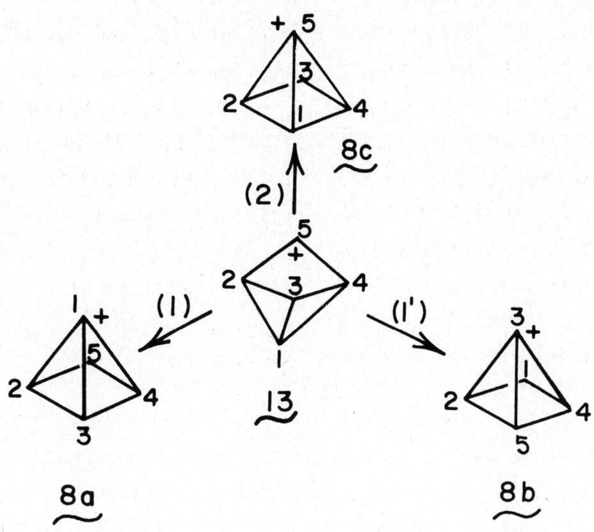

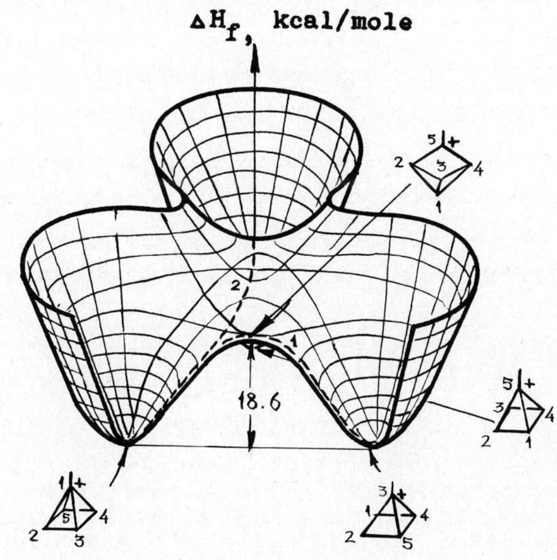

Fig.1. The MINDO/3 potential energy surface for the 13 → 8 rearrangement(ref.29)

only in a basal plane (path 1,1').

The stability of the C_{4v} structure 8 has been confirmed in a series of both semiempirical (ref.16,26,29-33) and ab initio (ref. 34,35) calculations. These indicated that conversion of 8 into more stable isomeric forms such as 14 (triplet) or 15,16 (singlet) should require high activation energies.

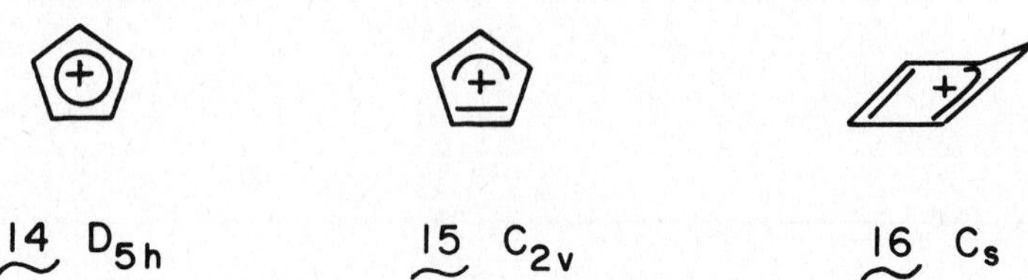

A detailed study of basis set effects and electron correlation contributions to stability differences between the isomeric forms 8 and 15 in their ground state electron configuration has been performed recently by Kohler and Lischka (ref.35). An unusually large stabilization effect by polarization functions and by electron correlation has been found (see Table 1) for the C_{4v} form of $(CH)_5^+$ 8 in relation to 15. This is in full agreement with earlier observations by Pople and Radom (ref.36,37) who pointed out that an inclusion of d orbitals and electron correlation corrections are of special importance for improving the calculated energies of cyclic ionic systems.

Table 1.
Relative stabilities for the $(CH)_5^+$ isomers (ref.35). STO-3G geometry.

Structure	C(7s3p)H(3s)	4-31G	C(7s3p1d)H(3s)	CEPA[a)]	MINDO/3[b)]
15[c)]	0	0	0	0	0
7	66.2	53.8	32.6	9	14.4

a)Electron correlation contributions are taken into account accordin to the CEPA-PNO scheme (R.Ahlrichs,H.Lischka,V.Staemmler and W.Kutzelnigg,J.Chem.Phys.,62 (1975) 1225)
b) ref.30
c)According to corrected MINDO/3 calculations (ref.35) the energy difference between structures 15 and 16 is equal to only 0.7 kcal/mo

At 1973 Hogeveen and Kwant reported the synthesis (ref.38) and ab initio calculation (ref.39) of a new pyramidal C_{5v} hydrocarbon structure - hexamethyl derivative of the $(CH)_6^{2+}$ dication 9 ($R=CH_3$).

A similar C_{5v} pyramidal form 17 has also been discovered in the MINDO-3 and ab initio STO-3G search of the $C_6H_5^+$ PES (ref.40,41). This structure is quite similar to the C_{4v} carbene 10.

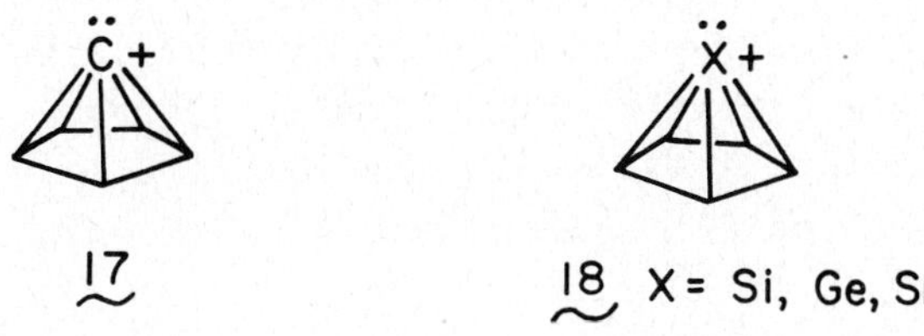

The analogous C_{5v} structures have also been suggested on the basis of semiempirical calculations (ref.41,42) for the ions 18 which were observed in the mass spectra of polynuclear metal carbonyl complexes (ref.43-45). Quite recently Jutzi (ref.46) has succeeded in preparing stable hexamethyl derivatives of 18 and has proved their C_{5v} structure 19 by direct X-ray diffraction measurements.

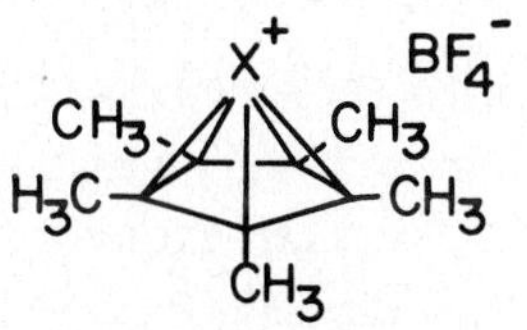

X^+ BF_4^- CH_3 CH_3 H_3C CH_3 CH_3

19a X=Ge
19b X=Sn

X

20a $X=CH^+$
20b $X=Sn^{++}$
20c $X=Pb^{++}$

On the other hand, the C_{6v} pyramidal structure 20a, one of the conceivable forms for the norbornadienyl cation (ref.47), has been shown by the MINDO/3 calculations (ref.48) not to belong to a local minimum on the $(CH)_7^+$ PES. However dications 20b (ref.49-51) and 20c (ref.52) do possess C_{6v} symmetry properties in a crystal state.

The pyramidal C_{4v} structure 10 (pyramidane has attracted our special attention. Although a tetracoordinated carbon atom in an apex possesses an extremely strained pyramidal bond configuration (ref.53), (approximate sp^6-hybridization) the strain energy per CC bond is equal to only 19 kcal/mol (MINDO/3 calculations (ref.18,55)). Both MINDO/3 and ab initio STO-3G and 4-31G (ref.56) calculations lead to the conclusion that structure 10 belongs to a sufficiently deep local minimum on the C_5H_4 PES. This allows to consider 10 as a realistic synthetical target.

Tricyclo [2.1.0.0^{2,5}]pentyliden carbene 21 was suggested as a suitable precursor of pyramidane 10 and some of its least motion transformations were calculated (ref.57) with use of the MINDO/3 reaction coordinate tecnique (ref.58). Figure 2 shows the cross-section of the 21 PES along the reaction path for 21 → 10 rearrangement as well as for some possible rival reactions.

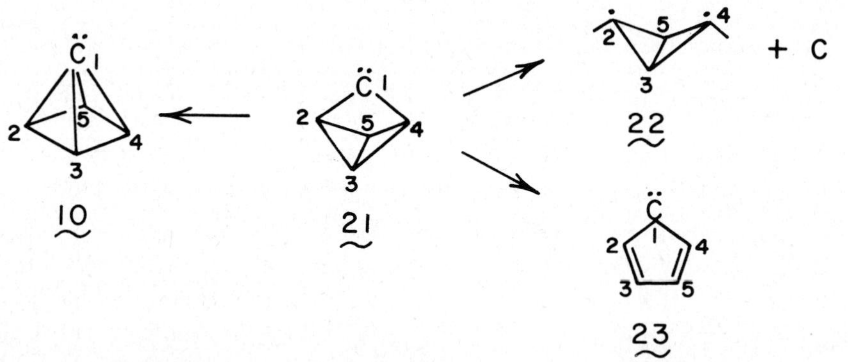

Both 21 → 10 and 21 → 23 rearrangements are thermally disallowed which corresponds to the high energy barriers (38 and 42 kcal/mol) calculated for these reactions. However photochemical reactions seem to be possible which is indicated in Figure 2 by the S_0 and T_1 surface crossings.

Taking into account the energy preference of the cyclopentadienylidene carbene structure 23 in comparison with the C_{4v} form 10 ($E_{relative}$ by MINDO/3 is 25 kcal/mol (ref.57) and by ab initio 4-31G is 52.5 kcal/mol (ref.59)), rearrangement 23 → 10 is hardly conceivable as a possible way to trap pyramidane. This explains a recent failture in attempts to observe the formation of the pyramidane bishomo derivative 25 via carbene 24 (ref.60).

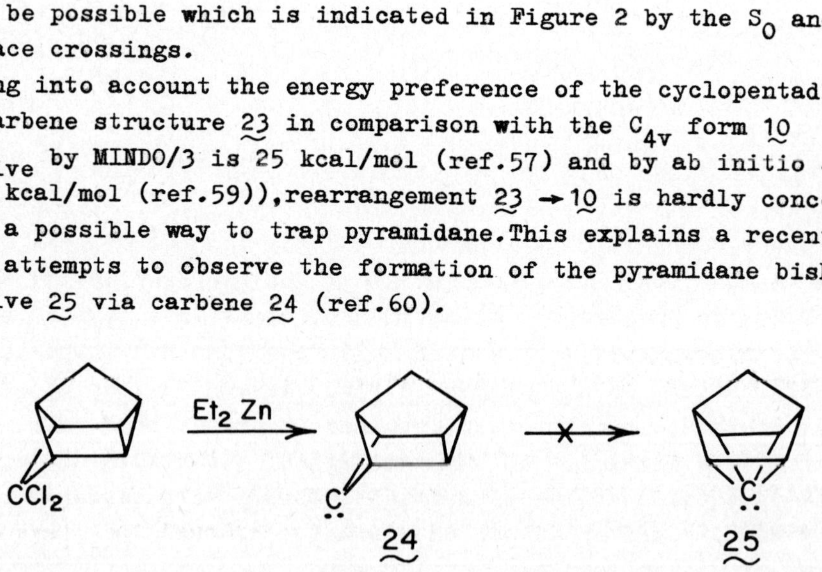

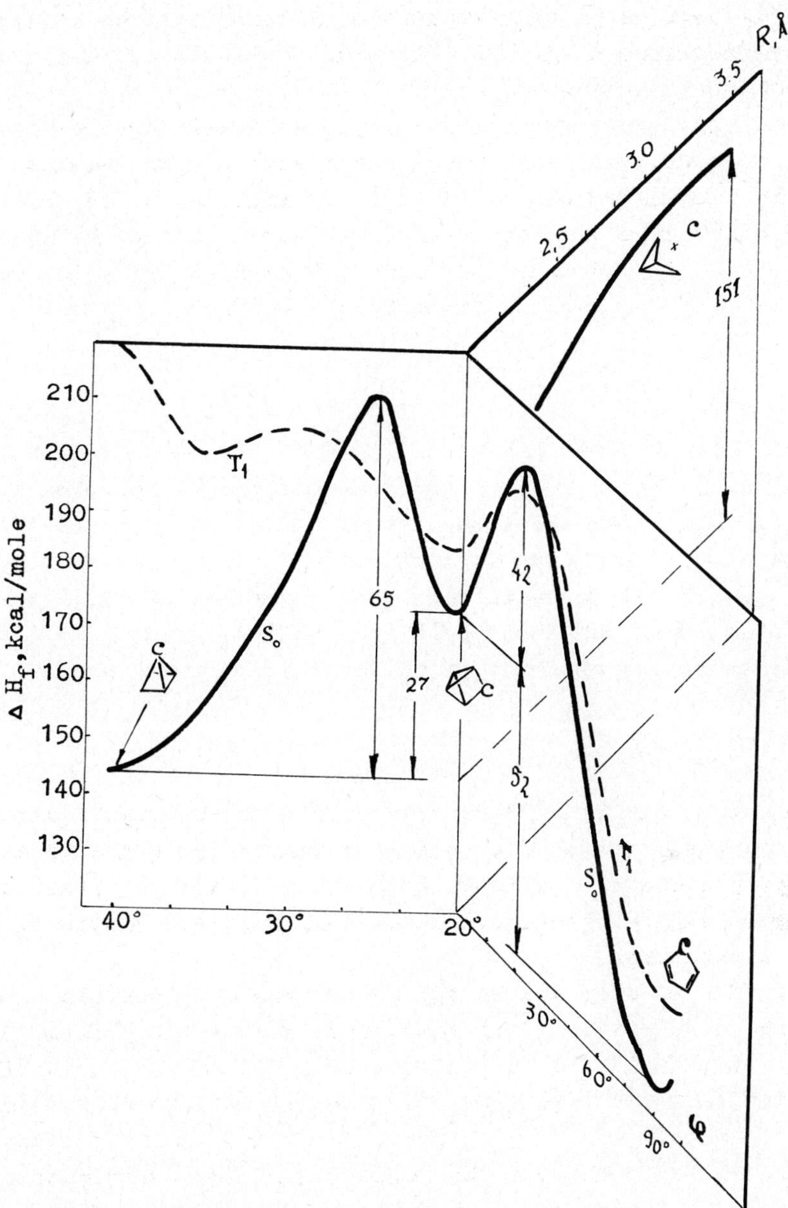

Fig.2. Minimum energy reaction paths for the $\underset{\sim}{21} \rightarrow \underset{\sim}{10}, \underset{\sim}{23}$ and $\underset{\approx}{22}+\underset{\sim}{C}$ conversions by MINDO/3 calculations (ref.58).

In the heterocyclic series suggestions concerning the formation of pyramidal structure $\underset{\sim}{26}$ were introduced as an explanation for the mechanism of the 2-phenyl thiophene to 3-phenyl thiophene isomerization (ref.61) and for the mechanism of the sulfur atom periphery migration in the Dewar thiophene 27 (R=CF_3) (ref.62).

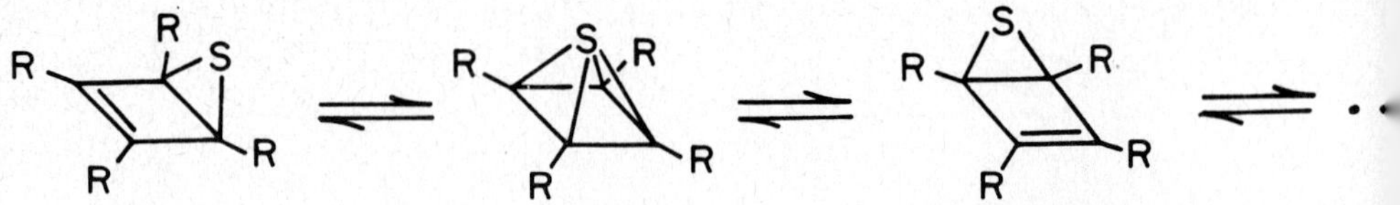

27a ⇌ 26 ⇌ 27b ⇌ ..

However both MINDO/3 and ab initio (ref.56) calculations lead to the conclusion that structure 26 (R=H) does not correspond to a local minimum and represent the top of a flat hill on the PES of the reaction. Similarly, the C_{4v} structure 28, which was proposed (ref. 64) as a possible intermediate in the fragmentation of thiophene ion to $C_2H_2S^{+\cdot}$ has been shown by a recent study (ref.65) of the threshold photoelectron spectrum of thiophen to be unreliable.

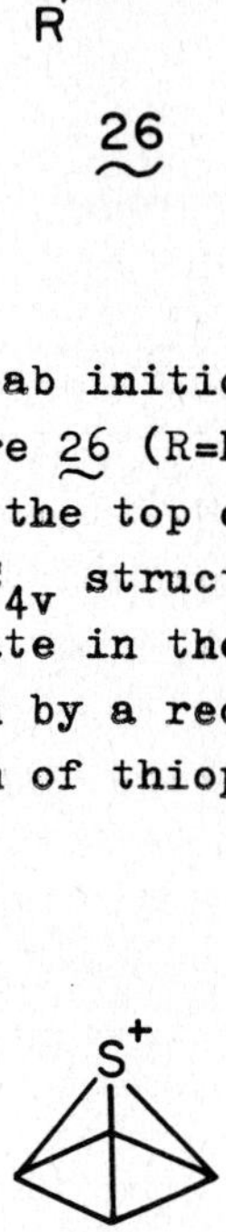

28

The following question arises from consideration of the data presented. How to explain qualitatively the stability of the pyramidal C_{nv} structures 7-10, 17-19, 20b, c, whereas some other structures of the same type (20a, 26, 28) cannot exist as stable species?

2.2. The 8 electron (8e) rule

In order to answer this question the fragment MO analysis in a simple form (ref.8,9) may be employed. All the pyramidal molecules and ions considered above are viewed as organic π-complexes 29 whic are formed by an interaction of the [n] annulene cycle with the apical atom X or bond XR.

Only π and σ -Walsh MO's of the [n] annulene cycles (see ref.66 for their clear pictorial presentation) may interact substantially with the apical center orbitals because of their sufficiently large overlapping and energy level closeness .Since the overlap (S_{ab}) with π -MO's is 3 to 5 times larger and since the orbital interaction energies are proportional to S^2_{ab} (ref.8,10,11,67),a qualitative analysis accounting for only π -MO apex orbital interactions seems to be enough for a good approximation.

Let us begin with pyramidal structure 29a to which compounds 7-10,17-20,26 and 28 belong. The resulting C_{nv} pyramidal structure MO's are derived by direct mixing of the fragment orbitals. The MO interaction diagram is given in Figure 3.

It reproduces in a slightly modified form all the common features of the several previous qualitative studies (ref.18,40,46,63,68) of some C_{nv} pyramidal systems.Mixing the lowest lying π -MO a_1 with a pair of sp_z hybridized AO's of an apical atom yields in the formation of three orbitals: strongly bonding,nearly nonbonding and strongly antibonding a_1 MO's of the C_{nv} pyramidal structure.The stabilization originates mainly from a mixing of the lowest [n] annulene e-MO's and the apical atom p-AO's resulting in a formation of the bonding 1e-MO's of the C_{nv} pyramidal structures.Taking into account that the a_1 MO in the basal cycle are always fully electron occupied,it becomes clear that the construction of stabilized C_{nv} pyramidal structures is possible only if the e-MO's of its basal fragment and the AO's of an apical center contain no more than six electrons.In an equivalent formulation,the total number of π-electrons in any size conjugated basal cycle and the valence electrons of an apical group (interstitial electrons) must not exceed eight (8e rule)(ref.63,68,69).

One can check immediately by a direct electron count,that for the stable C_{nv} pyramidal species 7-10 and 17-20b,c the number of interstitial electrons is indeed equal to eight while,structures 20a

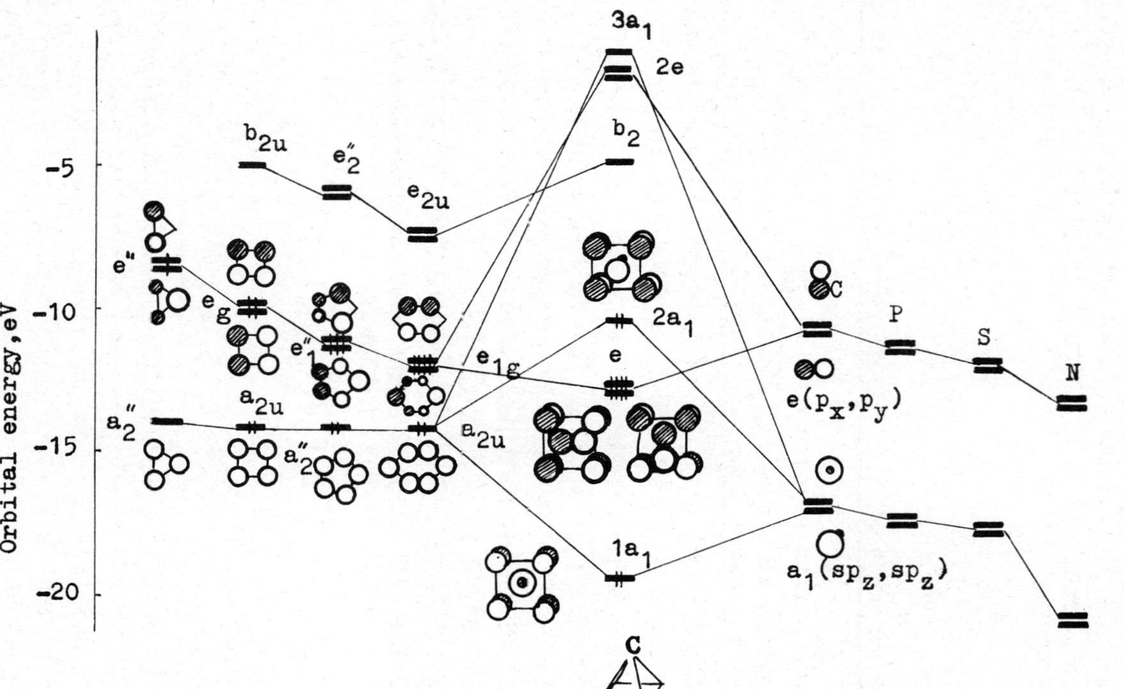

Fig.3. Orbital interaction diagram for fragment MO analysis C_{nv} pyramidal structures. Pyramidane 10 is chosen as an illustrative example. MO levels are calculated by the EHMO method. The sp-orbital energies are taken from H.Hinze, H.H.Jaffe, J.Am.Chem.Soc., 84 (1962) 540.

and 26 are ten electron species. For adopting the 8e rule requirement these molecules lower their symmetry by distortion to 27 in case of 26 and to fluctuating C_s structure 30 with switching off a π-bond in the case of the norbornadienyl cation 30a (ref.70).

30a ⇌ 30b

The orbital interaction diagrams for the pyramidal structure 29b are quite similar to those of 29a. This is demonstrated by Figure 4.

The nonbonding $2a_1$ level for 29a reduces to the low lying $1a_1$ level of 29b. As seen from Figure 4 the latter structure possesses only four bonding MO's that may be filled with no more than 8 electrons (8e rule). The 8e rule may be thought of as the tendency of an apical nontransition element center to form the rare gas electron shell.

Bearing the 8e rule in mind one can speculate about some unusual C_{nv} pyramidal molecules and ions 31-33 that meet this requirement.

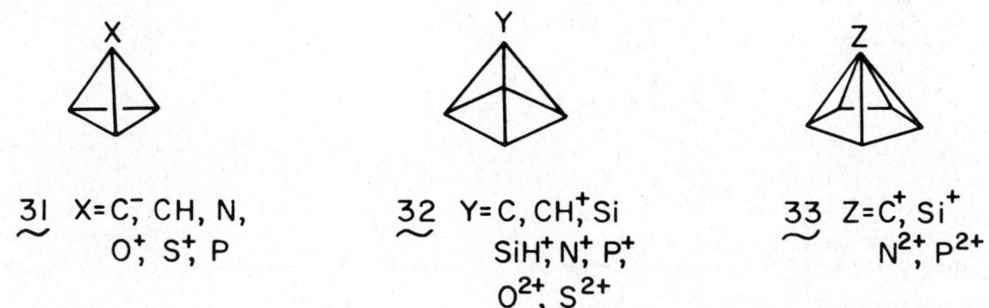

The MINDO/3 calculations (ref.63,68) confirmed that these C_{nv} structures do indeed correspond to true minima on their respective PES. On the other hand, various nonclassical polyhedral structures such as 34-36 which do not follow the 8e rule were shown to lie on hilltops of the PES's and to collaps to the less symmetrical forms 34a-36a.

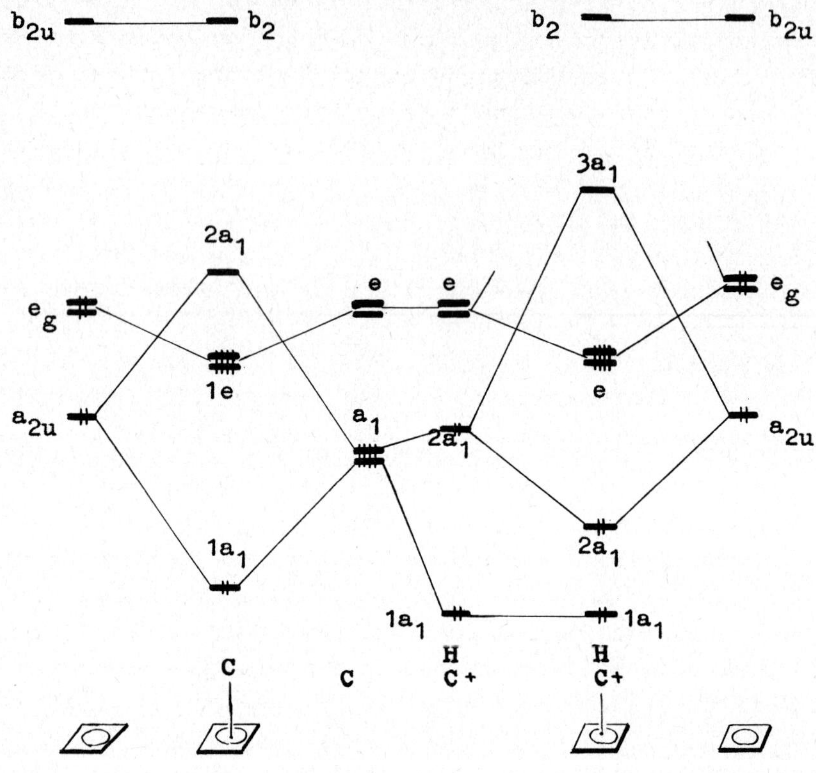

Fig.4. Interrelation between the fragment MO schemes for the structures 29a and 29b.

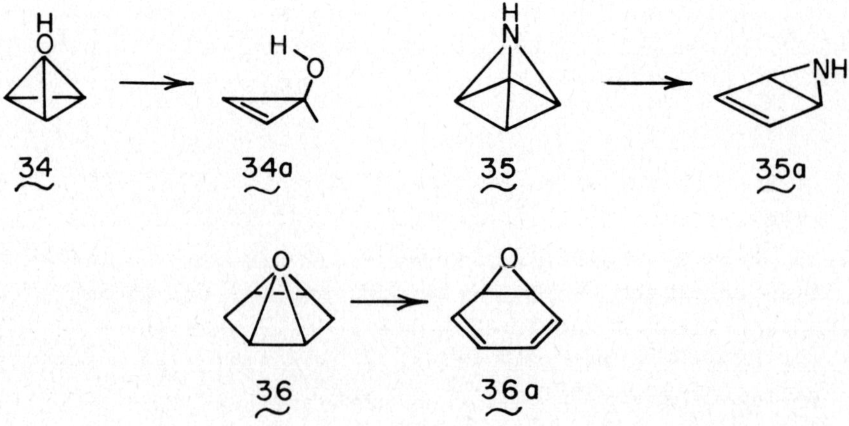

An implementation of the 8e rule requirement does not obviously guarantee a kinetic stability (sufficiently long life time) of the C_{nv} pyramidal structure.However,thorough ab initio (ref.71) and MINDO/3 (ref.72) calculations of tetrahedrane 7a have been performed and revealed its high kinetic stability.Analogous result has also been obtained for pyramidane 10 (ref.57) and azatetrahedrane 37 (ref.68,73) on the basis of MINDO/3 calculations of the reaction paths for the likely isomerizations and decompositions. The scheme given below includes (above arrows) the values (in kcal/mol) of the activation energies for such reactions of 37.These indicate the high kinetic stability of azatetrahedrane even in relation to the more stable isomers (azacyclobutadiene,3-azabicyclobutane [1.1.0]diyl and to the decomposure products.The values of the relative energies of the compounds in the scheme are given in Table 2.

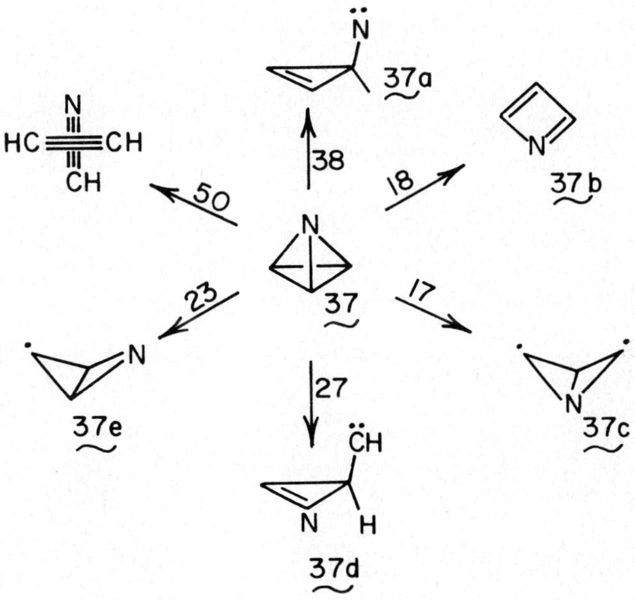

N
37a
N
HC≡CH
CH
38
18
37b
50
N
37
23
17
N
37e
27
ĊH
37c
N H
37d

The calculated geometries of several type 29a C_{nv} pyramidal structures (n=3-5) are given in Figure 5.

The bond formed by the apical centers (X) are 10-20% longer than typical single X-C bonds.Another interesting geometric peculiarity is the direction and magnitudes of the ring hydrogen tilts,especially the bending toward the apical group in the C_{4v} structures.This effect is associated with the geometric factors favoring the larger

Table 2.
Relative energies of the C_3H_3N isomers by MINDO/3 and ab initio calculations (ref.56,68,73)(kcal/mol)

Method	37	37a	37b	37c	37d	37e	HCCH+HCN
MINDO/3	0(H_f=114.3)	15.7	-29.0	6.0	14.0	-20.3	-24.5
4-31G[a]	0(-169.29545)[b]	-	-44.6	-	-	-	-92.5

a)H.B.Schlegel.Private communication to authors
b) in a.e.

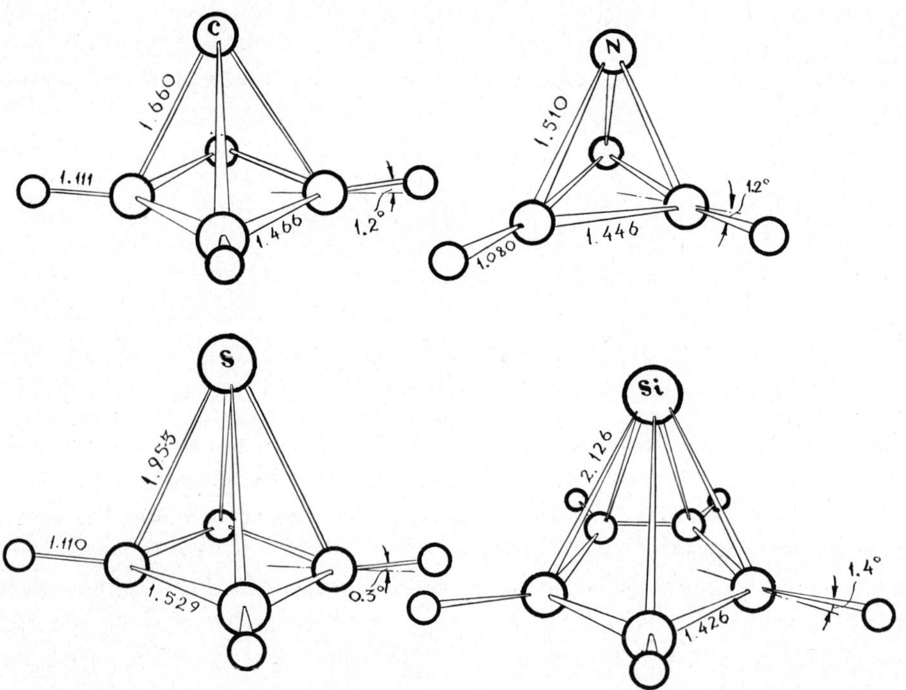

Fig.5. The ab initio STO-3G geometries of the type 29a C_{nv} pyramidal structures: azatetrahedrane (ref.56),thiophene dication (ref.56),and η^5-cyclopentadienyl-Si^+(ref.41).

overlap between the p-orbitals of the apical group and the e-set of the ring π-MO's (ref.41).

2.3. Pyramidal boranes and non-transition metal organometallic compounds.

Unlike the organic compounds pyramidal (nido) structures are quite typical for electron deficient boranes and carboranes (ref. 1,4,6,74). The special electron counting rules which take into consideration the total number of electrons holding their skeletons together have been worked out (ref.2-5). However the 8e rule may serve as an equivalent formulation. This follows directly from the early fragment MO analysis of the pyramidal nido-boranes (ref.1,75). Any borane tetraanions $B_nH_n^{4-}$, carborane dianions $C_2B_{n-2}H_n^{2-}$ and carboranes $C_4B_{n-4}H_n$, for example compounds 38, 39, do possess a pyramidal (nido) structure. Their correspondence to the 8e rule demands is easily checked.

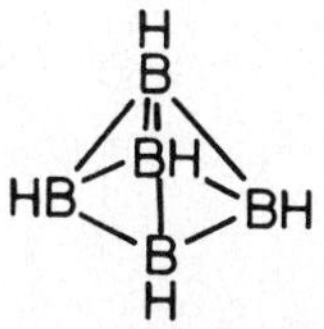

38

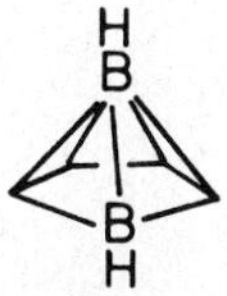

39

The C_{5v} pyramidal structures have been proved for a series of cyclopentadienyl berillium hydride derivatives 40, and a thorough theoretical analysis of the nature of the bonding in these compounds has been performed (ref.67,79). Beryllocene 41, an especially interesting compound, has been investigated with the use of various experimental techniques, including electron diffraction and the low temperature X-ray diffraction measurements (ref.77-80). In accordance with experimental data the semi-sandwich form 41b seems to be slightly preferable in comparison to the sandwich structure 41a. This is in conformity with ab initio STO-3G (ref.69) and the extended basis set (ref.79) calculations.

The pyramidal structure 42 has been established recently (ref.81) for the cation 42 which is isoelectronic to beryllocene 41b . Analogous C_{5v} structures 43 are found in the gas phase for cyclopentadienyltallium (ref..82) and indium (ref.83) (8 interstitial electrons) and calculated (ref.84,85) for cyclopentadienyl lithium (6interstitial electrons). In accordance with the 8e rule predictions, C_{7v} pyramidal form 44 for cycloheptatrienyl lithium may be suggested.

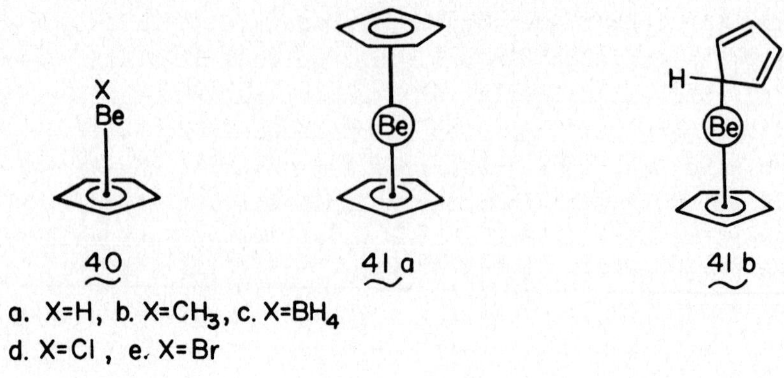

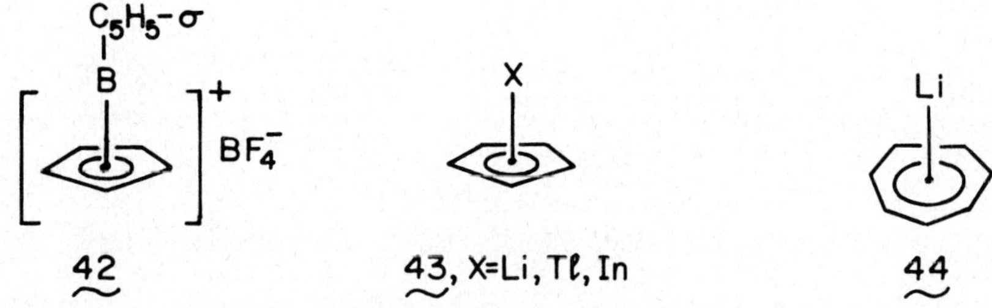

2.4. Homoconjugated cycles in a basal plane.

According to a theorem (ref.86) the principal symmetry properties of the π -MO set of a conjugated cycle are not influenced by any ruptures which are equivalent in a chemical sense to an incorporation of several insulating, for example CH_2 groups, to the conjugated cycle Hence the 8e rule may be extended to the pyramidal structures having polyene chains, isolated π-bonds or even isolated π-centers in a basal part of a molecule. This was exemplified by a recent study of the nonclassical isomer 45 of cyclopentyl cation 46.

Both MINDO/3 and ab initio calculations (ref.87) indicate structure 45 to be a local minimum on the $C_5H_9^+$ PES .It is however 30.8 kcal/mol (4-31G) energy higher than 46.

Bishomo-$(CH)_5^+$ cation 47 (ref.16,88) and a series of cations 48 (ref.89-91),49 (ref.92) have as well been prepared.Their high kinetic stability in relation to various rearrangements has been confirmed by MINDO/3 calculations for 47 (ref.16),48 (ref.33) and 50 (ref.93). By using the same method,structures 47-50 with various 4-electron groups istead of CH^+ and structure 51 have also been found to be local minima on their corresponding PES (ref.94).

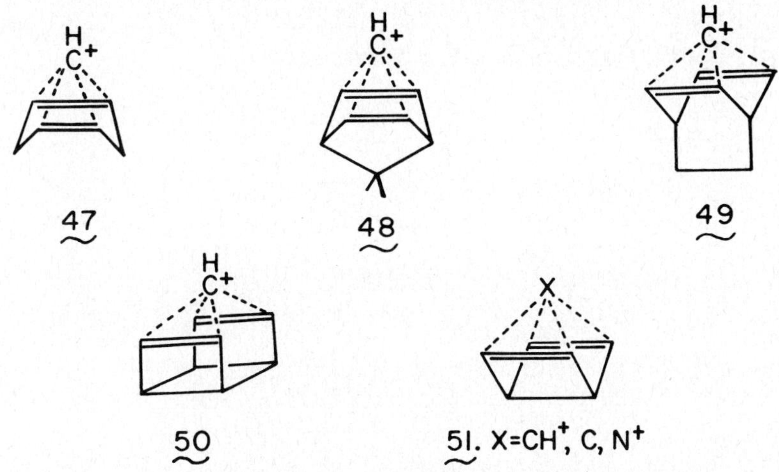

[4.4.4.4] Fenestrane 52 which may be considered as tetrahomopyramidane 53 has attracted special attention in association with prospects for the realization of structures containing planar tetracoordinate carbon (ref.14,95-98).Two conformations of [4.4.4.4] fenestrane with a strongly flattened tetrahedral 54a and pyramidal 54b configuration of the central carbon atom have been discovered in recent MINDO/3 and MNDO calculations (ref.99,100).The nature of the bonding in the latter is quite similar to that of pyramidane 10 (Figure 3).

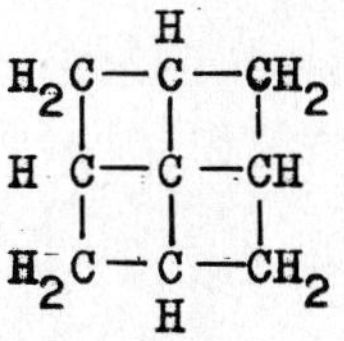

```
        H
H2C — C — CH2
 |    |    |
H C — C — CH
 |    |    |
H2C — C — CH2
        H
```

52

X

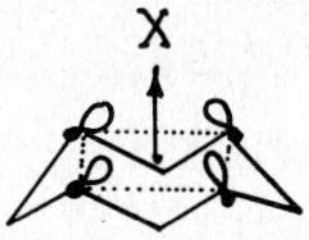

53

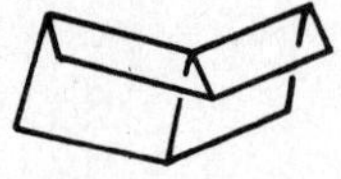

54a D_{2d} 54b C_{4v}

	54a	54b
MINDO/3	H_f=108.5 kcal/mol	137.8 kcal/mol
MNDO	H_f=150.6 kcal/mol	146.5 kcal/mol

It was suggested (ref.99) that the original determination (ref. 14,95) of the fenestrane (windowpan) structures may be extended to the n-gonal windows,so as to consider the type 55-57 compounds as [m.n.p] ,[m.n.p.q] ,[m.n.p.q.r] fenestranes.

55. X=CH, N 56. Y=CH^+, C, N^+, S^{2+} 57. Z=B, BeH, CH^{2+}

The 8e rule which was checked by MINDO/3 calculations (ref.99,101), works satisfactorily for these compounds.It should be noted that such a determination allows to consider tetrahedrane 7a as [3.3.3] fenestrane,i.e. the first member of the [m.n.p] fenestranes.Pyramidane 10 corresponds to [3.3.3.3] fenestrane,i.e. the first member of the type 56 compounds,and dication 9 represents [3.3.3.3.3] fenestrane 57.

3.SANDWICH STRUCTURES

Unlike widespread domain of the transition metal sandwich structures the analogous compounds of the first row elements are not yet known.The sandwich type structure 41a for beryllocene seems doubtful in light of recent experimental and theoretical studies.

The principal reason for the instability of nontransition element sandwiches 58 may be understood on the basis of the fragment MO analysis (ref.76). As Figure 6 shows only four bonding MO's can be formed by fragment orbitals of 58 both in D_{nd} and D_{nh} geometries.

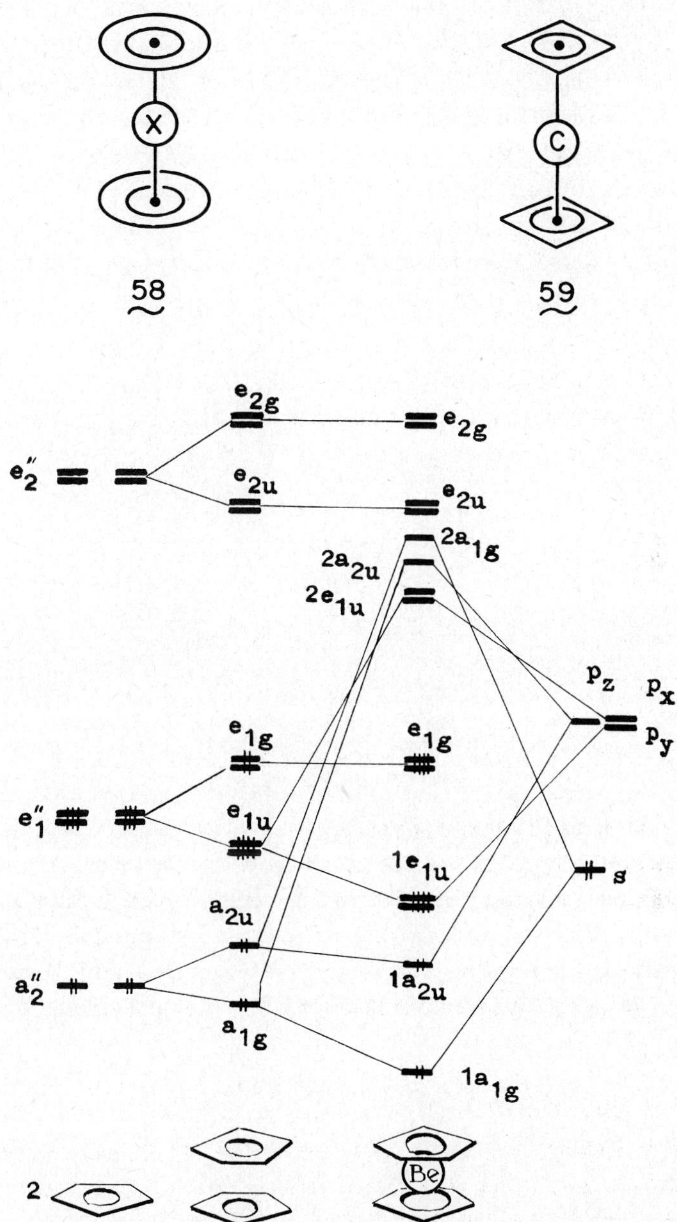

Fig.6. Orbital interaction diagram for the sandwich D_{5d} structure of beryllocene 41a (ref.69).a)MO of noninteracting cyclopentadienyl fragments;b)cyclopentadienyl splitting of energy levels of cyclopentadienyl moieties bringing together at an equilibrium distance in 41a; c) MO's of 41a.

Since the bis- η^5-cyclopentadienyl beryllium 41 structure possesses 12 interstitial electrons (valence electrons of central atom and π - electrons of conjugated cycles) it is unstable. For another 12-electron sandwich compound bis- η^4-cyclobutadienyl carbon 59 the strain energy per CC bond is extremely large (31 kcal/mol by MINDO/3 calculation (ref.102)). However compounds 60-62 (ref.76) and 63 (ref.102) following the 8e rule are expected to be stable. Actuall; the heat of formation calculated for carborane 63 is equal to -7 kcal/mol instead of 476 kcal/mol for 59.

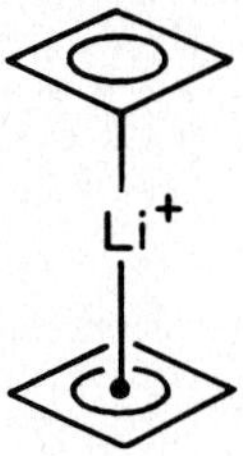

60 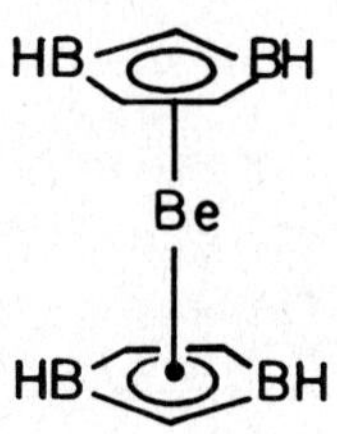 61 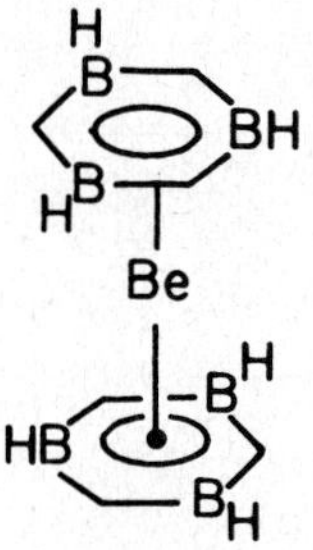 62 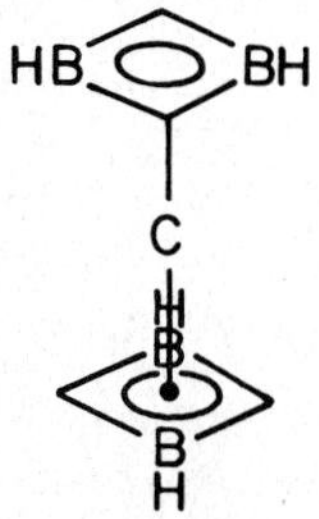 63

Schleyer (ref.76) has pointed out that maintenance of the 8e rule could not guarantee a stable sandwich structure if the e-set of π-MO's were strongly antibonding inside the cycle, see also ref.103 This is the case of η^3-cyclopropylidene compounds. Although 64a does follow the 8e rule, its most stable isomeric form is represented by the σ,σ -structure 64c. 64b takes the intermediate position.

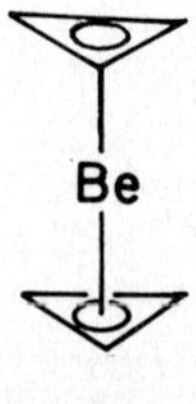

63a 64b 64c

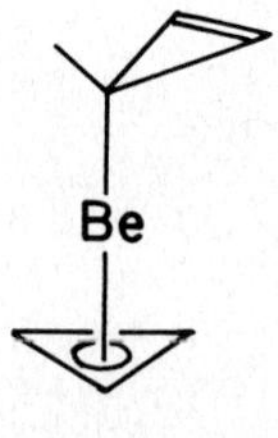

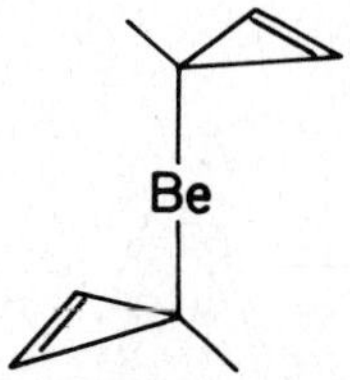

4. BIPYRAMIDAL STRUCTURES

Bipyramidal structures (inverse sandwiches) with a general structure 65 can be formally derived by the addition of a second apex to a pyramid.

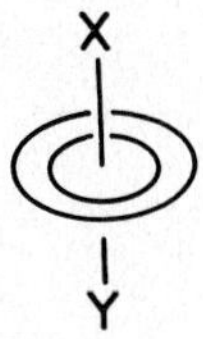

When considering the possible structures for $(CH)_5^+$ cation Stohrer and Hoffmann (ref.26) discussed the trigonal pyramid D_{3h} form 66 and found it to be less stable than the square pyramidal isomer 8. An energy difference was estimated by the MINDO/3 method to be 100 kcal/mol in favour of 8 (ref.30). Moreover the D_{3h} structure of both $(CH)_5^+$ cation and $(CH)_5^-$ anion do not belong to local minima on their PES's (ref.30,104). Similarly, the tetragonal bipyramid D_{4h} structure $(CH)_6$ 67, which may be considered as the benzene isomer, has been shown by ab initio calculations (ref.105,106) to be extremely unstable.

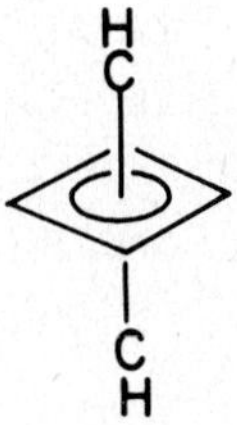

Instability of the bipyramidal structures 66 and 67 is not caused mainly by unusual coordination numbers of carbon in these compounds. Substitution of some CH vertexes in 66 and 67 for BH groups gives rise to well documented stable closo-carboranes $C_2B_{n-2}H_n$ (ref.2,4,6) for instance 68–70.

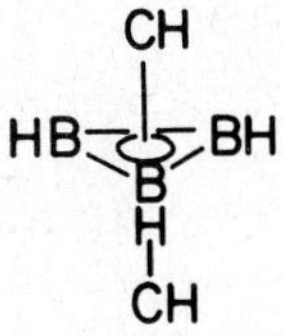

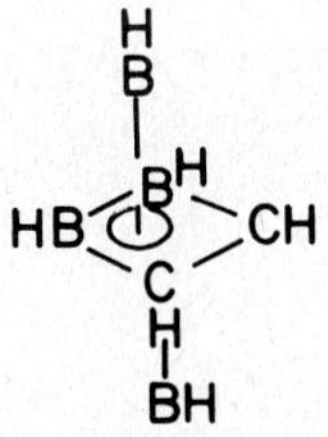

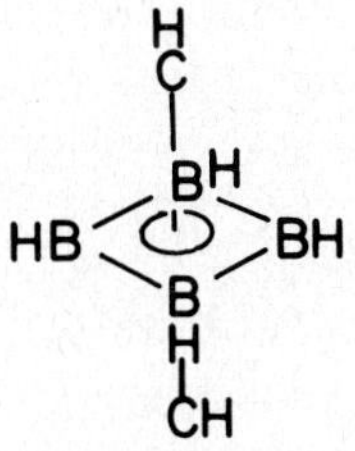

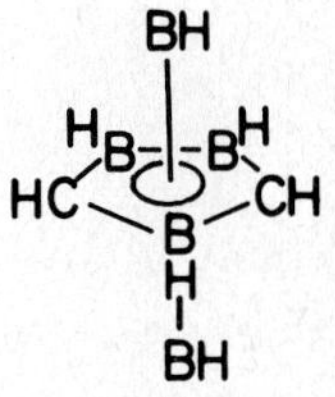

Let us consider the electron count requirements which bipyramidal structures **65** should also follow .As seen from Figure 7 only antisymmetric combinations of the apical center orbitals can interact with the π-MO's of the central plane cycle.The lowest π-MO a_{1u} forms upon mixing of $\sigma_u(s)$ and $\sigma_u(p_z)$ orbitals bonding ($1a_{1u}$), nonbonding ($2a_{1u}$) and antibonding MO's of **65** ,the first two being stabilized.The next two stabilized MO's of **65** originate from interaction of two π_g-orbitals and e_g-cyclic MO's.Therefore coupled with a $\sigma_g(s)$-orbital they form 5 stabilized MO's of the bipyramidal structure **65**.These orbitals can be populated obviously by no more than 10 electrons from π-MO's of a central cycle and valence electrons of apexes.This allows to formulate the 10 electron rule for stability of bipyramidal structures **65** (ref.68).An equivalent formulation (ref.107) which takes into consideration only strongly stabilized $1a_{1u}$ and e_g orbitals of **65** has also been given.

From the qualitative arguments of the 10e rule it follows that bipyramidal organic molecules or ions can hardly be discovered.Actually each CH^+ cap includes 4 electrons.Hence the central cycle should add no more than 2 interstitial electrons,i.e. the structure thus formed should have a great positive charge,for example,+3 in case of **66** and +4 for **67**.However this is quite unlikely even for electrostatic reasons (ref.108).At the same time,carboranes **68**-**71** and bipyramidal (closo) borane dianions $B_nH_n^{2-}$ (n=5-7)(ref.1,75) do obey the 10e rule requirement.The bipyramidal D_{4h} structure **72** has also been proved by ab initio calculation (ref.107) to be a likely candidate for a dilithiated diacetylene $C_4H_4Li_2$.A similar D_{4h} structure was also predicted for the Li_6 cluster (ref.109).

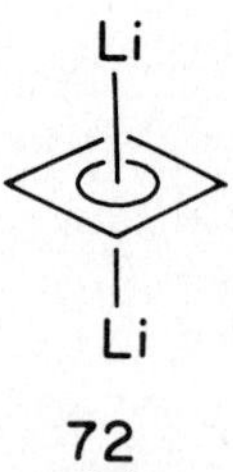

72

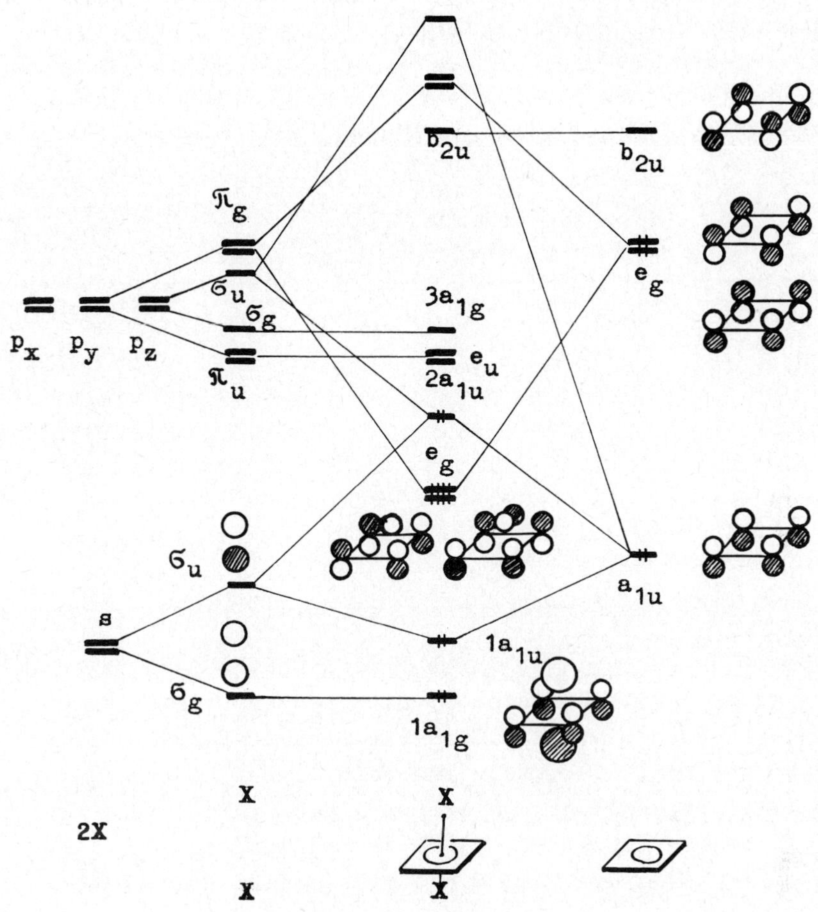

Fig.7. Interaction orbital diagram for the D_{nh} bipyramidal structures 65. D_{4h} form is chosen as an example.

5. THE ISOLOBAL ANALOGY PRINCIPLE AND THE ELECTRON COUNT RULES FOR THE ORGANOMETALLIC COMPOUNDS AND CLUSTERS

Recognition of the equivalence of certain transition metal fragments and groups comprising first row atoms that has been attained last time (ref.2,3,5,110) resulted in a formulation of a new valuable structural concept, isolobal analogy (ref.8,9,11-113). In accordance with Hoffmann(ref.113) two fragments can be termed isolobal if the number,symmetry properties,shapes and energies of their frontier orbitals are approximately the same.For example $Fe(CO)_3$ and CH^+ are isolobal since both contribute two vacant and symmetry identical orbitals and a pair of electrons(skeletal bonding electrons,ref.2) to structures in which they participate.Table 3 contains a brief summary of isolobal groups.

Table 3.
Isolobal and electron equivalent groups.

Nontransition element group	Valence electron	Skeletal bonding electrons	Transition metal fragment
CH, N, O^+	5	3	$Co(CO)_3$, Ni-(Cp- η^5), $Ir(CO)_3$
CH^+, C, BH	4	2	$Fe(CO)_3$, $Os(CO)_3$, Co-(Cp- η^5)
CH^{2+}, C^+, BeH	3	1	$Mn(CO)_3$, Fe-(Cp- η^5)
CH^{3+}	2	0	$Cr(CO)_3$, $Fe(CO)_2$ Cr-(C_6H_6- η^6)
CH_2	6	-	$Fe(CO)_4$
CH_3	7	-	$Mn(CO)_5$

Any chain,cyclic or even three-dimensional polyhedral structures constructed by coupling isolobal units,possess similar orbitals and as a consequence similar structural characteristics.The orbital pattern of the cyclic $Fe(CO)_{12}$ and $Fe(CO)_9$ trinuclear complexes is quite similar to those of cyclopropane and cyclopropényl (ref.114) and an electronic structure of the tetrahedral cluster 1 is similar to that of tetrahedrane 7a.Taking into account the isolobal analogy of the CH^+ and Co- η^5-(C_5H_5) units,the bipyramidal structure 5 can be regarded as obeying the 10e rule demand as well as pyramidal carbide cluster 4 follows the 8e rule.

Some other illustrative examples of organometallic cluster compounds considered together with the corresponding hydrocarbon compounds which were constructed from the isolobal units given below. For extensive review see also ref.2-5,9,111-119.

An important role of the isolobal analogy concept is determined by the possibility to elucidate with its help the similarities of electronic structure between various types of organic and transition element organometallic compounds.

The 8e and 10e rules derived with help of simple fragment MO analysis may be shown to be directly connected to some more widely accepted formulations (ref.2-5) of the skeletal bonding electron

Pyramidal Structures (8e rule)

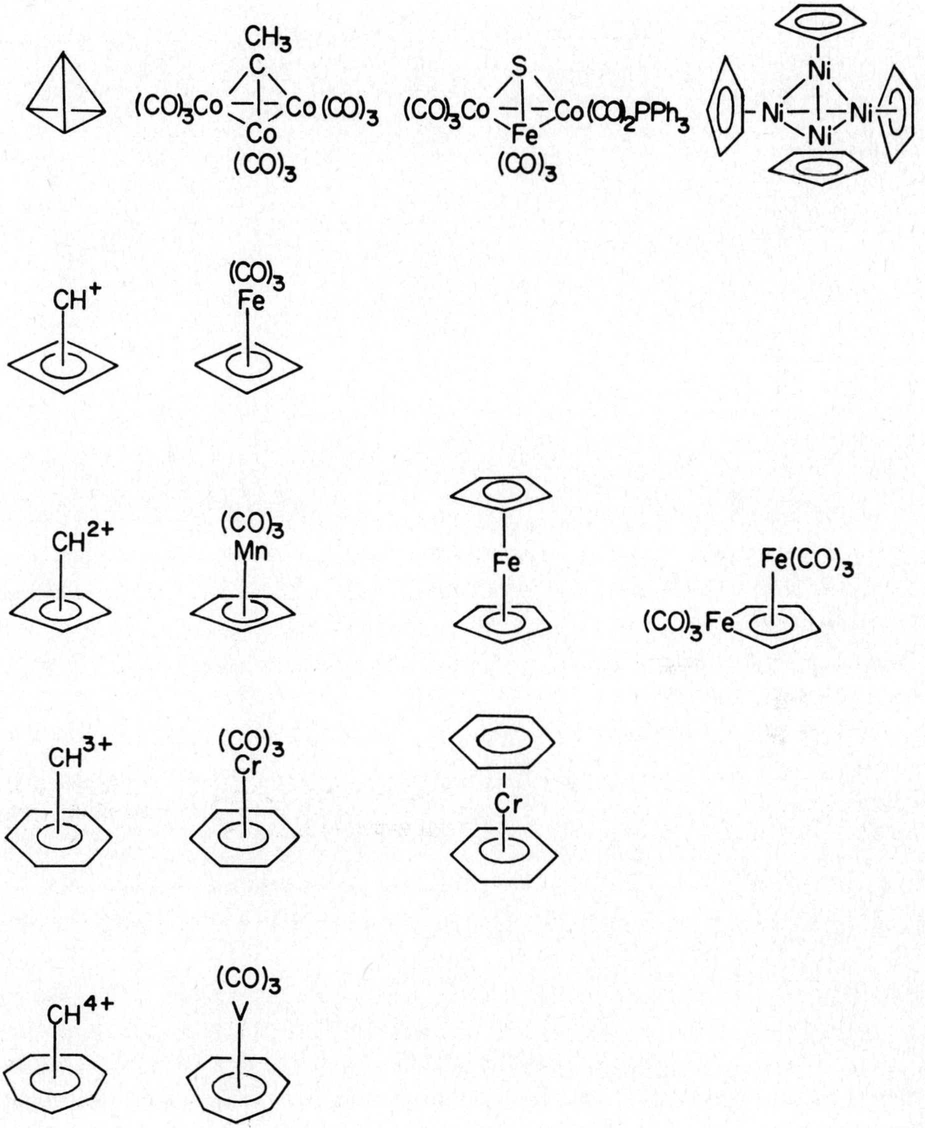

CH_3 C $(CO)_3Co$ $Co(CO)_3$ Co $(CO)_3$

S $(CO)_3Co$ $Co(CO)_2PPh_3$ Fe $(CO)_3$

Ni Ni Ni Ni

CH^+

$(CO)_3$ Fe

CH^{2+}

$(CO)_3$ Mn

Fe

$Fe(CO)_3$ $(CO)_3Fe$

CH^{3+}

$(CO)_3$ Cr

Cr

CH^{4+}

$(CO)_3$ V

Bipyramidal Structures (10e rule)

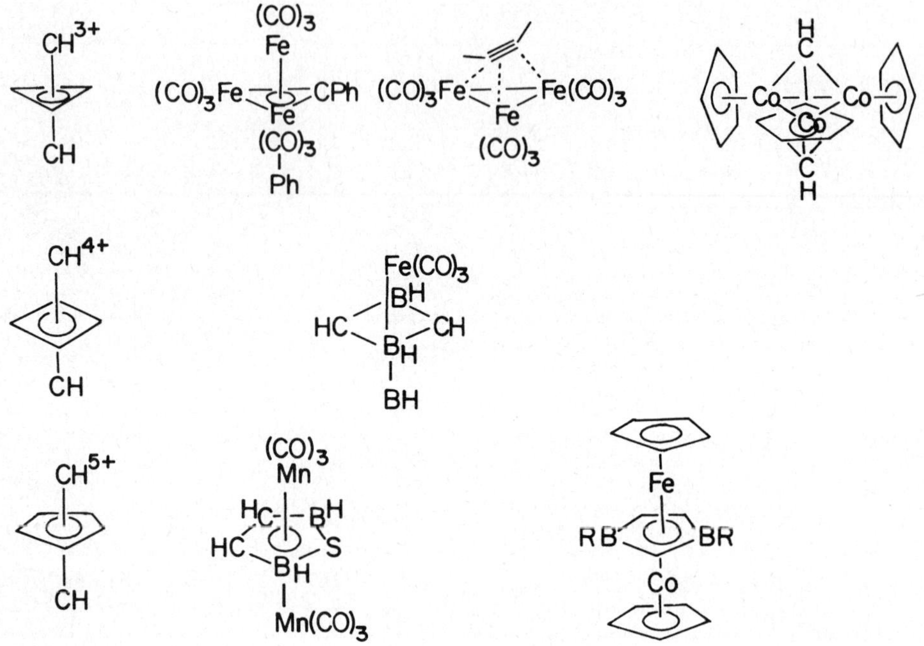

count for boron hydrides and metal carbonyl clusters. Actually it may be easily checked that 2(n-1) skeletal electrons from the whole (2n+4) of those required for the stable pyramidal nido-boranes $B_nH_n^{4-}$ belong to the inner σ-bonds of the basal cycle while 2 additional electrons occupy the σ-orbital of the apical atom exocyclic bond (or form the nonbonding electron pair on such an atom). Therefore the number of electrons binding the apical and basal fragments together is equal to $[(2n+4)-2(n-1)]+2=8$ (the 8e rule). For the bipyramidal closo-structures $B_nH_n^{2-}$ (2n+2 skeletal bonding electrons) the 10e rule may be derived in a similar way: $(2n+2)-2(n-2)+2+2=10$.

The electron count rules above described are topological in their nature. They elucidate the common character of principal features of the electronic and structural organization of organic, inorganic, and organometallic compounds. Amongst many other implications this is supported by the possibility to extend the well known (ref.2-4) structural correlations for the boron hydride derivative redox reactions.

$$\text{Closo} \underset{-2e}{\overset{2e}{\rightleftharpoons}} \text{Nido} \underset{-2e}{\overset{+2e}{\rightleftharpoons}} \text{Arachno}$$

to the organic polyhedral structures
(n-2)-gonal bipyramid(10e) $\underset{-2e}{\overset{+2e}{\rightleftharpoons}}$ (n-1)-gonal pyramid(8e) $\underset{-2e}{\overset{+2e}{\rightleftharpoons}}$ n-gonal cycle (6 π electrons).

A dication 9 ($R=CH_3$) preparation from the Dewar hexamethyl benzene (ref.17) serves as an example of the above type of conversion.

As Figure 8 shows the cage closing is expected on the basis of MINDO/3 calculations (ref.56) to accompany reactions in which the thiophen dication 73 would be generated.

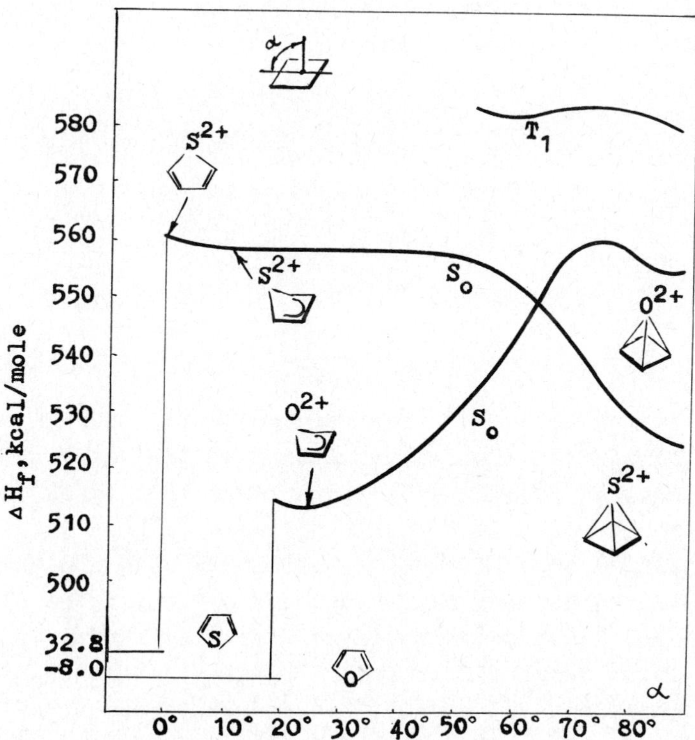

Fig.8. MINDO/3 minimum energy reaction paths for the 73 → 32 type dication transformations.

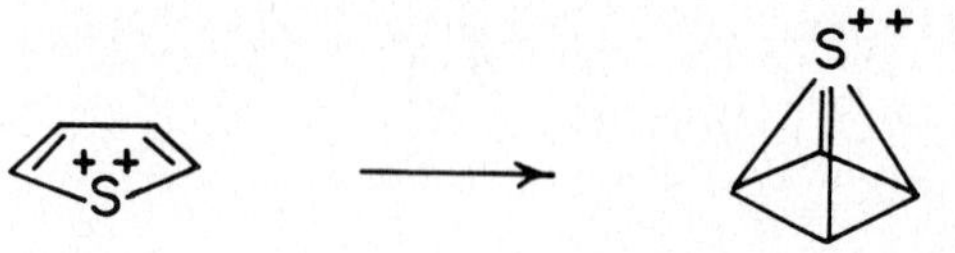

73 → 32, Y=S

6. [N] PRISMANES

The orbital interaction diagrams of the pyramidal and bipyramidal structures which are represented by Figure 3 and 7 demonstrate an important stabilizing role of the π-face bonding with various apical groups. The analogous $\pi_\sigma - \pi_\sigma$ type interactions have been shown (ref.120) to be responsible for the significant part of bonding forces tying two [n]annulene cycles together upon the [n]prismane formation.

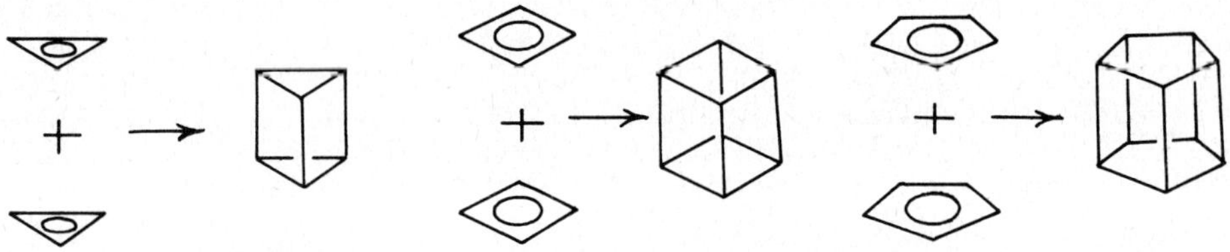

74 75 76

The main stabilizing effect holding two cycles together in [n]prismane originates from a strong splitting of the antibonding π-MO's of the annulene fragments at close range. This makes their in-phase combination be bonding MO of [n]prismane (Figure 9).

It should be pointed out that such an effect is attained only at very short distances between two [n]annulene cycles. At the distances of the Van der Waals contacts (more than 3 Å) the $p_\sigma - p_\sigma$ overlap is small which results in a small π^*-level splitting and the total repulsion effect (Figure 9). This was confirmed by molecular beam studies (ref.182) of benzene dimer and by ab initio (ref. 122) and semiempirical (ref.123) calculations as well.

The orbital interaction diagram in Figure 9 elucidates the reason for [n]prismane clusters to be appropriate structures for

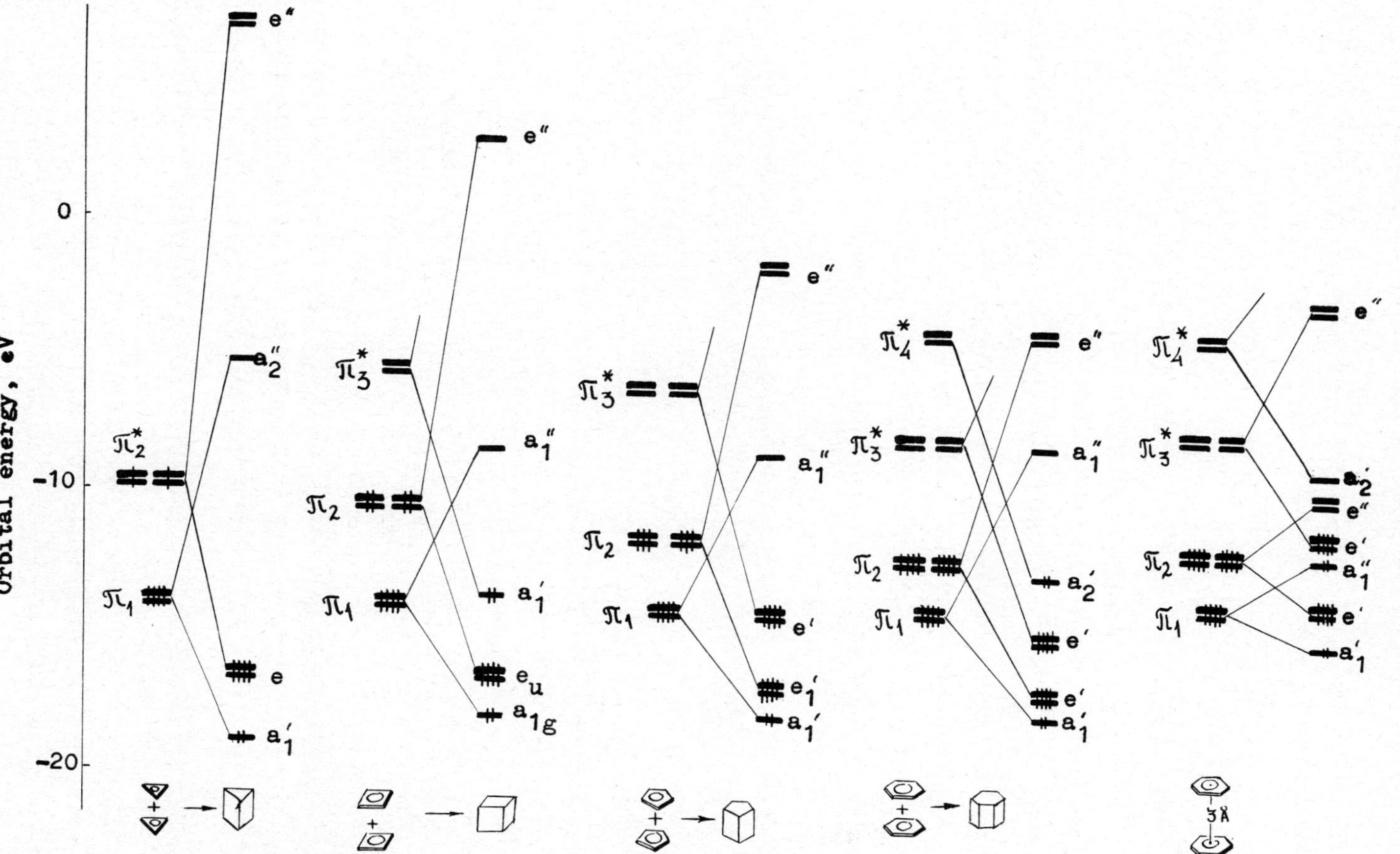

Fig. 9. MO's of [n] prismanes which are derived from the π_6- π_6 type interaction of [n] annulene π-orbitals (S_{p-p} =0.3). At the right splitting of the π-MO levels of two benzene upon their bringing together at the Van der Waals contact distance is shown ($S_{p_6-p_6}$=0.1). This dimer is unstable

only hydrocarbon $(CH)_{2n}$ compounds.D_{2d} Form of closo-borane $(BH)_8^{2-}$ is predicted to be almost 400 kcal/mol preferred to the $O_h(D_{4d})$ structure (ref.124).This is caused by the availability of only two electrons in $(BH)_8^{2-}$ which may be placed on the bonding MO's of the O_h structure.On the contrary the D_{nd} chalcogenide clusters S_8,Se_8, S_4N_4,Se_{10}^{2+} (ref.125) would have several electron occupied antibonding MO's.At the same time the cluster compound 6 is isoelectronic to cubane 75 that causes a similarity of their structures.

7. CONCLUSION

The simple orbital interaction scheme considered above seems to be an appropriate qualitative approach to understanding the nature of bonding and the prediction of novel nonclassical polyhedral organic molecules and ions.Its allows to formulate some common rules of an electron count which may be applied for both organic and organometallic clusters.This underlines the common character of the principal features of their electronic and structural organization.It should be noted finally that the orbital approach ,which gives special attention to the role of the $\pi_\sigma - \pi_\sigma$ interactions,has been recently extended with success even in the field organic and inorganic molecular crystals (ref.126).

8. REFERENCES

1 W.N.Lipscomb, Boron Hydrides,New York,Benjamin,1963.
2 K.Wade,Chem.Brit.,11 (1975) 177.
3 K.Wade,Adv.Inorg.Chem.Radiochem.,18 (1976) 1.
4 R.W.Rudolf,Accounts Chem.Res.,9 (1976) 446.
5 D.M.P.Mingos,Adv.Organometal.Chem.,15 (1977) 1.
6 R.Grimes,Carboranes,Academ.Press,New York,1970.
7.V.G.Albano,P.Ciani,M.Sansoni,D.Strumolo,B.T.Heaton,S.Martinengo, J.Am.Chem.Soc.,98 (1976) 5027.
8 R.Hoffmann,T.A.Albright,D.L.Thorn,Pure Appl.Chem.,50 (1978) 1.
9 R.Hoffmann,Science,
10 M.H.Whangbo,H.B.Schlegel,S.Wolfe,J.Am.Chem.Soc.,99 (1977) 1296
11 N.D.Epiotis,W.R.Cherry,S.Shaik,R.L.Yates,F.Bernardy,Structural Theory of Organic Chemistry,Topics Curr.Chem.,Berlin,Springer,197[illegible]
12 B.M.Gimarc,Molecular Structure and Bonding,New York,Academ.Press, 1979
13 H.C.Brown,Nonclassical Ion Problem,New York-London,Plenum Press, 1977
14 A.Greenberg,J.F.Liebman,Strained Organic Molecules,New York, Academ.Press,1978.
15 M.D.Newton,in Applications of Electronic Structure Theory ,ed. H.F.Schaefer,Modern Theoretical Chemistry,v.4,New York,Plenum Press,1978,223 p.
16 S.Masamune,Pure Appl.Chem.,44 (1975) 661.
17 H.Hogeveen,P.W.Kwant,Accounts Chem.Res.,8 (1975) 413
18 V.I.Minkin,R.M.Minyaev,I.I.Zacharov,V.I.Avdeev,Zh.Org.Khim., 14 (1978) 3.

19 N.S.Zefirov,A.S.Kozmin,A.B.Abramenkov,Usp.Khim.,37 (1978) 289.
20 G.Maier,S.Pfriem,U.Schaefer,R.Matusch,Angew.Chem., 90 (1978) 552.
21 E.Heilbronner,T.B.Jones,A.Krebs,G.Maier,K.Malsch,J.Rockligton, J.Schmelzer,J.Am.Chem.Soc.,102 (1980) 564.
22 J.D.Dill,A.Greenberg,J.F.Liebman,J.Am.Chem.Soc.,101 (1979) 6814.
23 H.Kollmar,J.Am.Chem.Soc.,102 (1980) 2617.
24 J.A.Kerr,Chem.Rev.,66 (1966) 465.
25 R.E.Williams,Inorg.Chem.,10 (1971) 210.
26 W.D.Stohrer,R.Hoffman,J.Am.Chem.Soc.,94 (1972) 1661.
27 S.Masamune,M.Sakai,H.Ona,J.Am.Chem.Soc.,94 (1972) 8955,8956.
28 V.I.Minkin,N.S.Zefirov,M.S.Korobov,N.V.Averina,A.M.Boganov, L.E.Nivorozhkin,Zh.Org.Khim.,submitted for publication.
29 V.I.Minkin,N.S.Zefirov,M.S.Korobov,R.M.Minyaev,N.V.Averina, A.M.Boganov,L.E.Nivorozhkin,J.Am.Chem.Soc.,submitted for publication.
30 M.J.S.Dewar,R.C.Haddon,J.Am.Chem.Soc.,95 (1973) 5836.
31 S. Yoneda,Z.Yoshida,Chem.Lett.,(1972) 607.
32 H.Kollmar,H.O.Smith,P.v.R.Schleyer,J.Am.Chem.Soc.,95 (1973) 5834.
33 C.W.Jefford,J.Mareda,H.Perlberger,U.Buerger,J.Am.Chem.Soc.,101 (1979) 1371.
34 W.J.Hehre,P.v.R.Schleyer,J.Am.Chem.Soc.,95 (1973) 5837.
35 H.Kohler,H.Lischka,J.Am.Chem.Soc.,101 (1979) 3479.
36 L.Radom,P.C.Hariharan,J.A.Pople,P.v.R.Schleyer,J.Am.Chem.Soc., 98 (1976) 10.
37 J.A.Pople,Int.J.MassSpectrom.Ion Phys.,19 (1976) 84.
38 H.Hogeveen,P.W.Kwant,Tetrahedron Lett.,(1973) 1665.
39 H.Hogeveen,P.W.Kwant,J.Postma,P.T.v.Duynen,Tetrahedron Lett., (1974) 4351.
40 W.A.M.Castenmiller,H.M.Buck,Rec.J.Roy.Neth.Chem.Soc.,96 (1977) 207.
41 K.Krogh-Jespersen,J.Chandrasekhar,P.v.R.Schleyer,J.Org.Chem., 45 (1980) 1608.
42 Yu.A.Borisov,Yu.S.Nekrasov,Izv.AN USSR (ser.Khim.) (1980) 1693.
43 B.J.Aylett,K.M.Colquhon, J.Chem.Res.,(1977) 148.
44 I.S.Nekrasov,D.V.Zagorevskii,V.F.Sizov,F.S.Denisov,J.Organomet. Chem.,97 (1975) 253.
45 Yu.S.Nekrasov,V.F.Sizov,D.V.Zagorevskii,Yu.A.Borisov,J.Organometal.Chem.,205 (1981) 157.
46 P.Jutzi,F.Kohl,P.Hofmann,C.Kruger,Y-H.Tsay,Chem.Ber.,113 (1980) 757.
47 S.Winstein,C.Ordonneau,J.Am.Chem.Soc.,80 (1960) 2084.
48 C.Cone,M.J.S.Dewar,D.J.Landman,J.Am.Chem.Soc.,99 (1977) 372.
49 H.Luth,E.L.Amma,J.Am.Chem.Soc.,91 (1969) 7515.
50 Th.A.Anel,E.L.Amma,J.Am.Chem.Soc.,100 (1978) 5941.
51 J.L.Jefferts,M.B.Hossain,K.C.Mollog,D.Helm,J.J.Zuckerman, Angew.Chem.,19 (1980)309.
52 A.G.Gash,P.F.Rodesiler,E.L.Amma,Inorg.Chem.,13 (1974) 2429.
53 V.I.Minkin,R.M.Minyaev,I.I.Zacharov,Chem.Commun.,(1977) 213.
54 M.Krogh-Jespersen,J.Chandrasekhar,E.Wurthwein,J.B.Collins, P.v.R.Schleyer,J.Am.Chem.Soc.,102 (1980) 2263.
55 V.I.Minkin,R.M.Minyaev,Zh.Org.Khim.,15 (1979) 225.
56 V.I.Minkin,R.M.Minyaev,V.I.Pavlov,to be submitted for publication.
57 R.M.Minyaev,V.I.Minkin,N.S.Zefirov,Yu.A.Zhdanov,Zh.Org.Khim., 15 (1979) 2009.
58 M.J.S.Dewar,Chem.Brit.,97 (1975) 11.
59 K.Krogh -Jespersen,P.v.R.Schleyer,unpublished results.
60 C.W.Lefford,V.de los Heros,Tetrahedron Lett.,21(1980) 913.
61 J.Skramstad,B.Smedsrad,Acta Chem.Scand.,B31 (1977) 625.
62 H.Kwart,T.George,J.Am.Chem.Soc.,99 (1977) 629.
63 V.I.Minkin,R.M.Minyaev,Zh.Org.Khim.,15 (1979) 1337.
64 H.Wynberg,R.M.Kellogg,H.van Driel,G.E.Berhuis,J.Am.Chem.Soc.,

89 (1967) 3501.
65 J.J.Butler,T.Baer,J.Am.Chem.Soc.,102 (1980) 6764.
66 W.L.Jorgensen,L.Salem,The Organic Chemist's Book of Orbitals, Academ.Press,New York,1973
67 L.Salem,J.Am.Chem.Soc., 90 (1968) 543.
68 V.I.Minkin,R.M.Minyaev,Izv.So Acad.Nauk USSR (ser.khim.) (1980) 87.
69 E.D.Jemmis,S.Alexandratos,P.v.R.Schleyer,A.Streitwieser,H.F.Schaeffer,J.Am.Chem.Soc.,100 (1978) 5695.
70 G.A.Olah,G.K.S.Prakash,T.N.Rawdah,D.Whittaker,J.C.Reco,J.Am.Chem. Soc.,101 (1979) 3935.
71 H.Kollmar,J.Am.Chem.Soc.,102 (1980) 2617.
72 R.C.Bingham,F.Carrion,M.J.S.Dewar,H.Kollmar,J.Am.Chem.Soc., 102 (1980) 21
73 R.M.Minyaev,V.I.Minkin,Theor.and Experiment.Khim.,16 (1980) 659.
74 E.L.Muetterties(ed.),Boron Hydride Chemistry,Acad.Press,New York, 1975.
75 R.Hoffmann,W.N.Lipscomb,J.Chem.Phys.,36 (1972) 2179.
76 J.B.Collins,P.v.R.Schleyer,Inorg.Chem.,16 (1977)152.
77 O.P.Charkin,Stability and Structure of Gase Inorganic Molecules, Nauka,Moscow,1980.
78 A.Allmenningen,A.Haaland,J.Lusztyk,J.Organomet.Chem.,170 (1979) 2
79 R.Gleiter,M.C.Bohm,A.Haaland,J.Lusztyk,R.Johansen,J.Organomet. Chem.,170 (1979) 285.
80 J.Lusztyk,K.B.Staroweiyski,J.Organomet.Chem.,170 (1979) 293.
81 P.Jutzi,A.Seufert,J.Organomet.Chem.,161 (1978) C5 p.
82 J.K.Tyler,A.P.Cox,J.Sheridan,Nature (London) 183 (1959) 1182.
83 S.Shibata,L.S.Bartell,R.M.Gavin,J.Chem.Phys.,41 (1964) 712.
84 S.Alexandratos,A.Streitwieser,H.F.Schaefer,J.Am.Chem.Soc., 98 (1976) 7959.
85 N.T.Anh,M.Elian,R.Hoffmann,J.Am.Chem.Soc.,100 (1978) 110.
86 M.J.Goldstein, R.Hoffmann,J.Am.Chem.Soc.,93 (1971) 6193.
87 W.Franke,H.Schwarz,H.Thies,J.Chandrasekhar,P.v.R.Schleyer, W.J.Hehre,M.Saunders,G.Walker,Angew.Chem.Int.Ed.,18 (1980)485.
88 S.Masamune,M.Sakai,A.V.Kemp-Jones,H.Ona,A.Venot,T.Nakashima, Angew.Chem.Int.Ed.,12 (1973) 769.
89 H.Hart,M.Kuzuya,J.Am.Chem.Soc.,96 (1974) 6536; 97 (1975) 2450.
90 A.V.Kemp-Jones,N.Nakamura,S.Masamune,Chem.Commun.,(1974) 109.
91 H.Hart,R.Willer,Tetrahedron Lett.,(1978) 4189.
92 R.H.Coates,E.R.Freetz,Tetrahedron Lett.,(1977) 1955.
93 W.L.Jorgensen,J.Am.Chem.Soc.,99 (1977) 4272.
94 V.I.Natanson,R.M.Minyaev,Theoret.and Experim.Khim.,17 (1981) 264.
95 V.Georgian,S.Saltzman,Tetrahedron Lett.,(1972) 4315.
96 K.B.Wiberg,G.B.Ellison,J.J.Wendoloski,J.Am.Chem.Soc.,98 (1976) 12
97 J.B.Collins,J.D.Dill,E.D.Jemmis,Y.Apelog,P.v.R.Schleyer,R.Seeger, J.A.Pople,J.Am.Chem.Soc.,98 (1976) 5419.
98 W.T.Hoeve,H.Wynberg,J.Org.Chem.,45 (1980) 2925,2930.
99 V.I.Minkin,R.M.Minyaev,V.I.Natanson,Zh.Org.Khim.,16 (1980) 673.
100 E.U.Wurthwein,J.Chandrasekhar,E.D.Jemmis,P.v.R.Schleyer, Tetrahedron Lett.,22 (1981) 843.
101 R.M.Minyaev,V.I.Natanson,Izv.SCNC (natural science)(1980) 55.
102 V.I.Minkin,R.M.Minyaev,Zh.Org.Khim.,15 (1979) 225.
103 O.Yu.Lopatko,N.M.Klimenko,M.E.Dyatkina,Zh.Str.Khim.,13 (1972) 11
104 R.E.Stanton,J.W.McIver,J.Am.Chem.Soc.,97 (1975) 3632.
105 M.D.Newton,J.M.Schulman,M.M.McManus,J.Am.Chem.Soc.,96 (1974) 17.
106 K.Meer,J.C.Mulder,Theor.Chim.Acta,41 (1976) 183.
107 A.J.Kos,P.v.R.Schleyer,J.Am.Chem.Soc.,102 (1980) 7928.
108 L.Radom,H.F.Schaeffer,J.Am.Chem.Soc.,99 (1977) 7522.
109 H.Beckman,J.Koutecky,P.Botschwina,W.Meyer,Chem.Phys.Lett., 119 (1979) 67.

110 A.S.Foust,M.S.Foster,L.F.Dahl,J.Am.Chem.Soc.,91 (1969)5631.
111 M.Elian,R.Hoffmann,Inorg.Chem.,14 (1975) 1058.
112 M.Elian,M.M.L.Chen,D.M.P.Mingos,R.Hoffmann,Inorg.Chem.,15 (1976) 1148.
113 J.M.Lauher,M.Elian,R.H.Summerville,R.Hoffmann,J.Am.Chem.Soc., 98 (1976) 3219.
114 B.E.R.Schilling,R.Hoffmann,J.Am.Chem.Soc.,101 (1979) 3546.
115 R.N.Grimes,Coord.Chem.Rev.,28 (1979) 47.
116 E.Band,E.L.Muetterties,Chem.Rev.,78 (1978) 639.
117 H.Werner,J.Organomet.Chem.,200 (1980) 335.
118 E.L.Muetterties,J.Organomet.Chem.,200 (1980) 177.
119 J.E.Ellis,J.Chem.Educ.,53 (1976) 1.
120 V.I.Minkin,R.M.Minyaev,Zh.Org.Khim.,17 (1981) 237.
121 J.M.Steed,T.A.Dixon,W.Klemperer,J.Chem.Phys.,70 (1979) 4940.
122 K.B.Lipkowitz,J.Am.Chem.Soc.,100 (1978) 7535.
123 D.B.Chestnut,P.E.S.Wormer,Theoret.Chem.Acta,20 (1971) 250.
124 D.A.Kleier,W.N.Lipscomb,Inorg.Chem.,18 (1979) 1312.
125 R.E.Gillespie,Chem.Soc.Rev.,8 (1979) 315.
126 J.K.Burdett,J.Am.Chem.Soc.,102 (1980) 450.

Molecular Structure and Conformation: Recent Advances,
I.G. Csizmadia (Ed.), *Progress in Theoretical Organic Chemistry*, Volume 3

CORRELATION ENERGY AS A STABILIZING FACTOR IN MOLECULAR STRUCTURE

M. A. ROBB

1. INTRODUCTION

During the past 5 years there have been dramatic advances in the computational aspects of the electron correlation problem. Most of the computational difficulties associated with the configuration interaction (CI) method [1-2] and the multi-configuration self consistent field (MC-SCF) method [3-6] have been resolved. We are approaching the situation where these types of calculations may be carried out as routinely as ordinary SCF calculations. However, in these methods there is much empiricism. This is associated with, for example, the truncation and choice of expansion in terms of configurations. Thus it still requires much experience to apply these very general methods to chemical problems and, more important, to interpret the results.

In a recent monograph, Carsky and Urban [7] have reviewed the complete field of ab-initio calculations to mid 1979. The reader of this excellent monograph should find that much of the empiricism associated with performing SCF (eg. basis set choice) or calculations that include electron correlation can now be replaced with very definite statements about which method works best under certain conditions. In their section on electron correlation, Carsky and Urban were concerned almost entirely with the case where the SCF method is already a good approximation (in the region of the equilibrium internuclear separation).

The objective of this article is quite different. We shall attempt to formulate a definition of electron correlation that is valid over a complete potential surface. We shall discuss the separation of the structure independent part of the electron correlation problem from the structure dependent part of the electron correlation problem using the language of the CI and MC-SCF methods. Finally using a few selected case studies drawn from the literature, we hope to give some physical insight into the nature of the structure dependent part of the correlation problem.

2. ELECTRON CORRELATION AND THE MO-CI METHOD

2.1 THE MO-CI FORMALISM

It is our intention in this section to present the CI method in a very general way so that we may use this formalism as a language to discuss the correlation problem itself. We shall omit all discussion of practical aspects of calculations at this stage and refer the reader to the literature (see the reveiw article [8] by I. Shavitt, for example).

Let us assume that we have at our disposal a set of orthogonal orbitals ϕ (which may have originated from an SCF calculation). Each orbital has a space part and a spin part so that we can write:

$$\phi_i = \phi_i \alpha$$

$$\bar{\phi}_i = \phi_i \beta$$

It is now possible to specify what we mean by a configuration by specifying a) the occupancies of the spatial orbitals (0,1,2) and b) the total spin 2S+1. From the notion of a configuration we can define configuration state functions (CSF) which are wavefunctions that describe a configuration. Since there are many ways of coupling together the spins of the individual electrons to produce the same total spin, each configuration will generate several CSF.

An example serves to clarify the point. Let us consider the case of 4 electrons and 4 spatial orbitals. There are three possible spin states (2S+1=1,3,5) with 20, 15 and 1 configuration state functions respectively. For the singlet configuration $(1)^2(2)^1(3)^1$ there is one CSF. On the other hand, for $(1)^1(2)^1(3)^1(4)^1$ there are 2 CSF corresponding to different ways of coupling the spins to a singlet. For our purposes, we use the term configuration to refer only to the spatial occupancy of the orbitals.

The CI method is based upon the expansion of the wavefunction Ψ_α for state α in CSF Φ_K

$$\Psi_\alpha = \sum_K C_{\alpha K} \Phi_K \quad 1$$

where the $C_{\alpha K}$ are determined using linear variation method (or perturbation theory). Thus one is led to the usual matrix eigenvalue method

$$H C_\alpha = E_\alpha C_\alpha \quad 2$$

with

$$H_{KL} = \langle \Phi_K | H | \Phi_L \rangle \quad 3$$

The problem of the evaluation of the matrix elements in equation 3 now completely solved [1,2]. The only practical problem is that the expansion length in equation 1 gets astronomical very rapidly for large numbers of orbitals. Thus one must truncate this expansion in some fashion.

Much of one's intuition about the nature of the electron correlation problem is associated with the truncation of the CI expansion 1. In addition, while the energy obtained with the complete CI expansion is independent of the choice of orbital basis (i.e. the energy is invariant to an orthogonal transformation of the orbital basis used to define the configurations) this is not true for a truncated expansion.

For the particular case where equation 1 is truncated at one term

$$\Psi_\alpha = \Phi_0 \quad 4$$

we have the single configuration SCF method (SC-SCF). Here the orbitals in Φ_0 are optimized to minimize $\langle \Phi_0 | H | \Phi_0 \rangle$. Alternatively, if we truncate equation 1 at some finite number of terms and optimize both the orbitals and expansion coefficients $C_{\alpha K}$ we have the so-called multi-configuration SCF (MC-SCF) method. In MC-SCF the orbitals are thus chosen to optimize

$$\langle \Psi_\alpha | H | \Psi_\alpha \rangle = \sum_K C_{\alpha K} \langle \phi_K | H | \phi_L \rangle C_{\alpha L} \quad 5$$

with $C_{\alpha k}$ fixed.

At this stage it is useful to define various truncated CI expansions for a subsequent discussion of the physical nature of electron correlation. Let us consider a subset of the configurations (which we shall call our reference space) of a full CI defined in some orbital basis that has been defined by an SCF or MC-SCF calculation. We shall divide our orbitals into core orbitals (which are doubly occupied in all reference configurations), valence orbitals (which have variable occupancies in the reference configurations) and virtual orbitals (which are unoccupied in the reference configurations). As a simple example, consider an MC-SCF wavefunction for the ^3P state of the carbon atom. Here the 1s and 2s orbitals have double occupancy whereas the 2p orbitals have variable occupancy. The remaining atomic orbitals would be classified as virtual. On the other hand, we could choose a reference set of CSF which consisted of $(2s)^2(2p)^2$, $(2s)^1(2p)^3$, and $(2s)^0(2p)^4$ configurations. The 2s orbital would now be classified as a valence orbital.

With these definitions we can now classify all the terms in the CI expansion by orbital occupancy. We shall use 3 paramaters to describe the various configuration sets, t_u, the number of electrons occupying virtual orbitals, t_c, the number of core holes (which is equal to the numbewr of electrons occupying core orbitals), and t_v, the number of electrons occupying valence orbitals. The reference set of configurations thus have $t_u=0$ and $t_v=N_v$ where N_v is the number of valence electrons. This type of definition will be useful in discussing the correlation energy for situations where the wavefunction can not be initially approximated by a single CSF. For a single SCF configuration we have $t_c=0$, $N_v=0$

As an example, for the carbon atom, if we choose the 2p orbitals to be valence orbitals and the 1s, 2s to be core orbitals, then the reference configurations have $t_c=0$, $t_v=3$, $N_v=3$, $t_u=0$ but a configuration

$$(1s)^2\ (2s)^1\ (2p)^3$$

has $t_c=1$, $t_v=N_v+1=3$, $t_u=0$, while

$$(1s)^2\ (2s)^1\ (2p)^1\ (3s)^2$$

has $t_c=1$, $t_v=N_v-1,=1$, $t_u=2$. Various possible CI expansions can be classified by the maximum value of t_c and t_u since t_v is normally allowed to have all possible values. Thus $t_c=2$ $t_u=2$ will be referred to as a double excitation CI. Further, if $N_v=0$ (i.e. we have only core orbitals such as might be derived from an SCF calculation) we have only one reference configuration and we refer to the method as single reference CI. However, if $N_v>1$ then we have multi-reference CI. Provided one chooses one's orbitals via MC-SCF methods, then an expansion up to $t_c=2$, $t_u=2$ contains all configurations that contribute at second order of perturbation theory as in SCF theory.

The preceeding description of truncated CI expansions is rather different than that normally associated with CI. Usually one refers to a doubly excited CSF as one that differs by two spin orbitals from a particular reference CSF. A configuration that is doubly substituted from one reference configuration may be quadruply excited from another reference configuration. However, the definitions stated in the preceeding paragraphs are based upon orbital populations, given an initial partition of the basis into active (core + valence) and inactive (virtual) orbitals. This type of definition is desirable if one is to attempt a physical discussion of electron correlation.

2.2 A DEFINITION OF ELECTRON CORRELATION USING THE LANGUAGE OF CI.

In this section we wish to discuss the physical interpretation of electron correlation using the language of CI. Our objective is to attempt to separate the dynamical (approximately structure independent) and non-dynamical (structure dependent) parts of the electron correlation energy. This separation is essential in discussing applications since as chemists we are interested in energy differences and thus the dynamical correlation, while very large in magnitude, may make only a very small contribution to chemical energy differences.

We begin by limiting ourselves to the case where the CI wavefunction is dominated by a single term, the SCF wavefunction Φ_0. This wavefunction will be a good approximation for closed shell molecular systems (with no near degeneracies) in the region of the equilibrium nuclear configuration. The correlation energy in this case is just the difference between the SCF energy (with an infinite basis) and the full CI energy. As is well known, this correlation energy has a simple interpretation. The coulomb and exchange operators which occur in the Hartree-Fock operator contain only the average effect of the electron repulsions of the remaining electrons. If we consider a two-electron example, the SCF wavefunction would have the form

$$\Phi_0 = |\phi_1(1)\ \bar{\phi}_1(1)| \qquad 6$$

The best possible wavefunction must have an explicit dependence on the interelectronic co-ordinates

$$\Psi = |\phi_1(1)\ \bar{\phi}_1(2)|\ \exp\ r_{12} \qquad 7$$

which we can write approximately as

$$\Psi = |\phi_1(1)\ \bar{\phi}_1(2)| + |\phi_1(1)\ \bar{\phi}_1(2)|\ r_{12} \qquad 8$$

In a CI expansion of Ψ we are thus attempting to expand $|\phi_1(1)\ \bar{\phi}_1(2)|\ r_{12}$ in a complete basis of CSF built from orbitals that are orthogonal to ϕ_1 (i.e. the virtual MO). For small changes in nuclear geometry, the effect of $|\phi_1(1)\ \bar{\phi}_1(2)|\ r_{12}$ will not change dramatically, since ϕ_1 will not change markedly. Similarly for many-electron systems, this type of electron correlation energy will be largely structure-independent provided the number of electron pairs and the spatial arrangement of electron pairs which are nearest neighbours is maintained (such as internal rotation).

Thus the physical concept of electron correlation is quite clear in those situations where the wavefunction is dominated by a single term, the SCF wavefunction. Here the correlation energy is largely structure-independent. This type of electron correlation is referred to as dynamical correlation. The situation is much more complicated in molecular systems which cannot be described by a single term SCF wavefunction. However, these situations, which occur during bond breaking, and making are the most interesting in quantum chemistry.

Let us begin a more general discussion of electron correlation with the H_2 molecule. In the region of the equilibrium nuclear configuration, we can write the wavefunction in terms of the SCF wavefunction

$$\Psi = \Phi_0 = (1\sigma_g)^2$$

However at infinite separation the $1\sigma_g$ and $1\sigma_u$ orbitals become degenerate and the correct SCF wavefunction is

$$\Psi = \Phi_0 - \Phi_1$$

with $\Phi_1 = (1\sigma_u)^2$ Thus in order to describe the molecule over the range from equilibrium to infinite separation at the SCF level, we must use an MC-SCF wavefunction of the form

$$\Psi = C_0\,\Phi_0 + C_1\,\Phi_1$$

It is clearly not meaningful in this situation to define electron correlation relative to the $(1\sigma_g)^2$ configuration over the full range of nuclear geometries. Thus we shall henceforth adopt the following definition of electron correlation energy.

> The total correlation energy is the energy difference between the energy obtained with an MC-SCF wavefunction that permits correct dissociation into SCF atoms (HFPD) and the full CI wavefunction.

Thus it is only appropriate to discuss the correlation energy of the H_2 molecule relative to a two-configuration space $1\sigma_g^2$, $1\sigma_u^2$.

If we call the $1\sigma_g$ and $1\sigma_u$ orbitals valence orbitals, then it remains to discuss the correlation energy recovered in the remainder of the expansion. There is one additional CSF that can be constructed from these two valence orbitals corresponding to the configuration $1\sigma_g^1\,1\sigma_u^1$. We shall refer to the complete set of CSF obtained by allowing all possible arrangements of the valence electrons among the valence orbitals (ie all the ($t_v=N_v$ CSF) as the internal or complete active space (CAS) wavefunction. The difference in energy between the HFPD wavefunction and the CAS wavefunction is the internal correlation energy. In the case of H_2 the configuration $1\sigma_g^1\,1\sigma_u^1$ does not contribute because it has a different symmetry. The internal correlation energy has no counterpart for a closed shell SCF reference function.

The set of CSF with $t_u=2$ play the same role as in single reference (SCF) CI namely, to incorporate the correct r_{12} dependence into the wavefunction. Thus configurations such as $2\sigma_g^2$, $1\pi^2$ etc. play the same role at the equilibrium internuclear separation and at infinite separation. The energy lowering which arises from these $t_u=2$ CSF is called external correlation.

The CSF corresponding to $t_u=1$ play a special role in multi-reference theories of electron correlation which we shall discuss in more detail in the next section. We shall use the term semi-internal correlation energy to denote the difference between the CAS wavefunction and the $t_u=1$ CSF. These $t_u=1$ configurations are important in multi-reference CI because (unlike single reference CI) the contribution of the singly excited CSF is not small.

The nature of semi-internal correlation can be best illustrated with an example using the F_2 molecule. Here the proper disociation configurations (HFPD) are identical to H_2, $3\sigma_g^2$ and $3\sigma_u^2$ with a core configuration $1\sigma_g^2 1\sigma_u^2 2\sigma_g^2 2\sigma_u^2 1\pi_u^4 1\pi_g^4$. The configuration $1\pi_u^3\ 3\sigma_g^1\ 3\sigma_u^1\ 2\pi_g^1$ (ie $t_c=1$, $t_u=1$) makes a very large contribution to the energy. This configuration corresponds to a single excitation ($3\sigma_g - 3\sigma_u$) in the valence (or internal) orbitals and a single excitation ($1\pi_u - 2\pi_g$) from the core orbitals into the virtual orbitals. Thus semi-internal correlation corresponds to the simultaneous excitation within the internal orbitals and into the virtual orbitals.

Let us now summarize our definitions of electron correlation in the case of multi-reference CI.

> The non-dynamical or internal correlation energy is the difference between the CAS-MCSCF (or internal CI) and an MC-SCF calculation which includes the smallest number of configurations required to discribe dissociation into SCF atoms (HFPD). The semi-internal correlation is the difference between this CAS-MCSCF energy and the energy of a multi-reference CI calculation with $t_u=1$. The remainder of the correlation energy is refered to as the external correlation ernergy and plays the same role as the dynamical correlation energy of systems that can be described by closed shell SCF wavefunctions.

It is the sum of the internal and semi-internal correlation energy that is of major importance to the quantum chemist since it is this part that contains most of the structure dependence. The neglect of the external correlation energy should have much the same effect as in the single reference SCF theory.

Let us end this section by giving a few historical references. The partition of the correlation energy into internal, semi-internal and external correlation was originally developed by Sinanoglu in the mid 1960's (see for example chapter IX in reference [9] and the papers reprinted therin for a full discussion). This type of partition was exploited by Schaefer and co-workers [10] and Goddard and his co-workers [11,12] in actual calculations from the end of the 1960's. The wavefunctions used had a wide variety if acronyms such as GVB, POL-CI etc. In this section we have defined these wavefunctions more precisely using the orbital population indices t_c, t_v, and t_u.

2.3 THEORETICAL ASPECTS OF THE CALCULATION OF INTERNAL AND SEMI-INTERNAL CORRELATION ENERGIES.

As we have just discussed, one requires a definition of electron correlation that is applicable for a manifold of excited states over the complete potential energy surface. In this case, the structure dependent parts of the electron correlation energy arise from the internal and semi-internal correlation. Thus we need to briefly discuss the main theoretical features of the methods that are used to calculate these effects. These two methods are MC-SCF (internal correlation) and first order CI (FO-CI) for the semi-internal part.

Let us suppose that we truncate our CI expansion with t_u=1 assuming that our reference space corresponds to a full internal CI (CAS MC-SCF). We shall write our CI wavefunction as

$$\Psi = \sum_A C_A \Phi_A + \sum_R C_R \Phi_R \qquad 9$$

where the CSF A have t_v=N_v, t_u=0 and the CSF R have t_v=N_v-1 or t_c=1 and t_u=1. If the orbitals occuring in the set A have been determined by an MC-SCF procedure, then the extended Brillouin theorem holds and we have

$$\langle \Psi_0 | H | \Psi_{\alpha\to r} \rangle = 0 \qquad 10$$

with

$$\Psi_0 = \sum_A C_A \Phi_A \qquad 11$$

$$\Psi_{\alpha\to r} = \sum_A C_A \Phi_A(\alpha\to r) \qquad 12$$

Here $A(\alpha \to r)$ is used to denote the CSF where the occupied orbital α (core or valence) is replaced by a virtual orbital r. Thus the MC-SCF wavefunction can be written as

$$\Psi^{MC-SCF} = \Psi_0 + \sum_{\alpha, r} X_{\alpha r} \Psi_{\alpha \to r} \qquad 13$$

where for optimum orbitals the $X_{\alpha r}$ go to zero. It is clear that the set of functions $\Psi_{\alpha \to r}$ span a subspace of the set R (since both sets have t_u=1). However, the set R will be much larger than the set $\Psi_{\alpha \to r}$. Thus one cannot expect the C_R to go to zero as the $X_{\alpha r}$ do (as in ordinary SCF).

If we assume the $\Psi_{\alpha \to r}$ are orthogonal (since they could be orthogonalized) then we have the approximate relationship

$$X_{\alpha r} = \sum_A \sum_R \langle \Phi_A(\alpha \to r) | \Phi_R \rangle C_A C_R \qquad 14$$

$$= D_{\alpha r}$$

where $D_{\alpha r}$ is the first order density matrix element between orbitals α and r (see ref. 3 for a full discussion). Thus if we perform a multi-reference CI the matrix elements of one-body Fock type operators between reference configuration A and configuration R with t_u=1 will be approximately zero. However, there will still be non-zero contributions from two-particle operators. In the limit where the reference set A consists only of a single SCF function, the matrix element between the reference function and any t_u=1 state can be expressed as a one-body Fock type operator and one has the usual Brillouin condition.

Let us now emphasize the importance of the preceeding discussion. The MC-SCF orbitals provide the optimum orbitals for the calculation of the internal (reference space) CAS correlation energy. Further, the contribution of the single excitations (t_u=1) for a multi-reference CI is not zero. It is this contribution which is referred to as semi-internal correlation and it has no counterpart in single reference SCF calculations.

Let us now return to our discussion of the FO-CI method corresponding to (t_u=1). It is now clear that the MC-SCF orbitals are the optimum ones to use and they also provide a definition of semi-internal correlation. We must now address ourselves to the case where the orbitals in FO-CI have not been optimized by MC-SCF. In this case there will be an additional contribution from t_u=1 CSF which is usually referred to as the polarization (or MC-SCF) contribution. If one wishes to attempt to describe several valence states simultaneously then one may wish to use some configurational average in performing the MC-SCF calculation. In this case we must take as our reference the energy corresponding to the eigenvalues of the CI problem containing only the reference configurations. The energy lowering obtained from the t_u=1 expansion now contains polarization and semi-internal correlation energies since these two aspects cannot now be separated.

The preceeding discussion leads us to consider two possible strategies for the computation of the correlation effects that are equally valid for all regions of a molecular potential energy surface. One may start with an MC-SCF wavefunction with the CAS configurations and then perform a CI calculation for the semi-internal (t_u=1) and external correlation (t_u=2). One may hope that the external correlation will be largely structure independent and that the t_u=1 contribution will dominate. On the other hand, one may hope that the semi-internal and external correlation may be dominated by the contribution of just a few of the virtual orbitals. In this case one would perform an MC-SCF calculation (since one has truncated the expansion) with a limited subset of the configurations which contribute to the semi-internal and external correlation energy. This latter procedure has recently begun to be used quite extensively.

2.4 SIZE EXTENSIVITY IN CI CALCULATIONS.

In the previous discussions we have assumed that contributions from t_u>2 CSF will be negligable. However, truncation of a CI wavefunction at t_u=2 leads to an incorrect dependence of the energy on the number of particles (see for example ref. 7, chapt.4). The inclusion of all CSF t_u=3,4 is impossible for all but the smallest systems. The various approximate methods have been formulated for including the effects of t_u=4 CSF in single reference double excitation CI calculations (the Davidson correction [13] and the CEPA method [14] are examples) have been quite successful.

In multi-reference CI these problems should be much less severe. Here one starts from a full internal CI (CAS) so the size extensivity errors which arise from the internal orbitals are rigourously eliminated. (Recently [15] various approximate methods have been developed for the multi-reference case as well). Perhaps the most important point to mention is that if one has a large contribution from t_u=4 CSF then it is a strong indication that one has not chosen one's reference space correctly.

3. SOME CASE STUDIES OF NON-DYNAMICAL CORRELATION

3.1 INTRODUCTORY REMARKS

It is our objective in this section to look closely at a few examples where correlation effects are known to be important for certain topological features of a potential energy surface. The effect of electron correlation on the properties of molecules near the equilibrium internuclear configuration has been exhaustively surveyed by Carsky and Urban [7] (see pp. 134–200). Thus we have chosen our examples solely on the criterion that there have been sufficient accurate calculations performed to illustrate some of the ideas of the previous sections.

3.2 THE F_2 MOLECULE

In many ways the F_2 molecule provides the perfect benchmark for quantum chemistry theories of electron correlation potential energy surfaces. For this molecule the Hartree–Fock method predicts that F_2 has a negative binding energy (with respect to two HF atoms) [16]. Further, since the closed shell SCF method predicts a dissociation into F^+–F^-, most theories of electron correlation (based upon SCF as a reference) break down at an internuclear separation of around 2A°. On the other hand, this molecule is small enough that it is possible to perform very high accuracy calculations with a wide variety of quantum chemistry methods using extensive gaussian basis sets which include polarization functions (see for example ref. 18–25).

Let us begin our discussion with a review of the role played by electron correlation in the region of the equilibrium geometry. Here the electronic configuration is

$$1\sigma_g^2\ 1\sigma_u^2\ 2\sigma_g^2\ 2\sigma_u^2\ 1\pi_u^4\ 1\pi_g^4\ 3\sigma_g^2$$

The value of r_e from SCF calculations [24] is 1.34 A° which improves to 1.38 A° with the addition of CI up to double replacements (t_c=2, t_u=2). However, the inclusion (approximately) of quadruply excited configurations (CEPA method) [24] (t_c=4, t_u=4) is required to bring r_e to 1.411 Å in close agreement with the experimental value (1.413 Å). The simple fact that the inclusion of quadruple excitations plays a very important role at the equilibrium geometry is indicative that we do not have the correct reference point at which to define external correlation in this case.

Let us therefore turn our attention to the calculation of the correlation energy of F_2 using HFPD as the reference point. It is useful to describe the wavefunction of F_2 for all values of the internuclear separation using both the language of MC-SCF theory and GVB (generalized valence bond) [25]. At infinite separation we have two F atoms in ^3P states with the electronic configuration $1s^2 2s^2 2p^5$. In terms of MC-SCF orbitals the wavefunction has the form (at infinite separation)

$$\text{core}\ 1\pi_u^4\ 1\pi_g^4\ [3\sigma_g^2 - 3\sigma_u^2]$$

whereas in GVB language, in terms of the HF atomic orbitals, we have

$$\text{core}\ (2p_\pi^A)^4 (2p_\sigma^A)^1 (2p_\sigma^B)^1 (2p_\pi^B)^4$$

We use the symbols A and B to distinguish the two nuclear centres. However, configurations such as

$$(2p_\pi^A)^3 (2p_\sigma^A)^2 (2p_\sigma^B)^1 (2p_\pi^B)^4$$

are degenerate with the previous configuration. The HFPD wavefunction must be a superposition of the $3\sigma_g^2$ and $3\sigma_u^2$ configurations. This corresponds to the identification of the $3\sigma_g$ and $3\sigma_u$ as valence orbitals (N_v=2). However, because of the degeneracy just noted, the $1\pi_u$ and $1\pi_g$ orbitals should also be included as active orbitals. Thus we should have a reference space with orbitals $1\pi_{g,u}$, $3\sigma_{g,u}$) which would correspond to 6 valence orbitals, and 10 valence electrons which would give 21 CSF. Further, in terms of a GVB wavefunction one would expect contributions from charge transfer configurations such as

$$(2s^A)^1 (2p_\pi^A)^4 (2p_\sigma^A)^2\ (2p_\sigma^B)^2 (2p_\pi^B)^4 (2s^B)^2$$

$$(2s^A)^2 (2p_\pi^A)^4 (2p_\sigma^A)^2 \quad (2p_\sigma^B)^1 (2p_\pi^B)^4 (2s^B)^1$$

Thus one might extend the reference space by including the $2\sigma_{g,u}$ orbitals to give a 36 CSF reference wavefunction.

The reference space may be thus constructed in a number of ways. The HFPD wavefunction has only two configurations built from the $3\sigma_{g,u}$ orbitals. If one includes the $1\pi_{g,u}$ or $2\sigma_{g,u}$ atomic orbitals as valence orbitals then one has 21 and 36 CSF. However these last two wavefunctions would also be generated from the first by taking the $1\pi_{u,g}$ and/or $2\sigma_{u,g}$ orbitals as core orbitals. The additional internal configurations correspond to $t_v=N_v-1$ and N_v-2 and $t_c=1$ and 2. The full 36 configuration wavefunction gives the internal correlation energy.

The results of some calculations that recover all or part of the internal correlation of F_2 are included in Table I. In the first row we present the original results of Das and Wahl [19] obtained with an extensive STO basis. The calculations in row 2 are those of Siegbahn et al. [22] using a gaussian basis with polarization functions, while those in rows 4 and 5 are our own calculations [26] (more heavily contracted). From the MC-SCF/2 calculations corresponding to the HFPD wavefunctions, it is apparent that one has only recovered about 40% of the binding energy (1.66 e.v. [23]). Further, the bond length is longer than the experiment (1.4118 Å [23]). The remainder of the internal correlation energy which is recovered in the MC-SCF/36 wavefunction does not change the situation significantly. However, once one includes internal correlation the curve is at least qualitatively correct (cf.the SCF result in Table I).

Let us now turn our attention to the semi-internal correlation. The semi-internal effect arises from CSF corresponding to $t_u=1$ and $t_v=N_v-1$ or $t_c=1$ where the reference space is the 36 CSF wavefunction just discussed. In F_2 the major contributions are expected to arise from CSF of the form

$$(1\pi_g)^3 (3\sigma_g)^1 (3\sigma_u)^1 (2\pi_u)^1$$

and corresponding configurations involving replacements of $1\pi_u$ by $2\pi_g$. This

Table I

Internal correlation energy of the F_2 molecule

Method	Total Energy/E_h r=1.42	D_e/e.v.	r_e/A°	ref.
MC-SCF/2	-198.8432	.68		19
MC-SCF/2	-198.8392	.63	1.479	22
MC-SCF/2	-198.8256	.60	1.482	26
MC-SCF/36	-198.8298	.72		26
SCF	-198.7496	-1.46		26

TABLE II

Semi-Internal Correlation Energy Effects for F_2

Method	D_e/e.v.	r_e/A°	ref.
MC-SCF/6	1.82	1.40	19
MC-SCF/14	1.75		26
FO-CIa	1.94		26
FO-CIb	1.85	1.418	21

TABLE III

Semi-Internal and External Correlation Effects for F_2

Method	D_e/e.v.	r_e/A°	ref.
MC-SCF/178	1.38		26
MC-SCF/2584	1.42		26
MC-SCF/684	1.69	1.421	23
CI/2R	1.20	1.437	22
CI/3Ra	1.58	1.417	23
CI/3Rb	1.21	1.436	23

CSF involves an excitation $3\sigma_g - 3\sigma_u$ in the reference space and an excitation of one of the electrons into the virtual space.

One may attempt to include part of the semi-internal correlation in an MC-SCF calculation by including the $2\pi_{u,g}$ orbitals in one's reference space. In Table 2 (first two rows) we present the results of such an attempt. The first result MC-SCF/6 is the six configuration wavefunction of Das and Wahl [F6]. This wavefunction consists of all the internal and semi-internal CSF (of the type discussed above) that contribute in $D_{\infty h}$ symmetry. The MC-SCF/14 calculation is from our own work [26] and spans the same CSF space as the Das-Wahl wavefunction. As one can see, one now overestimates the binding energy by only 8% and the bond length is much improved.

The full semi-internal correlation energy is obtained by including all the CSF with t_u=1 corresponding to the FO-CI wavefunction. In rows 3 and 4 of table II we present the results of Cartwright and Hay [21] (FO-CIb) and the results obtained in our own calculations in a similar basis [36] (FO-CIa). The main difference between rows 3 and 4 is the effect of orbital optimization in the MC-SCF. (Cartwright and Hay use SCF orbitals). Thus the semi-internal correlation energy results in a binding energy that is up to 17% too large.

One must finally look at the contribution of the external correlation energy. Some selected results are summarized in table III. As before, there are two possible strategies to recover the external correlation energy one may use MC-SCF with a subset of the virtual orbitals included in the CAS sapce (eg. the $2\pi_{g,u}$) or one may use multi-reference CI using the full (CAS) internal set of CSF as the reference space. The MC-SCF calculations MC-SCF/178 and MC-SCF/2584 are from our own work and include the t_c=1,2 t_u=1,2 CSF relative to the 2 configuration reference and the full CAS configuration set respectively . The MC-SCF/684 calculations are from the work of Siegbhan[23] and include only the $2\pi_u$ orbital from the virtual space. (These calculations use a very extensive gaussian basis with polarization functions).

The CI calculations CI/2R and CI/3R are also from the work of Siegbhan[23]. The calculation CI/2R included all single and double replacements of the two HFPD reference functions. Such a wavefunction does not contain all of the external correlation since configurations of the type

$$(1\pi_g)^3\,(1\pi_u)^3\,(3\sigma_g)^1\,(3\sigma_u)^1\,(2\pi_u)^2$$

corresponding to $t_c=2$ $t_v=2$ $t_u=2$ would be ommited since they would be triply excited with respect to the HFPD configurations. In the calculation CI/3Ra this defect is removed by the additional reference configuration

$$(1\pi_g)^3 (3\sigma_g)^1 (3\sigma_u)^1 (2\pi_u)^1$$

Also in the CI/3Ra calculation f functions have been added to the basis. This result is to be compared with CI/3Rb which did not contain f functions. Thus it is only when one uses a very complete CI exapansion in a full orbital basis does one begin to approach the experimental result.

On comparing the CI/2R and CI/3Rb results with CI/3Ra it is apparent that the calculation of the external correlation energy is extremely sensitive to basis set extension effects. Thus one has the same problems in the computation of the external correlation in multi-reference CI as in the single (SCF) reference case. However, if we now examine the MC-SCF results where only part of the external correlation is included one obtains better results with a much poorer Gaussian basis. The basis used for MCSCF/178 and MC-SCF/2584 is very similar to the basis used in CI/2R and CI/2Ra, yet the CI results are not in as good agreement with respect to experimental D_e. The main difference between MC-SCF/684 and MC-SCF/2584 is due to the fact that the former omits the $2\pi_g$ orbital from the reference space.

Let us now attempt to draw some careful conclusions from these results. The sum of the internal anmd semi-internal correlation tends to overestimate the binding energy but gives a good equilibrium bond length. The external correlation energy reduces the binding energy from this value.The computation of the external correlation energy is very sensitive to the effects of orbital optimization (cf. MC-SCF/2584 and CI/3Rb) and basis set extension (cf.CI/3Ra and CI/3Rb). One sees that if one does not allow orbital optimization in the computation of the external correlation energy then very large basis sets are necessary. On the other hand, an MC-SCF calculation which includes a part of the semi-internal and external correlation is insensitive to basis set extension (see ref. [23] for a full discussion of this problem). Thus from the detailed MC-SCF calculations presented in [23] it is apparent that one may recover that part of the external correlation energy that determines D_e by including only a small number of virtual orbitals in the reference space. The remainder of the external correlation energy is approximately structure independent.

If we compare the results of MC-SCF/178 and MC-SCF/2584 in table III we can also make some comment on size-extensivity. The MC-SCF/178 includes only t_u=2 configurations whereas the MC-SCF/2584 includes the full CAS set. It can be seen that the effect on D_e is very small . As discussed in section 2.4 this is to be expected in a multi- reference calculation.

Finally, if we return to the results obtained with the FO-CI wavefunctions (table II) it can be seen that the effects of orbital optimization may be very large here as well (cf.FO-CIa where one has used MC-SCF orbitals and FO-CIb where SCF orbials have been used). If we think of the FO-CI wave function in terms if HF AO (ie in valence bond terms) then the nature of thses effects can be clarified. As the molecule forms the single excitation (t_u=1) inter-atomic charge transfer configurations will begin to make very important contributions to the the binding energy. Optimization of the internal orbitals will cause them to distort to account for this charge transfer and the will become delocalized and diffuse. Because of this the intra-atomic correlation energy will decrease. Thus one has a delicate balance between orbital optimization effects and intra-atomic electron correlation. It would appear that the optimum approach would be build ones FO-CI wavefunction directly from the HF AO and not allow for orbital optimization. Then the intra-atomic correlation would be approximately constant. In this case the binding energy would have a polarization contribution contribution as well as a semi-internal contribution from the t_u=1 CSF. However, there are severe technical difficulties with this approach since the HF AO are not orthogonal except at infinite separation. Schaefer[20] has attempted calculations of this type but did not pursue them to high accuracy. In our own laboratory we have run some preliminary calculations with such wavefunctions using the same basis as for the FO-CIa calculations in table II. We obtained [27] a D_e of 1.46 e.v. and r_e=1.44 A^o.

3.3 THE HF MOLECULE

Many of the conclusions of the preceeding discussion on the F_2 molecule will apply to other diatomic molecules. However, we should briefly discuss some results of Dunning[28] on the HF molecule since they have performed a very careful analysis of the correlation effects on the ground state potential energy curve of this molecule.

While the ground state of HF can be described by the closed shell SCF wavefunction

TABLE IV

Binding Energy of HF ($^1\Sigma^+$) Computed with Various Methods (from ref.28)

Method	D_e/e.v.	R_e/A
SCF	4.27	.899
MC-SCF (HFPD)	4.89	.917
Internal CI	4.92	.917
FO-CI	5.72	.926
MC/CI	5.72	.920
Expt.	6.12	.916

$$(1\sigma)^2 (2\sigma)^2 (3\sigma)^2 (1\pi)^4$$

in the region of r_e. It is apparent that the HFPD wavefunction must have two configurations corresponding to

$$(1\sigma)^2 (2\sigma)^2 (3\sigma)^2 (1\pi)^4 - (1\sigma)^2 (2\sigma)^2 (4\sigma)^2 (1\pi)^4$$

at $r = \infty$. The full internal CI cold be generated by refering to the 2σ and 1π orbitals as core with t_c=0,1,2, t_v=2,1,0 and t_u=0. The FO-CI includes configurations with t_u=1 in addition this and the external correlation energy is recovered with a multi-reference (MR-CI) calculation with t_u=2.

Some results obtained from reference [28] are summarized in table IV. It can be seen that the major correction to the SCF/HFPD/internal CAS CI result comes from the FO-CI calculation (semi-internal correlation). The external correlation energy is very large in magnitude at r_e (3.90 e.v.) but changes by only .15e.v. at $r = \infty$ Further, the semi-internal correlation energy is converged with respect to basis set extension but the external correlation energy is not. Thus the accurate computation of the external correlation energy to the binding energy is a formidible task. However, most of the structure dependent part of the electron correlation appears to be recoverable at the FO-CI level (internal and semi-internal).

3.4 THE H_2O MOLECULE

The H_2O molecule provides a good prototype multiple surface system. There are many interesting topological features such as avoided intersections, saddle points and change of orbital character from valence to Rydberg. Several very accurate calculations are available on surfaces other than the $^1A'$ ground state. In this section we wish to examine a selection of these calculations with a view to discussion of correlation energy effects.

The ground $^1A'$ surface has been well studied in the region of the equilibrium geometry by Shavitt et al [29] using single reference CI methods. However, our interest is primarily in the way in which multi-reference theories of electron correlation may be used on this surface.

The ground state equilibrium electronic configuration of water may be written as

$$(1a_1)^2 (2a_1)^2 (1b_2)^2 (3a_1)^2 (1b_1)^2 (4a_1)^0 (2b_2)^0$$

where the $1b_2, 2b_2, 3a_1, 4a_1$, orbitals are combinations of the $0p_z$, $0p_x$ and H1s HF AO. In order to obtain dissociation into $O(^3P) + 2H(^2S)$ one requires all CSF which allow all possible occupancies of the $1b_2, 2b_2$ and $3a_1$, $4a_1$ orbitals (20 CSF in the absence of symmetry). The $1a_1$ (O1s) orbital can be assumed doubly occupied, so the full internal space should also include the $2a_1$ and $1b_1$ orbitals as well which would give rise to 105 CSF of singlet spin multiplicity. At infinite separation, in terms of localized orbitals one may write the wavefunction in terms of a perfect pairing configuration

$$^3P \; O\,(1s)^2 (2s)^2 (2p)^4 \qquad H\,(1s)^1 \quad H\,(1s)^1$$

In table V we have collected some data for the dissociation of H_2O into ground state atoms. If a potential surface is to be useful in studies of reaction dynamics it must reproduce the experimental binding energies. The SCF , MC-SCF and FO-CIa results are from our own work and use the modest STO-431G** basis while the FO-CIb calculation is from the work of Howard et al [30] and includes polarization and diffuse Rydberg type functions. It can be seen that the internal correlation energy contribution (MC-SCF/SCF) is 1.44 e.v., the semi-internal contribution is .74 e.v. and the remainder (ca.1.13 e.v.) must be allocated to external correlation energy (ignoring basis set extension effects). The FO-CIb wavefunction of Howard et al is also in good agreement with the single reference CI calculations of Shavitt et al [29] with respect to properties such as bond lengths and force constants. In addition this wavefunction reproduces the diatomic H-H and O-H potential energy curves quite well. Thus the FO-CI method which includes the internal and semi-internal correlation energy gives a good representation of the $^1A'$ surface. The contribution from the semi-internal correlation appears to be quite small since the wavefunction is well represented by a single CSF.

The surface for the reaction

TABLE V

Dissociation Energy for H_2O (into HF atoms)

Method	ΔE/e.v.
SCF	6.69
MC-SCF	8.13
FO-CIa	8.87
FO-CIb	8.94
Expt.	10.008

TABLE VI

Saddle Point Geometries and Barrier Heights for the Reaction $O(^3P) + H(^1\Sigma) \rightarrow OH(^3\Pi) + H(^2S)$

Method	r(O-H)	r(H-H)	ΔE/e.v.	ref.
SCF	1.67	1.10	1.53	29
MC-SCF	1.14	1.00	1.33	32
CI/1R	1.169	.908	1.06	33
CI/1R+4	1.146	.953	.835	33
FO-CI	1.15	.949	.604	30
FO-CI	1.23	.92	.541	32
Expt.			.511	

TABLE VII

Saddle Point Geometry and Barrier Height for the C_{2v} Insertion of $O(^3P) + H_2$ to give H_2O $(^3B_1)$

Method	r(O-H)	r(H-H)	ΔE/e.v.	ref.
SCF	1.010	1.209	3.40	33
CI/1R	1.012	1.28	2.47	33
CI/1R+4	1.013	1.31	2.23	33
FO-CI	1.05	1.34	1.84	30

$$O(^3P) + H_2(^1\Sigma_g^+) \rightarrow OH(^3\Pi) + H(^2S)$$

which corresponds to the $^3B_1$ excited state of H_2O is much more interesting since one may give some discussion of electron correlation from the results of recent calculations which involve quite different methodologies. Of particular interest is the saddle point which occurs as $O(^3P)$ approaches H_2 collinearily. Assuming the molecules collide along the z axis, the electronic configuration can be represented as

$$(1\sigma)^2 (2\sigma)^2 (1\pi_y)^2 (3\sigma)^2 (4\sigma)^1 (1\pi_x)^1 (5\sigma)^0$$

spin coupled to a triplet. The 3σ orbital is the H-H bond in the reactants and becomes the O-H bond in the products, while the 5σ orbital is the corrsponding σ^* orbital. The 4σ is the $2p_z$ orbital of the O atom in the reactants and becomes the H $1s$ orbital in the products. Thus there is no avoided crossing and one might expect that the open shell SCF should be a good initial approximation. However, the 3σ, 4σ and 5σ orbitals will change significantly over the reaction surface. Thus the internal correlation involving these orbitals will play an important role. One expects that the internal CSF such as

$$(3\sigma)^2 \rightarrow (5\sigma)^2$$

and

$$(3\sigma)^2 \rightarrow (3\sigma)^1 (5\sigma)^1$$

may play an important part in correlation effects. On the otherhand, when viewed from a valence bond point of view, the reactants correspond to a singlet coupling of the two H $1s$ orbitals and a triplet coupling of the O $2p_\sigma$ and $2p_\pi$ orbitals. This changes to a singlet coupling between the O$2p_\sigma$ and the H $1s$ and a triplet coupling of the O$2p_\pi$ and the H $1s$ electron of the departing H atom. Thus the internal space should involve at least the 3 σ orbitals O$2p_\sigma$ and the 2 H$1s$.

In table VI we have presented the results of some calculations on the saddle point geometry obtained using single and multi-reference methods. All the calculations used extensive Gaussian basis sets with polaization functions and diffuse functions. The MC-SCF calculation in table VI did not use the full internal space but only involved single and double replacements of the 3σ orbital with the 5σ orbital. The CI calculations were made with respect to a single configuration open shell SCF reference function and the CI+4 calculation included the effect of the quadruples using the CEPA method. Finally we present the results of two independent FO-CI calculations.

It can be seen that electron correlation has a very large effect on the saddle point geometry and the barrier height. Further, the FO-CI results (internal and semi-internal correlation) are in good agreement with experiment. In contrast, the single reference CI results still give a barrier which is too high and the effect of the quadruple excitations is quite large. In reference [33] it has been shown that only when one uses very large basis sets and attempts an extrapolation to infinite basis size does the single reference CI result for the barrier height begin to approach the experimental value.

Let us now attempt to draw some tentative conclusions from these results. It would appear that a multi-reference definition of electron correlation is essential. The external correlation energy obtained using a single reference function (open shell SCF) is slowly convergent and basis dependent. This is because configurations such as $3\sigma^2 \rightarrow 5\sigma^2$ are internal in the multi-reference case and external in the single reference case yet these configurations are important for discribing the barrier height. Further, a configuration such as

$$(1\pi_y)^1 (5\sigma)^2 (4\sigma)^2 (1\pi_x)^0 (2\pi_y)^1$$

is only a single excitation (t_c=1,t_u=1) in the multi-reference case but would be a quadruple excitation in the single reference case. Because of the large reorganization effects in the internal space one must include these effects from the outset as internal correlation. Only in this case is the external correlation largely structure independent.

Finally we must comment on the MC-SCF result in table VI. In this calculation only the 3σ and 5σ orbitals were allowed to have all possible occupancies. The full internal (CAS) MC-SCF would permit all possible occupancies of the 3σ, 4σ, 5σ, and 1π orbitals. A full internal space MC-SCF would undoubtably lower the barrier so that the semi-internal correlation would be less than that implied in table VI. However, it is important to note that the MC-SCF calculation gives a good result for the barrier position. This is important since energy gradient methods may be easily used with MC-SCF methods to determine saddle point geometries. One might then add semi-internal correlation effects to determine the height of the barrier more accurately.

Let us now turn briefly to the C_{2v} insertion reaction on the same surface ($^3B_1$). The ^3P ground state of the O atom is split into $^3A_2$, $^3B_2$ and $^3B_1$ and it is the latter that correlates with the lowest excited B state of H_2O with the configuration

$$(1a_1)^2 (2a_1)^2 (1b_2)^2 (3a_1)^2 (1b_1)^1 (4a_1(3s))^1$$

The $4a_1$ orbital has Rydberg character at equilibrium ground state geometry. In order to describe this surface one will require the same active internal orbitals as for the $^1A'$ surface. However, the $4a_1$ orbital will change from valence like to Rydberg like. Thus the optimization (via open shell SCF or MC-SCF) will be essential.

Some results for the barrier height and saddle point geometry are summarized in table VII. Again, one sees that the multi-reference calculation (internal and semi-internal correlation) gives a lower barrier than single reference CI based upon open shell SCF.

Let us now briefly summarize these results for H_2O. It is clear that to describe the surface manifold for H_2O one needs a reference space that allows all possible occupancies of those MO that correlate with the O $2p$ and H $1s$ AO. With this reference space one can describe many of the important features of the surface manifold at the FO-CI level (internal and semi-internal correlation). An attempt to use an open shell SCF reference function as the starting point for the computation of the external correlation energy leads to an expansion which is slowly convergent with respect basis set extension.

3.5 THE OZONE MOLECULE

The ozone molecule O_3 cannot be described qualitatively at the SCF level. Accurate SCF calculations[34] predict that the ground state is a triplet. The singlet state has a very large correlation energy due to its singlet bi-radical character. In addition there is much interest in a possible ring structure for this molecule which may lie below the dissociation limit. Over the past 5 years there have been several calculations performed at the double zeta plus polarization level (see for example [35-37] and references cited therin). These calculations give much insight into the nature of electron correlation in the case where the single configuration SCF cannot be used as a reference function.

The ground state of the open bent structure of the O_3 molecule has the following configuration of the valence orbitals (in C_{2v} symmetry)

$$(5a_1)^2 (3b_2)^2 (4b_2)^2 (6a_1)^2 (1b_1)^2 (1a_2)^2$$
$$(2b_1)^0 (7a_1)^0 (5b_2)^0$$

corresponding to 2 σ bond orbitals, 2 σ lone pair orbitals, 2 π orbitals (1 π lone pair), a π^* orbital, and 2 σ^* orbitals with 4 π electrons. The closed ring structure, on the otherhand, has the configuration

$$(5a_1)^2 (3b_2)^2 (4b_2)^2 (6a_1)^0 (1b_1)^2 (1a_2)^2$$
$$(2b_1)^2 (7a_1)^0 (5b_2)^0$$

with 6 π electrons. The $1a_2$ and $2b_1$ orbitals are very close in energy. One may localize these orbitals to give

$$\pi_{a/b} = 2b_1 \pm 1a_2$$

with the π lone pair as

$$\pi_c = 1b_1$$

Thus one may approximate the ground state electronic configuration using these localized orbitals as

$$(\pi_c)^2 (\pi_a)^1 (\pi_b)^1$$

which would correspond to the linear combination (in symmetry adapted orbitals)

$$(1a_2)^2 - (2b_1)^2$$

for a perfect bi-radical. Thus one must have at least two CSF to describe the bi-radicaloid nature of the ground state. For the transition from the ring to open structure one has a crossing of the $6a_1$ and $2b_1$ orbitals leading to an avoided intersection of the $1\,^1A_1$ and $2\,^1A_1$ potential energy curves. Thus in order to describe the open to ring structure transformation the internal space must include all the σ and π lone pair orbitals (since one is going from 4π electrons to 6π electrons). Finaly in order to describe the dissociation to O_2 and O one must include the σ orbitals $5a_1$ and $3b_2$ and the π orbitals $1a_2$ and $2b_1$ in the internal space.

Some calculations on the ring and open structures of O_3 are summarized in table VIII. Here ΔE referes to the difference between the ring and open geometries, ΔE^b to the barrier height, and ΔE^d to the dissociation energy to O_2 and O atom. The calculation MC-SCFa includes the HFPD CSF ($5a_1 3b_2 \rightarrow 7a_1 5b_2$) and the CSF needed to describe the bi-radicaloid nature of the ground state ($1a_2^2 \rightarrow 2b_1^2$). The calculation MC-SCFb contains all possible arrangements of the full internal space (9 orbital CAS MC-SCF). The internal CI calculation consisted of all possible double excitations (t_u=2) relative to a 2 configuation reference ($1a_2^2 - 2b_1^2$) using the orbitals from MC-SCFa. Thus the main difference between the internal CI and the MC-SCFb arises from the effect of orbital optimization which we shall discuss in more detail subsequently. The FO-CI calculation calculation includes all the t_u=1 CSF relative to the internal CI. The entries denoted CI/1R and CI/2R+4 are calculations which include all single and double excitations relative to the single SCF configuration. The last two entries in table VIII correspond to multi-reference CI calculations which include the effects of external correlation (t_u=2) however, they have been performed using quite different reference sets. The calculation MC-SCF/CIa used all single and double replacements from the two configurations which had the highest weight in a prelimanary MC-SCF calculation. In MC-SCF/CIb a single reference was used for the ring form and a two configuration reference for the open form. This proceedure (although representing the open form by a larger number of configations than the closed form) is acceptable since the ring structure is dominated by a single configuration.

The subtle differences in methods of calculation make it difficult to draw firm conclusions. However, the results are not at variance with previous case studies presented in this article. If we compare the internal CI and MC-SCFb results with the SCF and MC-SCFa (HFPD) calculations it can be seen that the internal correlation energy difference between the open and ring structures in very large indeed. As expected, the disociation energy is too small (expt 1.02

TABLE VIII

Calculations on the Ring and Open Structures of the Ozone Molecule (energies in e.v.)

Method	ΔE	ΔE^b	ΔE^d	ref.
SCF	.38			34
MC-SCFa	.52			34
MC-SCFb	1.42	2.54	.20	
Internal CI	1.31			34
FO-CI	1.20			34
CI/1R	.70			36
CI/1R+4	.95			36
MC-SCF/CIa	1.48	2.25		37
MC-SCF/CI+4	1.24	2.34		37
MC-SCF/CIb	1.22			34

TABLE IX

Ozone ($^1A_1 - ^3B_2$) Vertical Excitation Energies (from ref. 38 unless otherwise indicated)

Method	ΔE /e.v.
SCF	-3.13
π -MC-SCF	.80
GVB (MC-SCF)	-.19
$\sigma + \pi$ MC-SCFa	1.60
$\sigma + \pi$ MC-SCFb	1.75
FO-CI [34]	1.60

TABLE X

Effect of Internal Correlation on Molecular Orbitals (open structure) (from ref. 38 Energies in E_h)

Orbital source	Internal CI	MC-SCF
SCF	-224.3675	-224.4621
π MC-SCF	-224.3842	-224.4621
GVB (MC-SCF)	-224.4458	-224.4621
π + DS MC-SCF	-224.4620	-224.4621

e.v.). One might expect a large effect for semi-internal correlation for the open structure. However, for the ring and open structure the semi-internal correlation energy is 3.78 and 3.67 e.v. respectively and we have only a very small contribution to ΔE. If we compare the FO-CI results with MC-SCF/CIb we see that we have an almost negligible contribution from the external correlation energy as well. However, we see that the single reference CI calculations CI/1R and CI/1R+4 are in serious disagreement with the remaining calculations of the external correlation energy. This fact is merely an indication that one must have at least a two configuration reference function to define external correlation for the open structure. Thus it would appear that most of the correlation effect on the energy difference between the ring and open structures of O_3 is due to internal correlation energy.

The barrier between the ring and open forms of O_3 results from an avoided intersection of the $1\,^1A_1$ and $2\,^1A_1$ states with 4 and 6 π electrons. The most accurate calculations MC-SCF/CIa give a value of 2.34 e.v. for this barrier. Comparison with the CAS MC-SCFb calculations indicate that most of the correlation effects are internal in this case.

Since SCF theory predicts the $^3B_2$ state to be the lowest energy it is also of interest to examine the effect of electron correlation on the $^1A_1 - {}^3B_2$ energy separation. Some results are given in table IX. The calculations $\sigma+\pi$ MC-SCF correspond to a full internal CI in the 9 σ/π orbitals including (a) and excluding (b) d functions. The calculation which includes the d functions is in reasonable agreement with the experimental value of 1.70 e.v. Comparison with the FO-CI calculation shows that the effect of semi-internal correlation is quite small. The π MC-SCF calculation and GVB calculations demonstrate that one needs to use the full internal space. The π MC-SCF calculation includes only the π orbitals as the internal orbitals and the energy separation is far too small. In the GVB calculation the HFPD configurations are included as well.

Finally we should give some discussion of the effect of internal correlation on the orbitals themeselves. This is necessary because our definition of semi-intenal correlation is with respect to orbitals optimized with respect to an MC-SCF calculation with a full internal CI (CAS). However, in many calculations, the orbitals are obtained from simpler MC-SCF calculations (or GVB). In table X we give some results for a full internal CI (9 orbital σ/π) MC-SCF. In column 1 we give the choice of starting orbitals, in column 2 the full internal CI result obtained with these orbitals and in column 3 we give the result of the full internal (CAS) MC-SCF result. It can be seen that the SCF orbitals are very poor and the energy is lowered by 2.6 e.v. by orbital optimization. The π MC-SCF calculation includes the two dominant configurations yet the internal CI result is only slightly better than the SCF. In contrast the the GVB orbitals obtained from a calculation which includes the HFPD configurations ($\sigma \rightarrow \sigma^*$) as well, is in error by only .4e.v. relative to the CAS MC-SCF. Finally, a single and double excition CI relative to a full internal CI involving the π orbitals (ie t_c=1,2 and t_v=N_v-1,N_v-2) gives an excellent result. Thus the FO-CI results of table VIII, which used GVB orbitals should be very close to the optimum with respect to orbital variations.

Thus in the O_3 molecule we have an example where single configuration SCF gives results that are not even qualitatively correct. Further, the computation of correlation requires a multi-reference definition of electron correlation. In this case most of the important features of the surface can be studied by including internal correlation only.

3.6 THE CH-CH MOLECULE

As a final case study we should like to consider correlation effects on the dissociation of acetylene into two CH $^2\Pi$ radicals. Recent calculations by Siegbhan[39] represent the state of the art with respect to calculations of electron correlation and give a good indication of the future of such calculations on much larger molecules.

At the equilibrium internuclear separation the electronic configuration is iso-electronic with the N_2 molecule and can be written as

$$\text{core } (1\pi_u)^4 (3\sigma_g)^2 (3\sigma_u)^0 (1\pi_g)^0$$

The $3\sigma_u$ and $1\pi_g$ orbitals must be included as internal orbitals to descibe dissociation. Thus the full internal space has 6 orbitals and 6 electrons and gives 105 CSF of which only 52 contribute in C_{2v} symmetry. The core orbitals $1\sigma_{ug}$ and $2\sigma_{ug}$ can be assumed doubly occupied for all values of the C-C distance. However, such a wavefunction would not allow dissociation of the C-H bond.

For the collinear approach of two CH ($^2\Pi$) radicals there is an avoided intersection corresponding to a change of configuration from

$$(1\pi_u)^2 (3\sigma_g)^2 (3\sigma_u)^2$$

corresponding to two CH ($^2\Pi$) to

$$(1\pi_u)^4 (3\sigma_g)^2$$

which corresponds to two CH ($^4\Sigma^-$) radicals. This leads to a small barrier and suggests that the minimum energy path may involve a non-collinear approach.

There have been two quite detailed calculations on this reaction which have recently appeared in the literature. Some of these results are summarized in table XI. The internal CI calculation used open shell SCF orbitals for the two CH($^2\Pi$) radicals (a type of molecules in molecules valence bond calculation). The MC-SCF calculation included all CSF in the 6 orbital/6 electron CAS space and the MC-CI calculation included all t_u=1,2 CSF based upon the CAS internal space (all semi-internal and external correlation).

It can be seen that the value of D_e from the internal CI (valence bond) is rather poor. This is to be expected since the CH ($^2\Pi$) orbitals were used. In the region of r_e one would expect that the orbitals from the CH ($^4\Sigma^-$) would be more appropriate. In contrast, the MC-SCF gives an excellent result for D_e and there is only a slight improvement when the semi-internal/external correlation is added in the MC-CI calculation.

Now let us turn our attention briefly to the description of the avoided intersection. Here one has the curious result that the inclusion of semi-internal and external correlation actually increases the barrier (cf column 4 of table XI). At 2.8 Å there is a very sharp change in the wavefunction from two CH ($^2\Pi$) to two CH ($^4\Sigma^-$). Since the total correlation energy of CH ($^2\Pi$) would be greater than CH ($^4\Sigma^-$) due to the smaller number of electron pairs in the latter one will expect the barrier to increase. By bending the mollecule and thus allowing the initial formation of a single bond, the barrier has been shown to dissappear[39] at the MC-SCF level.

TABLE XI

CI calculations on CHCH
(distances in Å, energies in e.v.)

Method	r(C-C)	r(C-H)	D_e	ΔE^b	ref.
Internal CI	1.32		6.30	.33	40
MC-SCF	1.214	1.057	9.70	.42	39
MC-CI	1.208	1.061	10.00	.57	
Expt.	1.203	1.060	10.26		

4. CONCLUSIONS

In this work we have attempted to formulate a definition of electron correlation that is valid for all regions of a molecular potential energy surface. We have attempted to illustrate the usefulness of this definition with some selected case studies.

The internal correlation energy (CAS MC-SCF) is necessary to obtain results that are qualitivly correct over the potential surface. The semi-internal correlation, which has no counterpart in single configuration theory, contains the most important part of the structure dependence. The external correlation energy is very sensitive to basis set extension effects. It would appear that this convergence problem may be partly avoided by the partial inclusion of semi-internal and external correlation effects in an MC-SCF calculation.

Acknowledgements

The author is very greatful to Richard Eade and Zoe Slattery who performed many of the calculations reported in this article.

REFERENCES

1)J. Hinze,(ed),The Unitary Group for the Evaluation of Electronic Energy Matrix Elements,Lecture Notes in Chemistry Vol. 2, Springer-Verlag, 1981

2)J. Paldus, in H.Eyring and D. Henderson (ed) Theoretical Chemistry: Advances and Perspectives, Vol 2, Academic Press, New York, 1976, p.131

3)M. A. Robb and R. H. A. Eade, in I. G. Csizmadia and R. Daudel (ed)

Computational Theoretical Organic Chemistry Proceedings of the NATO ASI held at Menton, France, June 29-July 13, 1980, D. Reidel, Dordrecht, Holland, 1981

4)P. E. M. Sieghbahn, J.Almlof, A. Heiberg, and B. Roos, J. Chem. Phys. 74(1981)2384

5)H. J. Werner and W. Meyer, J. Chem. Phys. 73(1980)2342

6)D. L. Yeager and P. Jorgensen, J. Chem. Phys. 71(1979)755

7)P. Carsky and M.Urban, Ab Initio Calculations, Lecture Notes in Chemistry Vol 16, Springer-Verlag, Berlin, 1980

8)I. Shavitt, in H. F. Schaefer (ed), Methods of Electronic Structure Theory, Vol 3, Plenum,1977

9)O. Sinanoglu and K. A. Bruckner, Three Approaches to Electron Correlation in Molecules,Yale University Press, 1970

10)H. F. Schaefer, R. A. Klemm and F. E. Harris, J. Chem. Phys. 51(1969)4643

11)P. J. Hay, T. H. Dunning, W. A. Goddard, J. Chem. Phys. 62(1975)3912

12)P. J. Hay and T. H. Dunning, J. Chem. Phys. 64(1976)5077

13)S. R. Langhoff and E. R. Davidson, Int. J. Quantum Chem. 8(1974)61

14)W. Meyer, Int. J. Quantum Chem. S5(1971)341

15)S. Prime, C.Rees and M. A. Robb Mol. Phys. 44(1981)173

16)K. Hijikata, Rev. Mod. Phys. 32(1960)445

18)S. Fraga and B. J. Ransil, J. Chem. Phys. 36(1962)1127

19)G. Das and A. C. Wahl, J. Chem. Phys. 44(1966)87
Phys.Rev. Letters (1970)24
J. Chem. Phys. 44(1966)87

20)H. F. Schaefer, J. Chem. Phys. 52(1970)6241

21)D. C. Cartwright and P. J. Hay, J. Chem. Phys. 70(1979)3191

22)B. Jonsson, B. Roos, P. R. Taylor and P. E. M. Siegbahn, J. Chem. Phys. 74(1981)4566

23)M. R. A. Bloomberg, P. E. M. Siegbahn, Chem. Phys. Lett. 81(1981)4

24)R. Ahlrichs, H. Liska, B, Zurawski and W. Kutzelnigg J. Chem. Phys. 63(1975)4685

25) W. J. Hunt, P. J. Hay, and W. A. Goddard III, J. Chem. Phys. 57(1972)738

26)R. H. A. Eade and M. A. Robb (in preparation)

27)R. H. A. Eade, M. A. Robb and Z. Slattery (in preparation)

28)T. H. Dunning, J. Chem. Phys. 65(1976)3854

29)B. J. Rosenberg, W. C. Ermler and I. Shavitt J. Chem. Phys. 65(1976)4072

30)R. E. Howard, A. D. McLean and W. A. Lester, J. Chem. Phys. 71(1979)2412

31)R. F. W. Bader and R. A. Gangi, J. Amer. Chem. Soc. 93(1979)1831

32) S. P. Walch, T. H. Dunning and R. C. Raffenetti, J. Chem. Phys. 72(1980)406

33)R. Jaquett and V. Staemmler, Chem. Phys. 59(1981)373

34)P. J. Hay and T.H. Dunning, J. Chem. Phys. 67(1977)2290

35)C. Woodrow-Wilson and D. G. Hopper, J. Chem. Phys. 74(1981)595

36)R. R. Luchesse and H. F. Schaefer, J. Chem. Phys. 67(1977)848

37)G. Karlstrom, S. Engstrom and B. Jonsson, Chem. Phys. Lett. 57(1978)390

38)R. H. A. Eade, PhD. Thesis, University of London, 1981 (to be published)

39)P. E. M. Siegbahn, J. Chem. Phys. 75(1981)2314

40)M. Raimondi, M. Simonetta, And J. Gerrat, Chem. Phys. Lett. 77(1981)12

Molecular Structure and Conformation: Recent Advances,
I.G. Csizmadia (Ed.), *Progress in Theoretical Organic Chemistry*, Volume 3

ANALYTIC EQUATIONS FOR CONFORMATIONAL ENERGY SURFACES

M. R. Peterson and I. G. Csizmadia

ABSTRACT

The reliable location of minima and transition states is one of the more important uses of molecular orbital theory. An application of particular interest is conformational problems, both in the ground and lower excited states. In this work, surfaces (two rotating bonds) and hypersurfaces (three rotating bonds) were considered.

The approach is to represent a set of (ab initio) potential energy values, calculated at a grid of points, by an analytic least squares equation. The equations were most readily determined by an automatic stepwise regression technique which selected the best subset containing a given number of terms from a large number of candidate terms. All critical (stationary) points of the analytic equation were then rapidly and cheaply located.

The critical points were also determined by optimization directly on the ab initio potential energy hypersurface, utilizing the analytic first derivative of the energy (force), to ascertain the reliability and accuracy of the fitted equations. The molecular systems studied were n-butane (C_4H_{10}), formic acid (HCOOH), a protonated phosphoric amide ($H_2P(OH)NH_2^+$) and three ring-opened triplet state methyloxirane isomers (C_3H_6O), all at the STO-3G basis set level.

GLOSSARY

CI	configuration interaction
HF	Hartree-Fock
MO	molecular orbital
PES	potential energy surface (used in a general sense)
RHF	restricted Hartree-Fock SCF
RMS	root mean square
SCF	self-consistent field
SSCP	sums of squares and cross-products (matrix)
TS	transition state
UHF	unrestricted Hartree-Fock SCF
VA05AD	sum of squares optimization routine

1. INTRODUCTION

1.1 Qualitative Aspects of Surfaces and Hypersurfaces

The last decade has seen several important advances in ab initio molecular orbital (MO) theory and practice, including the widespread availability and application of automatic, ready-to-use programs like GAUSSIAN 70 [1], and the development of faster and more powerful computers, permitting MO calculations on systems of interest to organic chemists. Equally important have been the discovery of new techniques for rapid integral evaluation [2] and configuration interaction (CI) [3], allowing calculations of chemical accuracy to be carried out on molecules of ever-increasing size, and the implementation of energy derivative (force) programs [4] for precise determination of the geometries of both stable species and transition states.

MO calculations are almost always carried out within the Born-Oppenheimer approximation [5], by which the motion of the nuclei and electrons is decoupled as a consequence of the much larger nuclear masses, since the light electrons are assumed to adapt instantaneously to variations in the nuclear positions. Thus the total molecular energy is calculated for static nuclei as a function of their positions only, not their momenta. This energy is termed a potential energy, since it may be used as the potential energy function in the dynamic equations of nuclear motion.

The expression of the potential energy as a function of nuclear position leads to a potential energy surface (PES). These surfaces are at the foundations of most areas of chemistry, including reaction mechanisms, kinetics, spectroscopy and thermodynamics, and they are readily accessible via MO calculations. This work is a description of one approach to determining an analytic form for potential energy surfaces, and extracting some of the vast store-house of chemically important information contained in them, as applied to conformational problems in organic chemistry. This method will be compared with the determination of the same results directly by ab initio calculation without the intermediary of the analytic PES, with regards to accuracy and utility.

First, however, some of the varied uses of surfaces and hypersurfaces, particularly potential energy hypersurfaces, in chemistry will be briefly illustrated. The term 'surface' will usually denote a three-dimensional mathematical object, involving two independent variables (coordinates) and one dependent variable. 'Hypersurface' signifies that the dependent variable is a function of three (or more) independent variables, ie. a four- (or higher) dimensional object. Potential curves (two-dimensional objects) will not be considered explicitly, although the methods elaborated in the following are equally applicable to them.

The usual qualitative utility of any surface and hypersurface is to study the overall surface topology by visualization. This is accomplished most commonly by contour plots, or more recently by pseudo-three-dimensional plots. Some illustrations are the conformational PES for rotation and inversion processes in anions: $^{-}CH_2S(=O)H$ [6], CH_3^- [7], $^{-}CH_2SO_2H$ [8], $^{-}CH_2OH$ [9], and $^{-}CH_2NO_2$ [10].

Turning to reactive processes, three-dimensional contour plots for three states of trapezoidal H_4 have been discussed [11] in detail. The more conventional two-dimensional contour plots have been presented [12] for cross-sections of the CH_2+H_2 reaction hypersurface, and for multitudinous A+BC reaction surfaces [13]. Many of these studies have utilized spline interpolation to some extent, which has the advantage of passing through every point used to construct the surface.

The spline interpolation technique has also been applied [14] to the stereochemical behaviour of $HC(O^-)(OH)OCH_3$, the intermediate in the base-catalyzed hydrolysis of methyl formate. In addition, twenty-two critical points (stationary points) on the surface were located manually by inspection, the usual procedure for qualitative surfaces.

Another important application of contour plots and pseudo-three-dimensional plots is to electron density and difference density maps in crystallography [15-17]. Coppens [15] reviews many aspects of electron difference density maps, from both the experimental and theoretical standpoints, and includes many examples.

A related use [6] is contour maps of the overlap populations for the C-S and S=O bonds in $^{-}CH_2$-S(=O)H, as a function of rotation about the C-S bond and inversion at C.

A very recent innovation is two-dimensional NMR, which results in a pseudo-three-dimensional map where the resonance peaks are displayed (via a double Fourier transform) as a function of two chemical shifts for two different elements. For example, the ^{1}H and ^{11}B spectra were recorded [18] simultaneously for a carborane (2,4-$C_2B_5H_7$), which permits the facile resolution of peaks that overlap in the normal ^{1}H NMR spectrum.

1.2 Quantitative Aspects of Surfaces and Hypersurfaces

Obviously, any quantitative surface or hypersurface incorporates the qualitative topological features discussed in the previous section, but with the added ability to give an 'exact' numerical result once the values of all the independent variables are specified. This usually implies the existence of an analytic expression for computing the dependent variable as a function of the independent ones, but may involve the use of some interpolative technique.

The most widespread use of quantitative surfaces is in reaction dynamics, to calculate the potential energy for trajectory calculations. Many potential energy functions have been proposed, and interested readers are referred to the classic book "Molecular Reaction Dynamics" by Levine and Bernstein [13]. The only method of obtaining the PES discussed here is to fit an analytic equation to a series of ab initio points. The earliest use of this approach was for the $H^+ + H_2$ system [19], although the equation was not actually reported. Other systems studied, in rough chronological order, include: LiH_2^+ [20], H_3^+ [21], $CO_2 + H_2$ [22], HeH_2 [23], $(HF)_2$ [24], H_3 [25], HClBr [26], H_2F [26], H_3^+ [26], H_3 [27], H_2O [28], H_3 [29], and H_2F [29]. In several cases, cubic spline interpolation was utilized as part of the overall fit, which allows the calculation of the energy and its first derivative, required for the scattering calculations, but obviates the writing of the analytic equation in closed form. Many fits have incorporated constraints on barrier heights or positions, or both, to correct for deficiencies in the original ab initio points. It should be emphasized that high acurracy is required in all regions of the hypersurface, which almost always entails large CI calculations, as the single configuration (single determinant) SCF method rarely dissociates correctly. For example, AB dissociates to $A^+ + B^-$, not $A^{\cdot} + B^{\cdot}$. Several authors [21,24,25,27,28] have calculated the barrier position, the barrier height, the optimum complex geometry, the dissociation energy or the force constants from their hypersurface equation. In addition, many authors tested several equations for their suitability, and gave criteria for their choice of the 'best' fit.

A direct application of fitting analytic equations to ab initio energy values is the determination of force constants, and subsequently the normal mode frequencies, in a vibrational analysis. The advantage of this approach is that both the harmonic and higher-order anharmonic (cubic, quartic) force constants are quite simply related to the linear coefficients of the analytic equation. The technique has been applied to the complete vibrational analysis of water [30-33], which involves a hypersurface, and to the potential surface of linear HF_2^- [34]. Three authors [32-34] have determined the energies of excited vibrational states, and Ermler and Kern [30] even averaged several one-electron properties (multipole moments, field gradients, etc.) over the zero-point vibrational motion calculated from their hypersurface. They, along with Dunning et al [31], determined the theoretical equilibrium geometry of water from their Hartree-Fock hypersurfaces, while Hennig et al [33] used a CI hypersurface. The latter authors also calculated the geometry of the linear transition state and the barrier to inversion, but not from the fitted hypersurface. In addition to the region of coordinate space near the stable minimum geometry, the potential hypersurface of Sorbie and Murrell [32] allows the water molecule to dissociate, permitting dynamical calculations concomitant with the accurate computation of spectroscopic properties.

Bridging the gap between reactive and conformational problems, and between physical and organic chemistry, is the addition of $H^{\cdot}$ to ethylene, as both the hydrogen atom addition and migration pathways were fitted [35] by a hypersurface equation incorporating both ab initio and experimental data. The critical point geometries were computed from the hypersurface and compared with those optimized directly. Also, the force constants were evaluated from the equation at various points along the minimum energy pathways, and a vibrational analysis was performed for the minimum structure.

Turning to purely conformational processes, several surfaces have been fitted to the ab initio points, though chiefly to examine the surface topology via pseudo-three-dimensional plots, as the analytic equation was not subjected to further analysis. Examples where the equation used was actually stated include $CH_2{=}OH^+$ [36], $CH_2{=}SH^+$ [36], and $\cdot CH_2SH$ [37]. The same technique has been used for $^-CH_2OH$ [38], $^-CH_2CH_3$ [38], thiathiophthene [39], $CH_2{=}CH_2SH^+$ [40], $CH_2{=}CH_2OH^+$ [40], $CH_2{=}CH_2NH_2^+$ [40], and $F_2C{=}S$ [41], but without reporting the intermediate equation. Surfaces for the nuclear and electronic components of the total potential energy for the rotation-inversion of $^-CH_2NO_2$ have also been examined [42].

Analytic double rotor conformational PES were reported a decade ago for the methyl group rotations in propane [43,44], 2-methylpropene [44], and cis- and trans-2-butene [44]. The general form of the equation [44], a truncated Fourier series in the two dihedral angles θ_1 and θ_2, is shown in (1).

$$\begin{aligned} E(\theta_1,\theta_2) = {} & 1/2V_3(2-\cos 3\theta_1-\cos 3\theta_2) \\ & + 1/2V_6(2-\cos 6\theta_1-\cos 6\theta_2) \\ & + 1/4V_3'(1-\cos 3\theta_1)(1-\cos 3\theta_2) \\ & + 1/4V_3''\sin 3\theta_1 \sin 3\theta_2 \end{aligned} \quad (1)$$

The analytic expression was fitted to only five [43] or six [44] ab initio points for each molecule due to the high molecular symmetry, and the linear least squares coefficients (V_i) were interpreted directly as barrier heights. This is possible because all critical points occur for 0 and 60° torsional angles, resulting in much cancellation of terms in (1). For example, the barrier to rotation of a single methyl group by 60° is E(60,0)-E(0,0), or simply V_3. Note that the energy of the global minimum structure, the (0,0) conformation, was taken as zero to eliminate the constant term from the potential energy expression (1).

A final example illustrates that while much effort has been concentrated on potential energy surfaces and hypersurfaces, other possibilities do exist. Curtis, Langhoff and Carney [45] have reported an analytic fit to the dipole moment hypersurface of ozone, for a series of basis sets and degrees of wave-function sophistication. Theoretical infrared intensities were computed using dipole moment derivatives obtained from their hypersurface equation.

As is apparent from this review, the quantitative utility of potential energy surfaces and hypersurfaces has been centered in the area of reaction dynamics, and, to a lesser extent, on vibrational analysis. These and many other aspects of potential energy surfaces, such as the influence of symmetry rules, have been covered in an excellent review article by Murrell [46].

A neglected application of fitted equations is in the area of quantitative hypersurface topology: the accurate determination of critical point geometries concomitant with the visualization of the PES by contour or pseudo-three-dimensional plots. The advantage of such an approach lies in the extraction of significantly more information* from the identical set of points as would be used to generate the qualitative diagrams, though a substantially more accurate analytic equation would almost certainly be required. A method of achieving increased accuracy will be elaborated in section 2.2.

The determination of the critical point geometries from an analytic equation need not be efficient, as the evaluation of the molecular energy requires a negligible amount of computer time. However, the hypersurface equation can be used as a testing ground for various optimization procedures before they are utilized directly within the ab initio framework. Of particular importance is the efficient yet accurate location of first order saddle points, corresponding to chemical transition states. A more esoteric topological aspect is the postional and energetic relationships between the minima, the first order saddle point and the second and higher order saddle points, all of which may be represented by a single analytic equation.

The second derivatives of the analytic equation are related to the harmonic force constants, and higher order anharmonic contributions are also readily evaluated, especially if the analytic functional form is chosen to be readily differentiable. This feature is also important if the hypersurface is to be the subject of dynamical (trajectory) calculations. Note in passing that the first derivative of any energy hypersurface is the geometry: the points where the gradient hypersurfaces are all zero are termed the critical or stationary points, corresponding to minima (stable chemical species), saddle points of all orders (transition states) and maxima of the original energy hypersurface.

The feasibility of using analytic equations to obtain the geometries of ALL critical points contained on the hypersurface will be examined, as exemplified on the following systems drawn from organic chemistry: internal rotation in n-butane, rotation and inversion in formic acid, rotation and inversion in a protonated phosphoric amide ($PH_2(OH)NH_2^+$), and the triplet state conformational behaviour of three isomers of ring-opened methyloxirane (propylene oxide, $CH_3\overline{CHCH_2O}$). The critical point geometries and energies will be compared with those determined directly from the ab initio PES, to ascertain the reliability of the procedure.

---

*The wealth of data available from an analytic fit to a relatively small number of points was emphasized and used to great advantage in ref. [33].

2. COMPUTATIONAL METHODS

2.1 Generation of the Data Points

Within the Born-Oppenheimer approximation, the calculation of the total molecular energy by ab initio methods requires the specification of the nuclear geometry, the basis set for each atom, and the Hamiltonian operator. These three components are discussed in turn.

The experimental geometry, taken from x-ray diffraction or microwave studies, is a popular choice for all levels of calculation. However, this may pose difficulties, as calculations are often performed on experimentally unknown model systems, where substituents thought unimportant to the process under investigation are replaced by smaller groups (for example, methyl or phenyl groups by H, or Br by F). A partial solution is the use of standard geometries [47,48] based on experimental bond types and lengths, and valence angles. Difficulties still arise, though, for atoms not included in the tables, notably transition metal elements, and for ring structures, with the exception of compounds incorporating intact benzene or cyclohexane rings.

Clearly the most systematic procedure is to optimize the geometry within the computational method employed. This ensures that comparisons are consistently made at critical (stationary) points of the PES, while the experimental or standard geometry generally corresponds to a completely arbitrary point on the hypersurface. Geometry optimization is not possible for most semi-empirical MO methods, with the exception of MINDO/3 [49], MNDO [49], and PRDDO [50], as they usually give unrealistic geometries [50]. For ab initio calculations, complete geometry optimization is not yet feasible for larger molecules (containing more than five non-hydrogen atoms) in the absence of high symmetry, or virtually free computer time.

For conformational studies involving internal rotation, the cheapest approach is to simply change the desired dihedral angles, fixing the remaining geometrical parameters at their initial values. This is the rigid rotor model. The relaxed rotor alternative involves the partial, or even complete, optimization of other parameters, especially the valence angles adjacent to the rotating bond. Either of these modes may be combined with experimental, standard or optimized initial geometries.

Payne and Allen have reviewed [51] the merits of both the rigid and relaxed rotor approaches. They expect that Hartree-Fock calculations should accurately reflect even the small geometrical distortions that occur during internal rotation, indicating that the relaxed model should yield physically reasonable results. However, the fair success of the rigid rotor model is due in part at least to the partial cancellation of the distortional corrections by the zero point energy corrections. In particular, they advise that the latter contributions be analyzed for inversion barriers, where large bond length distortions (0.03 bohr) in the transition state may be found [51].

All calculations reported in this work have been performed with the standard STO-3G gaussian-type orbital basis set [52] introduced by Pople. The errors incurred by the use of minimal basis sets, in particular the very popular STO-3G basis, are highly systematic and well documented [49-51,53]. Specifically, the calculated bond lengths are mostly too long, with the exception of CC lengths, and valence angles too small. The average deviations are 0.03 Å and 4° respectively [53]. However, dihedral angles, important for conformational studies, are reproduced 'most satisfactorily' [53] with the notorious exception of peroxides which require d orbitals to obtain the correct results. Even minimal basis sets are capable of reproducing the subtle changes upon internal rotation [51], and general trends for large series of molecules [53].

More reliable energy differences, though, are obtained at more sophisticated levels of theory [50,53], often using the STO-3G optimized geometries. Isodesmic reactions, in which the numbers and types of bonds are not changed, are treated fairly accurately by minimal basis set calculations [49,50]. Other observations include a bias towards small strained rings [50], including non-classical carbonium ions [49]. In addition, activation energies for H transfer reactions are overestimated [49], ring-puckering is underestimated [49], and barriers to inversion are overestimated [51].

On the whole, STO-3G geometry determinations are reasonably accurate, giving, for the most part, systematic deviations from experiment. It appears that minimal basis set calculations may not always yield reliable results for processes which involve a change in the relative orientations of bonds on the same atom (rehybridization), since the basis set lacks sufficient flexibility [54]. Although internal rotation about what are formally single bonds is an isodesmic process, and usually involves no hybridization change, results from STO-3G barrier calculations are mixed [51,54]. Few general trends can be discerned, but the accuracy of the barrier heights seems to deteriorate as the hetero-atom content of the molecule increases.

Despite these potential drawbacks, there are several reasons favouring minimal basis set calculations. Perhaps the most important aspect of all is that sophisticated calculations are not required to ascertain the feasibility and reliability of representing the original (ab initio) data by a hypersurface equation. In fact, energies from large basis set or CI calculations could be handled with equal ease in exactly the same manner as will be described in the next sections, although the reliability testing by comparison with direct optimization would be a <u>significantly</u> more expensive undertaking.

Since the calculations reported in Chapter 3 comprised the first systematic coverage for each of the systems studied, it was logical and appropriate to commence with a simple level of theory, as there is no fundamental requirement for sophisticated computational methods. For n-butane and formic acid, the hypersurface and direct optimizations results may be compared with other STO-3G calculations which examined much smaller domains of the respective potential energy hypersurfaces.

Computer time considerations also arose, since a grid of fifty to one hundred ab initio points was used for each hypersurface fitted. These points were always, and should always be, chosen as to cover the entire coordinate space in a uniform manner, not concentrating all the points near the minima and projected saddle points.

The fundamental feature of closed shell Hartree-Fock theory is the use of a single Slater determinant wave function based on a set of doubly occupied molecular orbitals, which are in turn expanded by standard methods [55-57] as a linear combination of atomic orbitals. The n-butane, formic acid and protonated phosphoric amide data points were computed within this framework.

For open shell systems, such as triplet states, there are two Hartree-Fock approaches: the Unrestricted Hartree-Fock (UHF) [58] and the Restricted Hartree-Fock (RHF) [59] methods. They differ principly in the manner in which the electrons are partitioned: in the UHF formalism, the alpha and beta spin manifolds are separated (different orbitals for different spins), while the doubly occupied (closed shell) orbitals are distinguished from the singly occupied (open shell) orbitals in the RHF formalism.

All calculations for the triplet methyloxirane isomers were performed by solving the UHF equations as outlined by Pople and Nesbet [58]. The chief disadvantage of UHF calculations is that the wavefunction is not usually an eigenfunction of the total spin operator, S^2, and thus does not strictly correspond to a pure electronic state [53]. Rather, they may be "contaminated" by contributions from states of higher multiplicity; for example, the triplet wavefunction may acquire some quintet character. As a matter of standard procedure, the expectation value of S^2 should always be checked for UHF calculations. In every case, for all the triplet methyloxirane isomers, the containimation by states of higher multiplicity was quite small (less than 0.5%).

All of the SCF calculations used in the construction of the analytic hypersurface equations were performed with the GAUSSIAN 70 program package [1], which includes an efficient two-electron integral routine [60], and solves either the closed shell [55] or UHF open shell [58] SCF equations. The points for the n-butane and formic acid hypersurfaces, and the phosphoric amide surface, were computed on the IBM 370/165 II at the University of Toronto Computer Center, and the methyloxirane calculations were carried out on the CDC 6500 and 6600 computers at the Université Libre de Bruxelles.

2.2 Hypersurface Fitting

The general form of the analytic surface or hypersurface equation is

$$E(q) = \sum_{i=1}^{m} c_i t_i(q) \tag{2}$$

where $t_i(q)$ is the i'th term of the expansion, and q is the vector of the n independent coordinates (variables):

$$q = (q_1, q_2, \ldots, q_n) \tag{3}$$

The linear coefficients c_i are determined by standard least squares techniques [61].

For a surface equation, each term is a product of two functions f_1 and f_2:

$$t_i(q_1,q_2) = f_{1i}(q_1)f_{2i}(q_2) \tag{4}$$

which are each a function of just one of the independent variables. The terms for hypersurface equations,

$$t_i(q_1,q_2,q_3) = f_{1i}(q_1)f_{2i}(q_2)f_{3i}(q_3) \tag{5}$$

are expressed similarly. This constitutes a constraint on the analytic equation, but most of the possible terms can be written in this form. For example, $\cos(q_1+q_2)$ can be expressed as $\cos(q_1)\cos(q_2)-\sin(q_1)\sin(q_2)$, which is of the form (4).

The functions $f_{ji}(q_j)$ and all derivatives $\partial^r f_{ji}/\partial q_j^r$ must be continuous to be physically reasonable, as the potential energy is always smoothly varying. For rotational coordinates, they must also suit the periodicity, if any. The overall equation must also be adapted to the symmetry of the molecule; this is described in Chapter 3 for each equation fitted.

Concerning the choice of the functions $f_{ji}(q_j)$, rotational coordinates are usually represented by truncated Fourier series (sin bq_j, cos bq_j). The use of Fourier analysis for potential curves has been reviewed [51] recently. Aperiodic coordinates may be represented by polynomials q_j^b, although some care must be taken, as extraneous critical points may result from high order polynomials [28]. Examination of the surface, or cross-sections of hypersurfaces, has been recommended [25,28,44]. The present authors concur completely, and no irregularities were discovered in any of the fits reported in Chapter 3.

Functions and the entire methodology for dissociative coordinates have been discussed widely [20,21,23-25,27-29,32,35,44,62] in conjunction with analytic equations used for scattering calculations. Such functions are not required for the conformational processes discussed in this paper.

To aid in the description of large but highly localized maxima, Gaussian functions $\exp(-b(q-a)^2)$ have been used. They have been adapted to periodic coordinates by reducing (q-a) by addition or subtraction of multiples of 2π

such that $-\pi\leq(q-a)\leq\pi$, and requiring that the exponent b be sufficiently large to reduce the exponential to zero for $|q-a|=\pi$. This latter requirement ensures the continuity of the derivatives.

The non-linear parameters in the analytic equation, eg. a and b in $\exp(-b(q-a)^2)$, may be optimized provided the linear coefficients c_i are redetermined for each new set of the non-linear parameters. This is best accomplished by sum of squares optimization techniques, like the VA05 method discussed in section 2.3.

The first step in fitting an analytic equation is to select a grid which covers fairly uniformly the desired coordinate space, and to evaluate the energy at those points at the chosen level of theory. Next, the functions for each of the independent coordinates q_j are chosen, and combined to form the set of candidate equation terms by (4) and (5). The task remaining is the selection of the specific terms from the set to be included in the analytic equation.

A potential curve involves one independent variable q_1; if there are five functions $f_{1i}(q_1)$, selecting the best combination of the five terms could be readily accomplished by simply performing all possible regressions, that is, trying all the possibilities. A total of 31 linear least squares fits would be required, as the number of equations (N_e) containing R terms out of a total of T terms is

$$N_e(R) = \frac{T!}{R!(T-R)!} \tag{6}$$

For a surface, if each independent variable could be represented by five functions $f_{ji}(q_j)$, there is a maximum of twenty-five possible surface terms of the form (4). Assuming that an equation containing ten of those terms would be found to be satisfactory, there are 25!/10!15! = 3,268,760 combinations to be examined. While this is clearly beyond manual capabilities, a computer could perform all the required calculations in a few minutes.

For a hypersurface containing three independent variables, a similar example would require approximately 1.3×10^{26} least squares fits, even if a satisfactory equation was somehow known to contain only twenty-five of the 125 total terms. Even if computers could perform the regression analyses at the rate of one million per second, 4.1×10^{10} centuries would be required to complete them. Clearly, examining all possible regressions is not feasible, and manual procedures will be unsatisfactory. The latter point is amply illustrated by the n-butane hypersurface (section 3.1).

Several regression techniques are given in the classic text [61] by Draper and Smith: forward selection, backward elimination and stepwise regression. The forward selection method starts with no terms in the equation, and at each step inserts a new term into the equation. The term selected is the one most highly correlated with the residuals (fitting errors) remaining from the previous step.

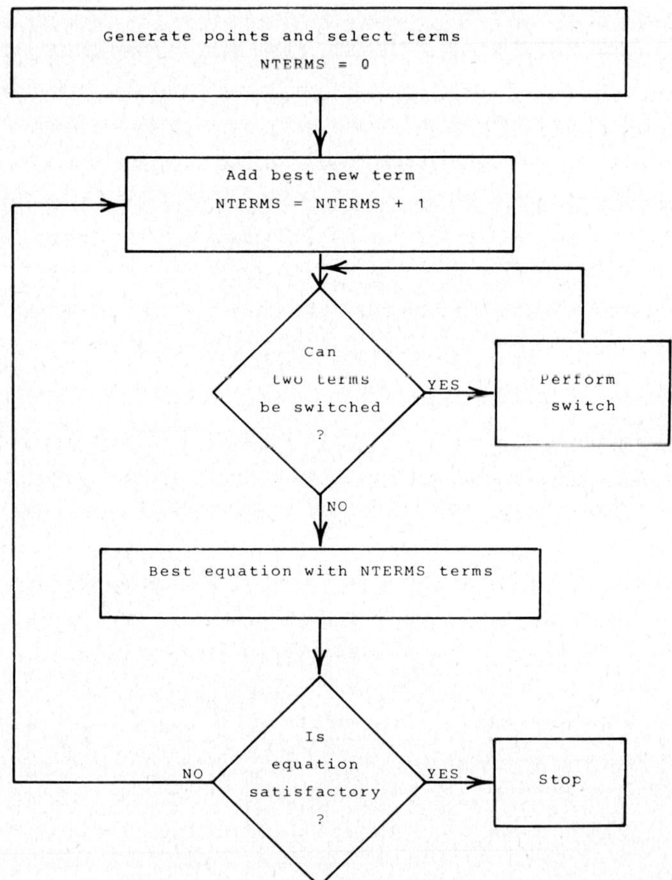

Figure 1. Flowchart for the optimization of the equation.

The process is halted when the latest term added does not effect a significant reduction in the sum of squared residuals, compared with previous terms and a preselected significance level.

A similar technique is used for backward elimination, except that all terms are initially included in the equation. At each step, the least important term remaining in the equation is removed, provided its significance is less than some preselected significance level.

Both of these methods are much more economical than the all-regressions approach. However, they both fail to explore the ramifications the introduction (or elimination in the backwards case) of a term has on the importance of other terms in (out of) the equation. This deficiency is overcome by the stepwise regression method, which is flow-charted in Figure 1.

The procedure commences with the correlation matrix, which is derived from the matrix $U=X'X$ of uncorrected sums of squares and cross-products (SSCP), where the element X_{ji} is the value of the i'th term $t_i(q)$ evaluated at the j'th point to be used in the fit and X' is the transpose of X. The column vector of energy values is appended to the matrix X, and will be identified by the subscript E. After correction of X for the means to give X_c, the corrected SSCP matrix (or covariance matrix) $V=X_c'X_c$ is reduced to the correlation matrix R by

$$R_{ij} = V_{ij}/(V_{ii}V_{jj})^{1/2} \tag{7}$$

Thus all terms are treated on an equal footing (note that the diagonal elements R_{ii} are unity due to the scaling (7)). The elements of the diagonal dominant correlation matrix are now all of the same order of magnitude, leading to minimum accumulation of round-off error in the ensuing regression analysis.

The initial equation contains no terms. The best term to add [61,63] is the one most correlated with the residual errors; it gives the maximum value of $R_{iE}R_{iE}/R_{ii}$. This term is duly incorporated, and the correlation matrix is adjusted for the inclusion of the new term [61], say term k. Row k becomes

$$R^*_{kj} = R_{kj}/P \tag{8}$$

and the remaining rows are obtained from

$$R^*_{ij} = R_{ij} - R_{ik}R_{kj}/P \tag{9}$$

where the superscript asterisk refers to the new R matrix, and the pivot element P is R_{kk}. R^{-1} (which will become the inverse of R), initially a unit matrix, is also updated in the same manner.

The next part of the procedure differs from the forward selection method discussed earlier: each term included in the equation is tested against each term excluded from the equation to determine if a better fit would result if the roles of those two terms were switched (ie., the included term is replaced by the excluded term). The best switch [63] maximizes the value of

$$\frac{R_{ii}^{-1}R_{jE}R_{jE} - 2R_{jE}R_{ji}^{-1}R_{iE}^{-1} - R_{jj}R_{iE}^{-1}R_{iE}^{-1}}{R_{ii}^{-1}R_{jj} + R_{ji}^{-1}R_{ji}^{-1}} \tag{10}$$

where i refers to an included term and j to an excluded term. Equation (10) assumes a much simpler form if the correlation between terms i and j, R_{ji}^{-1}, is zero:

$$\frac{R_{jE}R_{jE}}{R_{jj}} - \frac{R_{iE}^{-1}R_{iE}^{-1}}{R_{ii}^{-1}} \tag{11}$$

The first term represents the reduction in the error sum of squares to be realized upon the inclusion of term j, and the second term accounts for the increase due to the removal of term i. Note also that the improvement in the fit (if any) can be predicted without actually performing an inversion of the correlation matrix.

The switching procedure allows terms which become insignificant due to other terms added in later steps to be replaced, unlike the forward selection procedure in which a term is always retained once entered.

After a switch is made, all possible interchanges are evaluated again, until no further improvement is possible. This is then taken as the best equation containing a given number of terms, but there is no guarantee of this assumption, except to evaluate all possible regressions. If this equation is not satisfactory, the best excluded term is added to the equation and all possible replacements are tested again, as indicated in Figure 1.

Three criteria are used to terminate the looping process: first, the equation reduces the average error (not the root mean square error) of the fit below a preset cutoff, often 0.01 or 0.02 kcal/mol. The second criterion is that the maximum number of terms (NSTOP) to be included has been reached. Lastly, there is no further improvement possible, although NSTOP terms have not been included, nor the average error reduced sufficiently. These latter conditions usually indicate that the candidate set of equation terms lacked sufficient flexibility, and should be expanded.

The stepwise regression method just described reduces the error of the fit by the maximum amount at each addition or switch. This is equivalent to the maximum increase in R^2, the square of the multiple correlation coefficient:

$$R^2 = \frac{\text{total sum of squares - error sum of squares}}{\text{total sum of squares}} \tag{12}$$

where all sums of squares have been corrected for the means. Hence this method is termed MAXR2 [63].

A second alternative, very similar to the preceeding, is the MINR2 [63] method. Here, instead of selecting the addition or replacement that maximizes the increase in R^2, the one minimizing the increase in, but not decreasing, R^2 is chosen. This results in many more switches being performed with a given number of included terms; that is, the examination of many more possible equations - a more complete search, but still not approaching the computation of all regressions.

The MAXR2 and MINR2 procedures should arrive at the same series of best equations containing one, two, three, etc. terms, and did so for all examples tried.

The only remaining decision is the choice of the number of terms included in the equation - the regression program automatically determines the best equation containing any given number of terms. This choice is also governed by three major criteria: first, the average error should be sufficiently small, as discussed previously, to ensure accuracy. Each term in the equation should be significant statistically at the 10% level (ie. each term has less than a 10% chance of being zero for the given data). **Lastly,** the equation is checked visually by examining contour and pseudo-three-dimensional plots for irregularities.

Other stepwise regression procedures are given in Draper and Smith [65] and by Furnival and Wilson [64], and references therein. The latter paper also gives an efficient method of performing all possible regressions.

Except for the two butane hypersurfaces, all analytic equations fitted by the automatic stepwise regression technique were determined using the STEPWISE procedure of SAS76 [63]. As will be seen in Chapter 3, the procedure can yield good results; hence, the STEPWISE procedure [63] was rewritten in FORTRAN (it is originally in PL/1 and IBM Assembler Language) and was used to fit the second butane hypersurface (the first equation was determined entirely manually). The entire stepwise regression procedure requires only a few minutes, even on a minicomputer.

2.3 Hypersurface Analysis

2.3.1 Optimization of critical point geometries. The critical or stationary points of a function $f(q)$ are the points where the first derivatives with respect to each independent variable $\{q_i\}$ are all simultaneously zero:

$$g_i = \partial f(q)/\partial q_i = 0 \tag{13}$$

For potential energy hypersurfaces the function is total energy and $\{q_i\}$ are the internal coordinates of the molecule. At such points, the number of negative eigenvalues (principal directions or axes) of the second derivatives matrix (Hessian), is the order or index of the critical point. Minima have order 0, or no negative eigenvalues; thus the function increases in every direction from such a point. Maxima have order n, where n is the number of independent variables q_i; the function decreases in every direction from these points.

The minima of PES correspond to either stable molecules or intermediates. The geometries of stable species can often be accurately measured experimentally by spectroscopic or diffraction techniques, and their relative energies may also be determined. The critical points of order 1, termed saddle points, have Hessian matrices with only one negative eigenvalue. The lowest energy first order saddle point connecting two adjacent minima corresponds to the chemical transition state (TS). That the TS force constant matrix may have only one negative eigenvalue has been emphasized and proved by Murrell and Laidler [65].

Transition states are crucial to kinetics and mechanism, for their structures and energies determine both the outcome and rate of chemical reactions. The force constants account for isotopic substitution effects. Absolute rate theory, built around the transition state, relates kinetics and thermodynamics. However, the elusive TS has yet to be clearly observed, as reacting molecules simply do not remain in the neighbourhood of the PES saddle point long enough for accurate measurements [66,67]. Hence, the accurate calculation of TS structures and energies can, and probably will, be one of the most important and productive long term applications of MO theory.

Before turning to specific techniques, a few general comments: strictly, the order of the critical point should be determined through a Wilson GF matrix vibrational analysis [68,69], using a complete set of internal coordinates. The number of negative eigenvalues of the GF matrix, which correspond to imaginary vibrational frequencies, is the index of the critical point [46]. The positive eigenvalues will be interpretable as frequencies of normal modes of vibration only when mass weighted coordinates are used, as in the GF analysis. Otherwise, the magnitudes of the eigenvalues and the directions of the principal axes (the eigenvectors) have no physical meaning [67].

The discussion to this point has bypassed the possibility of a zero eigenvalue occuring at a critical point, implying no energy dependence for some internal motion. It has been argued by several authors [65,67] that such cases (e.g. three valleys meeting at a "monkey saddle" point) do not in fact occur in chemical systems, and that the supposed degenerate point is in reality either a shallow maximum or minimum.

The determination of saddle points has received little attention from numerical analysts; in fact, optimization methods are usually constructed to explicitly avoid such points. It appears that chemistry may be unique in that both saddle points and minima (of PES) play important roles, as transition structures and stable species respectively.

A traditional technique to locate saddle points has been to choose one geometrical coordinate as the independent variable, then optimize the remaining geometrical parameters, this process being repeated for a series of values for the independent variable. Thus it was hoped to generate a "reaction path", and the saddle point was taken as the maximum energy point along the path. Both of these expectations are incorrect [46,70-72]. In addition, the path is strongly dependent on the type of coordinates used; as seen previously, mass weighted coordinates are the only appropriate choice. Note that the critical points themselves are uniquely defined, independent of the coordinate system utilized.

The first reliable method is due to McIver and Komornicki [71], and involves minimizing the sum of squared gradient components:

$$S_g = \sum_{i=1}^{n} g_i^2 \tag{14}$$

The function S_g is clearly zero at a critical point, and positive elsewhere. This technique has been applied to several chemical systems [71,73,74]. One of the chief problems associated with this method are the possible appearance of non-zero local minima of S_g, which are not critical points as S_g must be zero at such points. A second feature is that the method can converge to any type of critical point - minimum, maximum or saddle point of any order. Thus the second derivatives matrix (Hessian) must be evaluated and diagonalized to complete the characterization of the critical point - this should be done routinely in any TS determination, unless the algorithm used guarantees the order of the critical point found. This ability of locating critical points of arbitrary order is used to advantage in this work, since a complete topological map of all the stationary points was desired. However, this can be a drawback to the technique if only first order saddle points are desired. In addition, the method cannot be used to generate reaction paths. The choice of a starting geometry is not generally a problem, as (stereo)chemical intuition is usually sufficiently accurate for convergence to the desired critical point.

Halgran and Lipscomb have developed [72,75] a synchronous transit method for determining both an approximate path and the saddle point structure. Their procedure first uses linear interpolation in ordinary (not mass weighted) cartesian coordinates between the reactant and product geometries. This linear synchronous transit (LST) path is the initial estimate of the reaction path. The total energy is evaluated at a series of points along the LST, and the maximum interpolated. An orthogonal optimization is then carried out, perpendicular to the LST, always maintaining the same relative (fractional) path position, leading to a lower bound on the saddle point energy. A new quadratic synchronous transit (QST) path is constructed through the two minima and the orthogonally optimized point. A series of points are calculated along the QST, and the predicted maximum energy point on this path is taken as the saddle point. Generally, the process is terminated at the QST level, as the two energy estimates are usually quite close.

The advantages of the technique are that the energy gradient is not required, convergence to a first order saddle point is efficient, and an approximate reaction coordinate is obtained. One disadvantage is that the saddle point geometry and energy are not uniquely determined, only bounded, although this is not often a serious drawback. Of course, it cannot be used to locate higher order saddle points.

In some cases, the orthogonally optimized structure is actually a local minimum (reaction intermediate) along the QST, although it was a maximum on the LST. When this occurs, two new LST paths are constructed from the intermediate point to the reactant and product geometries separately, the maxima are refined by orthogonal optimization, and two new QST paths constructed to locate the two saddle points. In other cases, more elaborate strategies were required [72].

A third method, the X method [76,77], was developed based on the properties of the hyperbolic paraboloid

$$z = f(x,y) = \frac{x^2}{a^2} - \frac{y^2}{b^2} = \left(\frac{x}{a} - \frac{y}{b}\right)\left(\frac{x}{a} + \frac{y}{b}\right) \qquad (15)$$

Thus there are two lines of constant z (z=0)

$$\frac{x}{a} = \frac{y}{b}, \quad \frac{x}{a} = -\frac{y}{b} \qquad (16)$$

that lie completely on the surface, and cross in an X at the saddle point (0,0). Details of the method are not discussed here, as preliminary testing on analytic surface equations demonstrated that the technique converged too slowly for ab initio use.

Recently, another approach to locating saddle points and minimum energy paths appeared [78], using a constrained simplex optimization procedure. Basically, the method involves the constrained minimization of the energy on successive hyperspheres surrounding both the minima simultaneously. The sequence of points so generated is then an approximate minimum energy pathway, and the maximum of this path is taken as the saddle point. The method does not require the energy gradient, and convergence to a first order saddle point may be verified by optimization in the subspace orthogonal to the one-dimensional energy path (c.f. the orthogonal optimization in the LST/QST method). Should this point prove to be a local minimum, the total path is divided into two segments, as in the LST method, and a new minimum energy pathway found for each.

A final method is direct force relaxation. This method utilizes a force constant matrix containing one negative eigenvalue to converge to a saddle point, through the Newton-Raphson relationship

$$X_{opt} = -H^{-1}g \qquad (17)$$

where H is the Hessian matrix and g is the gradient vector. Although very few details have been published, the method is thought more efficient than all others [79], and may become the method of choice when more details and a practical algorithm are available.

All critical points reported in this work, whether determined from analytic equations or by direct ab initio optimization, were determined by minimizing S_g, defined by (14). This is equivalent to solving the set of nonlinear equations

$$g_i = 0 \tag{18}$$

as the number of terms in the sum squares (n) is the number of geometrical parameters being optimized. Since (18) already involves the energy gradient, a method requiring explicit derivatives of g_i, i.e. the second derivatives of the energy, is not generally available for ab initio work, but second derivatives are of course readily available from the analytic equations. The best derivative-free algorithm [80] is NS01AD [81] when the number of terms (m) equals the number of independent variables (n) in the sum of squares

$$S = \sum_{i=1}^{m} h_i^2(x_1, x_2, \ldots, x_n) \tag{19}$$

and VA05AD [82] when m>n. While the present case falls into the former category, the more general routine VA05AD, already in use for other least squares otpimization applications in this laboratory, was selected, as both utilize the same algorithm [81], the NS01AD routine being slightly simplified due to the equality of m and n.

The algorithm [81] is a complex hybrid of the Newton-Raphson and steepest descents methods, related to the Levenberg/Marquardt procedure. In fact, the new directions and step lengths predicted by both methods are evaluated, then an appropriate linear combination chosen; the Newton-Raphson step aims for rapid convergence to a stationary point of S, while the bias towards the steepest descents direction attempts to ensure ultimate convergence to a minimum of S.

The method requires an initial estimate of the Jacobian matrix, which is evaluated by finite differences, whose elements J_{ij} are defined by

$$J_{ij} = \partial h_i/\partial x_j \tag{20}$$

In the present case, where $S=S_g$, $h_i=g_i$ and $x_j=q_j$,

$$J_{ij} = \partial g_i/\partial q_j = (\partial E/\partial q_i)/\partial q_j = \partial^2 E/\partial q_i \partial q_j \tag{21}$$

which is simply the energy second derivatives matrix, or Hessian. The initial Jacobian may thus be diagonalized and the number of negative eigenvalues used to complete the characterization of the critical point being sought -- the method always converges to the critical point nearest in space to the initial point, and should have the same index.

2.3.2 Energy gradient of ab initio wavefunctions. Routine geometry optimization by force (the force is the negative of the gradient or first derivative) methods is rapidly becoming a standard procedure in laboratories employing ab initio MO theory. Pulay pioneered [4] the use of the analytic first derivative of the energy with respect to the nuclear coordinates, as applied to both molecular geometries and force constants.

For closed-shell systems, the total SCF energy is [83]

$$E = V_N + 2 \sum_{m,n} D_{mn} h_{mn} + \sum_{m,n,o,p} P_{mnop} (mn|op) \tag{22}$$

where V_N is the nuclear repulsion energy, h_{mn} is a one-electron integral and (mm|op) is a two-electron repulsion integral (the integrals are over the atomic orbital basis η). The elements of the density matrix D are defined by

$$D_{mn} = \sum_{i}^{occ} C_{mi} C_{ni} \tag{23}$$

where the sum is over the occupied molecular orbitals, and

$$P_{mnop} = 2D_{mn}D_{op} - 1/2(D_{mo}D_{np} + D_{mp}D_{no}) \tag{24}$$

The first derivative of the total energy with respect to a variable A is then [83]

$$\frac{\partial E}{\partial A} = \frac{\partial V_N}{\partial A} + 2 \sum_{m,n} D_{mn} \frac{\partial h_{mn}}{\partial A} + \sum_{m,n,o,p} P_{mnop} \frac{\partial (mn|op)}{\partial A} - 2 \sum_{m,n} D'_{mn} \frac{\partial S_{mn}}{\partial A} \tag{25}$$

where

$$D'_{mn} = \sum_{i}^{occ} C_{mi} e_i C_{ni} \tag{26}$$

S is the overlap matrix and e is the vector of orbital energies. The first term of (25) is the nuclear-nuclear repulsion contribution, the second includes both the nuclear-electron attraction and kinetic energy contributions, and the third term is the electron-electron repulsion contribution. The last term arises from the changes in the MO coefficients as the basis functions rigidly follow their respective nuclei (the basis functions are always assumed to be assigned to an atom, even if the 'atom' has no mass or charge, i.e. a dummy atom). Further details may be found in several articles [4,74,83-87]. The force has been generalized to open-shell wavefunctions, both UHF [74,85] and RHF [86], and to CI wavefunctions [87].

One major advantage of calculating the gradient analytically is the enhanced numerical accuracy [84] compared with numerical differentiation of the energy. In addition, a single gradient evaluation yields the derivatives for all the atomic cartesian coordinates simultaneously, yet requires only about three times the computational effort compared with the entire SCF procedure [48,70,74,84,86]. To obtain the same information from just energy calculations,

3N+1 (N is the number of nuclei) SCF points would be required.

The availability of analytic energy gradients permits the efficient determination of the molecular geometries of stable species utilizing powerful gradient optimization methods, which are not discussed here, or to solve the gradient equations (section 2.3.1) to locate critical points of all orders. However, it is highly preferable to optimize the geometry using internal coordinates (bond lengths, and valence and dihedral angles) rather than cartesian coordinates, as the former parameters are coupled to a much lesser extent (the force constant matrix is much closer to a diagonal matrix), leading to faster convergence [84]. In addition, PES are most readily defined by internal coordinates. However, the gradient is much easier to evaluate [4,84] for the cartesian coordinates (A is x, y or z in equation (25)).

The transformation of the cartesian gradient vector (g_x) to the internal coordinate gradient vector (g_q) is [84]

$$g_q = (BmB')^{-1}Bmg_x \tag{27}$$

B is the well-known matrix from the Wilson GF vibrational analysis [68] that relates the internal and cartesian coordinates (q and x respectively) to first order:

$$q = Bx \tag{28}$$

The matrix m is arbitrary, and, as suggested by Pulay [4,84], has been taken as a unit matrix without encountering difficulties during the inversion of BmB'.

The force has been used extensively for complete geometry optimization (refs. 48,70,74,83,84,86,88-90 and refs. therein), and for the computation of quadratic, cubic and quartic force constants [83,84,90] by differences of the analytic gradients. Quadratic (harmonic) force constants were discussed briefly in section 2.3.1, but it is important to note that the diagonal force constants are systematically overestimated by Hartree-Fock SCF calculations [83,84]. Minimal basis sets yield especially poor results for quadratic force constants, but their performance improves for higher order force constants [83].

All gradient evaluations required during the direct optimization of the critical point geometries reported in this work were performed by the Schlegel FORCE program [91], which is compatible with GAUSSIAN 70 [1]. The B matrix, required in (27) for the transformation of the force to internal coordinates, was evaluated by a series of subroutines [69] embodying formulae developed [68,92] for the following internal coordinates: bond stretch, valence angle bend, bond torsion and linear angle bend.

All gradient optimizations were performed using the MONSTERGAUSS program system [93], running on the University of Toronto Chemistry Department SEL 32/75 minicomputer. The program incorporates, among many other options, the integral package (for s, p and d gaussian functions), the closed-shell SCF and the UHF open-shell SCF routines from GAUSSIAN 76 [94], the FORCE program [91] and the VA05AD gradient optimization method (see section 2.3.1). It was designed especially for the automatic optimization of the geometry for both closed-shell and open-shell molecules, and also incorporates other more efficient methods of optimization designed to locate only the minima of ab initio PES.

3. APPLICATIONS TO CHEMICAL SYSTEMS

3.1 n-Butane

n-Butane, the simplest hydrocarbon capable of existing in two stable conformations, anti (A) and gauche (G), was the first conformational hypersurface determined [95]. The triple-rotor hypersurface treated all the C-C torsional coordinates, with the (0°, 0°, 0°) conformer corresponding to (syn, eclipsed, eclipsed) as shown below.

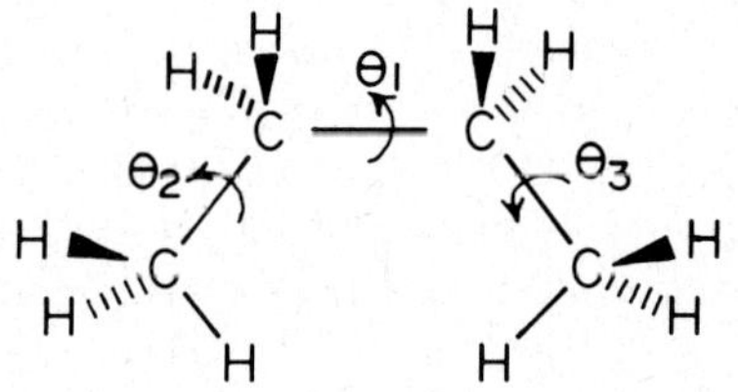

(0, 0, 0)

As may be seen in Table 1 of ref. 95, the molecule has been well studied experimentally and theoretically, by both empirical molecular mechanics and ab initio MO methods. However, all previous ab initio work, and even some of the much less costly empirical studies, had fixed the methyl groups in the staggered conformation (i.e. not allowed for methyl rotation), and assumed that the saddle points for the A (anti) to G (gauche) conversion occurred at the θ_1=120° eclipsed conformation. Some workers have also assumed the gauche conformer had a torsional angle θ_1 of 60° despite accumulated experimental and theoretical data to the contrary. A conformational hypersurface permitted ready examination of the effect the coupling of the three rotors has on the various structures, energy differences and barrier heights.

Hendrickson, in the course of his molecular mechanics calculations on the conformations of cycloalkanes, reported [96] that the gauche minimum occurred at (63°, 55°, 55°). The A→G saddle point was found at the point predicted by simple stereochemical arguments about eclipsed and staggered bonds, namely (120°, 60°, 60°). The bond lengths were fixed at 1.533 and 1.09 Å for C-C and C-H, respectively, for all conformers studied. His barrier height values, however, are lower than the experimental values.

Scheraga and co-workers determined [97] the critical points of the n-butane hypersurface, using rigid rotation. The gauche conformer was located at (65°, 52°, 52°), and the A→G saddle point at (121°, 60°, 60°). They obtained a large G→G barrier of nearly 14 kcal/mol, compared to the experimental value of 6 kcal/mol.

Bartell calculated [98,99] the gauche minimum to be at 66.8°, 56.7°, 56.7°), optimizing all internal coordinates. All of these empirical results indicate that there is appreciable methyl group torsion as the methyl groups interact in the G conformer, but the A→G saddle point is very near the eclipsed conformation (120°, 60°, 60°) that many previous workers have assumed.

A total of 63 unique (i.e. not symmetrically related) points on the hypersurface were chosen by independently varying θ_2 and θ_3 from 0 to 90° in 30° increments, for θ_1 values ranging from 0 to 180°, again in 30° steps. A point was also taken at (70°, 60°, 60°) near the anticipated gauche minimum. All calculations were performed using the minimal STO-3G basis set [52], for rigid rotation at the experimental geometry [100]. This geometry was used, with rigid rotation, for all previous ab initio studies on n-butane where a standard geometry was not used. The geometries were not optimized for each point so that differences in the determination of the critical points from the hypersurface, which permitted completely independent torsional freedom, could be ascertained. Partial geometry optimization would probably improve the results but would obscure the comparison with previous ab initio work.

These points were then fitted in a least squares sense by a thirty-four term analytic equation of the standard form (2), each term being a product (5) of three functions:

$$t_i(\theta_1,\theta_2,\theta_3) = f_{1i}(\theta_1)f_{2i}(\theta_2)f_{3i}(\theta_3) \qquad (29)$$

where θ_1, θ_2 and θ_3 are the torsional angles (measured in radians). The terms to be included were selected entirely manually.

The average error of this fit, labelled fit I and detailed in Table 1, was 0.21 kcal/mol, which represents a relative error of 0.47% since the calculated SCF energy difference between the global maximum and global minimum was 45.0 kcal/mol.

All equations tabulated in this work are given in the form utilized in Table 1. The left-hand column is the set of functions used for each independent hypersurface coordinate (coordinate 1 is θ_1, coordinate 2 is θ_2 and coordinate 3 is θ_3). Each function is numbered sequentially for each coordinate. Thus the first function for all three variables is simply the constant 1.0, while the seventh function for θ_1 is $\cos 6\theta_1$.

The ninth θ_1 function is an example of a "repeating Gaussian", $\exp(-2\theta_1^2)$, where θ_1 is constrained to lie between $-\pi$ and π by addition or subtraction of multiples of 2π. This type of function is very useful for describing large, localized maxima for conformational coordinates, which must repeat at least every 2π radians. Of course, the traditional Gaussian function does not have the correct periodic behaviour for use with rotational variables.

Table 1. Hypersurface equation I for n-butane

FUNCTIONS FOR COORDINATE 1:

1	1.0
2	COS 1X
3	COS 2X
4	COS 3X
5	COS 4X
6	COS 5X
7	COS 6X
8	COS 7X
9	EXP(-2.00000 X ** 2)
10	EXP(-2.75000 X ** 2)
11	EXP(-4.00000 (X - 22.91831) ** 2)
12	EXP(-4.00000 (X + 22.91831) ** 2)
13	EXP(-2.25000 (X - 45.83662) ** 2)
14	EXP(-2.25000 (X + 45.83662) ** 2)
15	EXP(-4.00000 (X - 57.29578) ** 2)
16	EXP(-4.00000 (X + 57.29578) ** 2)
17	EXP(-2.00000 (3X) ** 2)

FUNCTIONS FOR COORDINATE 2:

1	1.0
2	SIN 3(X - 22.91831)
3	SIN -3(X + 22.91831)
4	COS 3X
5	COS 3(X - 46.98254)
6	COS 3(X + 46.98254)
7	COS 3(X - 49.84733)
8	COS 3(X + 49.84733)
9	EXP(-1.00000 (3X) ** 2)
10	EXP(-1.50000 (3X) ** 2)
11	EXP(-2.50000 (3X) ** 2)

FUNCTIONS FOR COORDINATE 3:

1	1.0
2	SIN 3(X - 22.91831)
3	SIN -3(X + 22.91831)
4	COS 3X
5	COS 3(X - 46.98254)
6	COS 3(X + 46.98254)
7	COS 3(X - 49.84733)
8	COS 3(X + 49.84733)
9	EXP(-1.00000 (3X) ** 2)
10	EXP(-1.50000 (3X) ** 2)
11	EXP(-2.50000 (3X) ** 2)

TERMS IN SURFACE EQUATION:[a]

TERM	C	X1	X2	X3
1	5.625137	1	1	1
2	14.063079	9	1	1
3	1.835729	10	11	11
4	.553187	17	9	9
5	-.139316	2	1	1
6	2.220172	4	1	1
7	.077008	8	1	1
8	.352337	1	4	4
9	.439450	7	10	10
10	-1.215114	3	4	4
11	-1.127706	4	4	4
12	.447148	5	4	4
13	1.693128	6	4	4
14	.870575	7	4	4
15	1.451550	1	4	1
16	1.451550	1	1	4
17	-.707753	2	4	1
18	-.707753	2	1	4
19	.780997	4	4	1
20	.780997	4	1	4
21	.871383	5	4	1
22	.871383	5	1	4
23	.869224	6	4	1
24	.869224	6	1	4
25	.320839	7	4	1
26	.320839	7	1	4
27	-11.089787	15	2	2
28	-11.089787	16	3	3
29	8.768772	13	7	7
30	8.768772	14	8	8
31	-5.592748	11	5	1
32	-5.592748	11	1	5
33	-5.592748	12	6	1
34	-5.592748	12	1	6

[a]These coefficients calculate the energy in kcal/mol relative to the ab initio energy of the (180°, 60°, 60°) conformer.

These functions, $f_{1i}(\theta_1)$, $f_{2i}(\theta_2)$ and $f_{3i}(\theta_3)$, are combined to form the analytic equation as shown in the right-hand column of Table 1. Each of the thirty-four terms is a product of a linear coefficient C, determined from the least squares fit, a function of θ_1 (the function number is given in the column headed X1), a function of θ_2 (from column X2), and a function of θ_3 (from column X3). Thus the first term is 5.625137(function 1 for θ_1)(function 1 for θ_2)(function 1 for θ_3). However, function 1 for each θ_i is 1.0; thus term 1 is simply the constant, 5.625137. The tenth term is -1.215114(function 3 for θ_1)(function 4 for θ_2)(function 4 for θ_3), or $-1.215114\cos 2\theta_1 \cos 3\theta_2 \cos 3\theta_3$.

The equation consists of terms for independent rotation about each of the bonds (terms 2, 5-7, 15 and 16) and terms which account for the correlation among the various torsions (terms 3, 4, 8-14, 17-34). The equation may also be broken down into terms which are symmetric with respect to rotation (that is θ_i and $-\theta_i$ are not distinguished) and those that are antisymmetric. The first twenty-six terms fall into the former category, while the last eight terms (27-34) contribute to the hypersurface asymmetry.

There are several restrictions placed on the form of the equation by the symmetry of n-butane:

(1) The conformations (θ_1, θ_2, θ_3) and (θ_1, θ_3, θ_2) are enantiomeric, so the equation must be invariant to the interchange of θ_2 and θ_3. In other words, for all values of θ_1, the hypersurface must have a diagonal mirror plane cutting the (θ_2,θ_3) subspace.

(2) The conformations (θ_1, θ_2, θ_3) and ($-\theta_1$, $-\theta_2$, $-\theta_3$) are equivalent, forcing inversion symmetry about the origin (0°, 0°, 0°). The point (180°, 0°, 0°) is therefore also a center of inversion.

(3) The syn (θ_1=0°) and anti (θ_1=180°) conformers may have higher symmetry than other conformations of n-butane when either of θ_2 or θ_3 is 0 or 60°: if θ_3, say, is either eclipsed or staggered, (θ_1, θ_2, θ_3) is equivalent to (θ_1, $-\theta_2$, θ_3). One must always be wary of special conformations with higher symmetry, and design the asymmetric equation terms accordingly (the symmetric terms will always be correct).

(4) The equation must be periodic in θ_1, θ_2 and θ_3 with periods of 360, 120 and 120° respectively.

This last requirement was readily met by using a Fourier expansion, but due to the sharp maxima and large energies calculated for syn conformers (θ_1 near 0°), repeating Gaussians functions proved to give a more compact description of the hypersurface. These functions are used in terms 2-4, 9 and 27-34.

Property (1), the equivalence of θ_2 and θ_3, was realized in two ways, the simplest of which was to use a product $f_{2i}f_{3i}$ where the functions f_{2i} and f_{3i} are identical. This may be seen in Table 1 for functions 1-14 and 27-30. Alternatively, two terms could be added as in $C_i f_{1i} f_{2i} + C_i f_{1i} f_{3i}$ where f_{2i} is identical with f_{3i}. This may be seen for the pairs of terms 15,16 through 25,26, 31,32 and 33,34.

The centers of inversion, property (2), were already incorporated into terms 1-26 since $\cos\theta = \cos(-\theta)$ and all the repeating Gaussians were centered at $\theta = 0°$. While deviation of θ_1 from the ideal staggered and gauche angles has been accounted for by the Fourier terms 5-7, 9-14 and 17-26, no such allowance has been made for the methyl rotors. Thus all θ_2 and θ_3 minima will occur at 60° for all values of θ_1.

The last eight terms in Table 1 allow for deviations in the (θ_2,θ_3) subspace. Note that the effects of these terms are localized in θ_1 coordinate space by the large negative exponents of the θ_1 repeating Gaussians functions. Pairs of terms 27,28 and 29,30 are required to maintain the inversion centers, and the phase of the sine function in term 28 must be reversed to that of 27 when the direction of rotation is inverted. The pair of terms 31,32 is required by property (1) and the pair 33,34 by property (2), resulting in four terms with the same linear coefficient being generated by using term 31.

In total there are twenty-three different terms, the remaining eleven terms being required to give the hypersurface the correct symmetry properties. There are fifteen nonlinear parameters in Table 1 that were partially optimized manually. The twenty-three linear coefficients were determined by least squares regression.

Two cross-sections of the hypersurface are shown in Figures 2 and 3, as pseudo-three-dimensional plots. One (Figure 2) gives $E(\theta_1,\theta_2)$ with θ_3 held staggered, while the other (Figure 3) displays the dependence of the potential energy on the methyl group torsional angles for the anti conformer. Figure 4 shows the cross-section for central C-C bond rotation with both methyl groups held staggered, and Figure 5 depicts the methyl group rotational potential curves for the gauche and anti conformers.

Since the publication of the analytic equation I, the more automatic stepwise regression procedure, described in section 2.2, was developed. It was of some interest to ascertain whether that procedure could produce a better equation than that found entirely manually.

The first step was to select the functions to be used to represent each independent variable: in this case, trigonometric functions (sine and cosine), and repeating Gaussians. Since the ab initio points were calculated at a regular (30°) grid, a maximum value of n_i in $\sin n_i\theta_i$ and $\cos n_i\theta_i$ could be found: $\sin 6(30m)$ is always zero, for m integral, limiting sine terms to $\sin 5\theta_i$. Since

θ_2 and θ_3 must have three-fold periodicity, sin$3\theta_2$ and sin$3\theta_3$ were the only possible sine functions for those coordinates. For the cosine functions, cos 7(30m) equals cos 5(30m), limiting n_i to 6 or less.

The next step was the determination of the hypersurface symmetry, which was discussed previously for n-butane. The use of molecular models is generally most helpful here, especially in complicated cases such as this one.

Figure 2. Cross-section $E(\theta_1,\theta_2)$ of the n-butane hypersurface (θ_3=60°). To view these plots with the perspective in which they were drawn, the eye should be a few centimeters above the plane of the paper and about 20 cm from the plot origin (0,0).

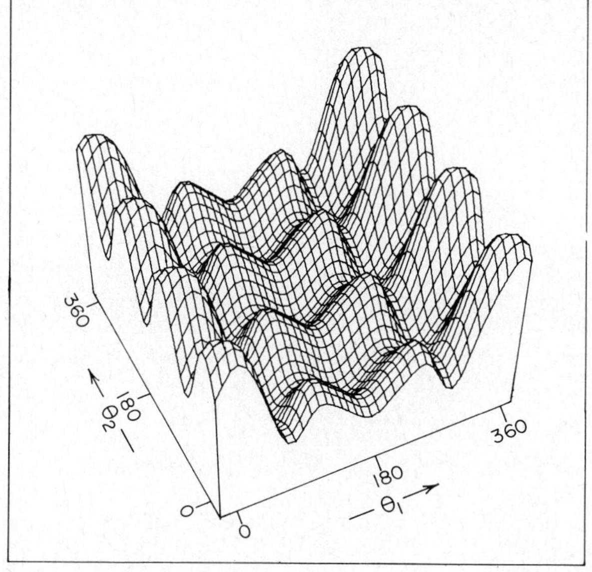

This figure taken from ref. [95] with permission.

The set of candidate equation terms was then generated. The most obvious were the completely symmetric terms

$$\cos n_1\theta_1 \cos n_2\theta_2 \cos n_3\theta_3 \tag{30}$$

which have all the desired properties listed previously, if $n_2=n_3$. For $n_2 \neq n_3$, property (1), the interchange of θ_2 and θ_3, is violated. The most general method to determine the correct composite term is to take the product (30) and subject it to all the desired equivalence relationships. For example, property (1) gave

$$\cos n_1\theta_1 \cos n_2\theta_3 \cos n_3\theta_2 \tag{31}$$

Figure 3. Cross-section $E(\theta_2,\theta_3)$ of the n-butane hypersurface (θ_1=180°).

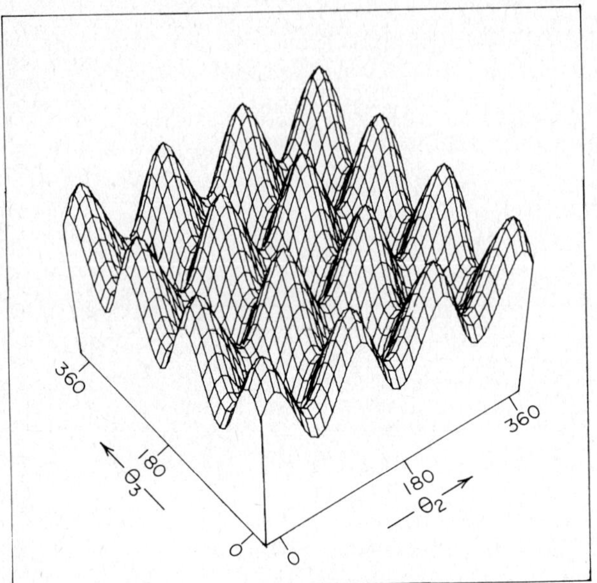

This figure taken from ref. [95] with permission.

Figure 4. Potential energy for central C-C bond rotation in n-butane (θ_2=θ_3=60°).

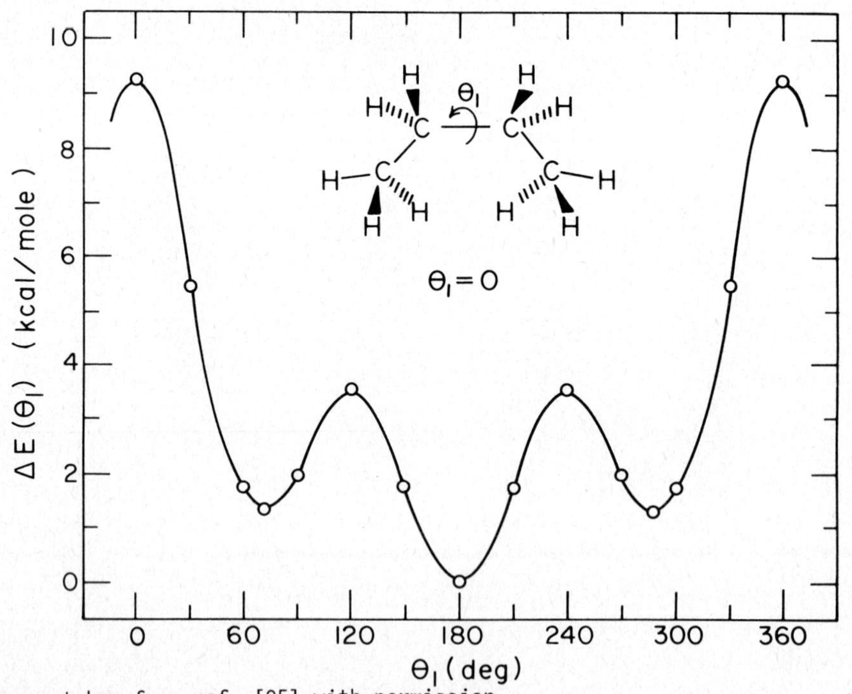

This figure taken from ref. [95] with permission.

Figure 5. Potential energy curves for methyl group torsion in n-butane (θ_3=60°).

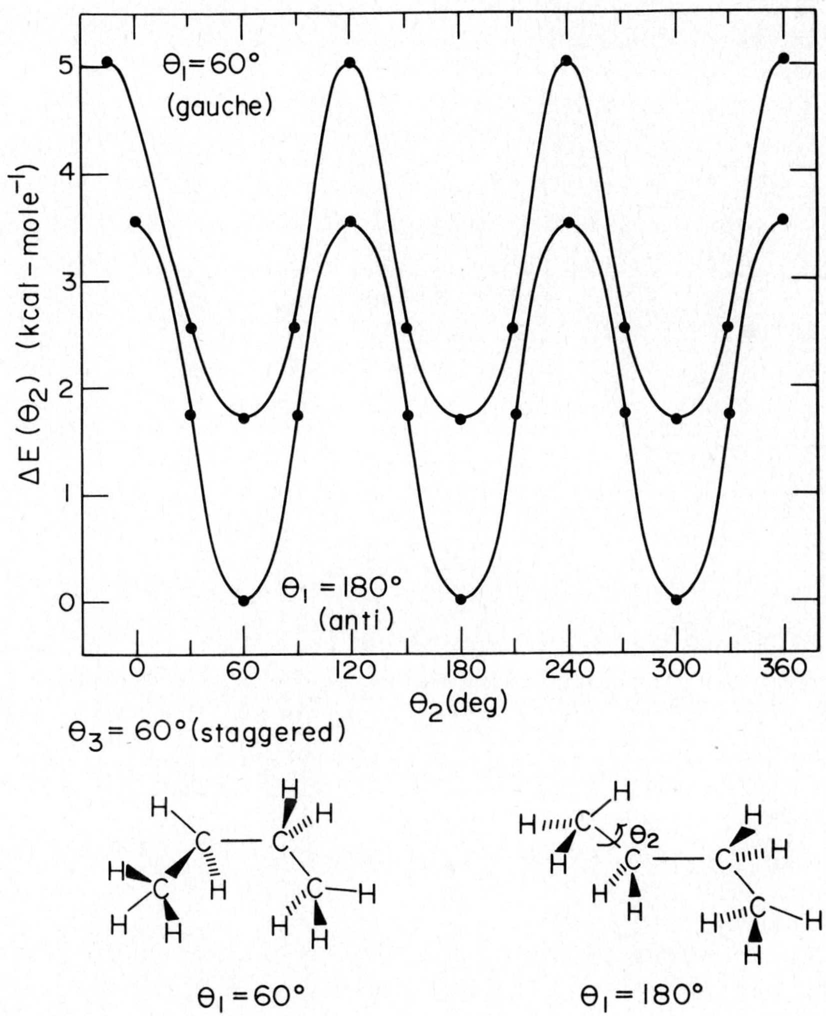

This figure taken from ref. [95] with permission.

while property (2), inversion through the origin, gave

$$\cos n_1\theta_1 \cos n_2\theta_2 \cos n_3\theta_3 \tag{32}$$

since

$$\cos(-\theta) = \cos\theta \tag{33}$$

Thus

$$\cos n_1\theta_1 (\cos n_2\theta_2 \cos n_3\theta_3 + \cos n_3\theta_2 \cos n_2\theta_3) \tag{34}$$

satisfied all the desired properties, provided n_2 and n_3 were integral multiples of 3. In (30) and (34), the cos $n_1\theta_1$ function was also replaced by a repeating Gaussian centered at θ_1=0°, leading to more candidate terms.

The first twenty-nine candidate terms were derived from

$$f_1(\theta_1) \cos n\theta_2 \cos n\theta_3 \tag{35}$$

where f_1 is cos $n_1\theta_1$ (for n_1=0,1,2,3,4,5,6) and exp $(B_1(A_1\theta_1)^2)$ for the pairs (A_1,B_1) = (1.0,-2.0), (1.0,-2.75) and (3.0,-2.0). The values of n in (35) were 0, 3 and 6. The term n_1=n=0 was omitted since it is simply a constant. Terms 30 through 59 followed from (34), with the cos $n_1\theta_1$ function replaced by the $f_1(\theta_1)$ functions given above for (35). The pairs (0,3), (0,6) and (3,6) were used for (n_2,n_3) in equation (34).

Replacing the cos $n_2\theta_2$ cos $n_3\theta_3$ portion of (35) by repeating Gaussians led to thirty more terms of the form

$$f_1(\theta_1) \exp(B(3\theta_2)^2) \exp(B(3\theta_3)^2) \tag{36}$$

with the same set of f_1 functions as for (35), and B = -1.00, -1.75 and -2.50.

When the repeating Gaussian in (35) was moved away from the origin, the term

$$\exp(B_1(\theta_1-A_1)^2) \cos n\theta_2 \cos n\theta_3 \tag{37}$$

resulted. This term did not obey property (2), inversion, which gave

$$\exp(B_1(-\theta_1-A_1)^2) \cos(-n\theta_2) \cos(-n\theta_3) \tag{38a}$$

$$= \exp(B_1(\theta_1+A_1)^2) \cos n\theta_2 \cos n\theta_3 \tag{38b}$$

Thus the sum

$$[\exp(B_1(\theta_1-A_1)^2)+\exp(B_1(\theta_1+A_1)^2)] \cos n\theta_2 \cos n\theta_3 \tag{39}$$

has the correct symmetry, and was used as candidate terms 90 through 101, with A_1 taking on the values 15.0, 30.0, 45.0, 60.0, 75.0 and 90.0, B_1 = -4.0 and n = 3 and 6.

The proof of property (3), the special symmetry requirement, for such terms is straightforward for θ_1=0°, since the repeating Gaussians are then both equal to $\exp(B_1(A_1)^2)$, reducing the term (39) to

$$2 \exp(B_1(A_1)^2) \cos n\theta_2 \cos n\theta_3 \tag{40}$$

which is clearly symmetric about θ_2=0° for θ_3 = 0 and 60°. In fact, (40) is symmetric in θ_2 for all values of θ_3.

For θ_1=180°, it is necessary to show the equivalence of the two exponentials, so they can be factored out as in (40). This reduces to demonstrating

$$(180-A_1)^2 = (180+A_1)^2 \tag{41}$$

which has two solutions:

$$180 - A_1 = 180 + A_1 + 360m \tag{42}$$

$$180 - A_1 = -(180 + A_1) + 360m \tag{43}$$

keeping in mind that these are repeating Gaussians, whose argument $\theta_1 - A_1$ is reduced to lie in the range $-\pi$ to π. The solution to (42) is

$$A_1 = -180m \tag{44}$$

which is clearly unsatisfactory. However, the solution to (43) is

$$m = 1 \tag{45}$$

which is independent of A_1. Thus the repeating Gaussians are equivalent for θ_1=180°, and can be factored out of (39), leading to

$$2 \exp(B_1(180-A_1)^2) \cos n\theta_2 \cos n\theta_3 \tag{46}$$

which is symmetric in the θ_2 subspace for all values of θ_3, as before.

Another type of term utilizing the off-center θ_1 repeating Gaussian was

$$[\exp(B_1(\theta_1-A_1)^2)+\exp(B_1(\theta_1+A_1)^2)][\cos n_2\theta_2 \cos n_3\theta_3 + \cos n_3\theta_2 \cos n_2\theta_3] \tag{47}$$

which satisfies the required properties. The proof of property (3) follows readily from those given above. The values of A_1 and B_1 were those used in (39), while the (n_2,n_3) pairs were (0,3), (0,6) and (3,6) as in (34).

Turning to terms catering to the hypersurface asymmetry, sine functions can be utilized for θ_2 and θ_3 as a product $\sin3\theta_2\sin3\theta_3$, which obeys the four desired properties, as does any product

$$f_1(\theta_1)\sin3\theta_2\sin3\theta_3 \tag{48}$$

where f_1 is symmetric in θ_1. In this case, property (3) results because $\sin3\theta_1$ is zero for both θ_i = 0 and 60°, regardless of the value of θ_1.

The next seven candidate terms arose from (48), f_1 being $\cos n_1\theta_1$ for n_1 ranging from 0 to 6. Utilizing repeating Gaussians for f_1, a sum of two components was necessary because

$$\exp(B(\theta_1-A)^2)\sin3\theta_2\sin3\theta_3 \tag{49}$$

became

$$\exp(B(\theta_1+A)^2)\sin3\theta_2\sin3\theta_3 \tag{50}$$

upon application of property (2), inversion. Terms 127 to 132 involve the sum (49)+(50), with a = 15.0, 30.0, 45.0, 60.0, 75.0 and 90.0, and B = -4.0.

The next group of ten candidate terms have the form

$$\sin n_1\theta_1(\sin n_2\theta_2\cos n_3\theta_3+ \cos n_3\theta_2\sin n_2\theta_3) \tag{51}$$

with n_1 ranging from 1 to 5, n_2=3 and n_3 = 3 or 6. The summation was required because the single component

$$\sin n_1\theta_1 \sin n_2\theta_2 \cos n_3\theta_3 \tag{52}$$

while satisfying property (2), inversion, became

$$\sin n_1\theta_1 \sin n_2\theta_3 \cos n_3\theta_2 \tag{53}$$

upon application of property (1), interchange of θ_2 and θ_3. Therefore, (52) and (53) must be combined to form (51).

The next twenty-four terms are asymmetric in the (θ_2,θ_3) subspace, but with the effect localized in the θ_1 space by the use of repeating Gaussians. The basic form is

$$\exp(B_1(\theta_1-A_1)^2)\sin3(\theta_2-A)\sin3(\theta_3-A) \tag{54}$$

which obeys the interchange property (1), but became

$$\exp(B_1(-\theta_1-A_1)^2)\sin3(-\theta_2-A)\sin3(-\theta_3-A) \tag{55a}$$

$$= \exp(B_1(\theta_1+A_1)^2)\sin3(\theta_2+A)\sin3(\theta_3+A) \tag{55b}$$

upon inversion, property (2), since

$$\sin(-\theta) = -\sin\theta \tag{56}$$

Thus the candidate terms were the sum of (54) and (55b), for A = 10.0, 20.0, 40.0 and 50.0°. The values of A_1 were 15.0, 30.0, 45.0, 60.0, 75.0 and 90.0°, with B_1=-4.0.

Note that for A = 0 and 30°, the functional dependence for the methyl group rotations becomes pure $\sin3\theta_i$ and $\cos3\theta_i$ respectively; both of these terms have already been included.

The proof of property (3), the special symmetry, follows from that given earlier, by showing the equivalence of the repeating Gaussians for θ_1 = 0 and 180°, and factoring out that portion of the term:

$$f_1(\theta_1)[\sin3(\theta_2-A)\sin3(\theta_3-A)+\sin3(\theta_2+A)\sin3(\theta_3+A)] \tag{57}$$

This expression can then be expanded, using

$$\sin(x+y) = \sin x \cos y + \cos x \sin y \tag{58}$$

to, dropping the constant $f_1(\theta_1)$ function,

$$[\sin 3\theta_2 \cos(-3A)+\cos 3\theta_2 \sin(-3A)][\sin 3\theta_3 \cos(-3A)+\cos 3\theta_3 \sin(-3A)]+[\sin 3\theta_2 \cos 3A+ \cos 3\theta_2 \sin 3A][\sin 3\theta_3 \cos 3A+\cos 3\theta_3 \sin 3A] \tag{59}$$

Expanding and collecting terms results in

$$2(\sin 3\theta_2 \sin 3\theta_3 \cos^2 3A + \cos 3\theta_2 \cos 3\theta_3 \sin^2 3A) \tag{60}$$

For θ_2 or θ_3 either eclipsed or staggered, the sine component is zero, leaving only the required symmetric (cosine) dependence on the methyl group torsional angles.

The final thirty candidate terms arose from the simple product

$$\exp(B_1(\theta_1-A_1)^2)\sin 3(\theta_2-A) \tag{61}$$

Property (1), interchange, applied to (61) gave

$$\exp(B_1(\theta_1-A_1)^2)\sin 3(\theta_3-A) \tag{62}$$

resulting in a sum of two components:

$$\exp(B_1(\theta_1-A_1)^2)\ [\sin 3(\theta_2-A)+\sin 3(\theta_3-A)] \tag{63}$$

Application of property (2), inversion, to (63) gave

$$\exp(B_1(\theta_1+A_1)^2)[-\sin 3(\theta_2+A)-\sin 3(\theta_3+A)] \tag{64}$$

from (56). Thus the final terms, numbered 167 to 196, were a sum of (63) and (64); each term is comprised of four components, two added and two subtracted. The values of A_1, B_1 and A were as for the previous set of terms, except that A = 0° was retained (A = 30°, resulting in $\cos 3\theta_i$, has already been included though).

The stepwise regression analysis was carried out with these 196 candidate terms, and sixty-five ab initio points (two grid points had been omitted in the previous fit). The equation containing twenty-three terms plus a constant was selected, as adding another term gave that term a 17% probability of being zero. The terms included in the fit chosen, hereafter termed fit II, are detailed in Table 2. Note that each term has been expanded to its components. The least significant included term is the pair (41,42) which has an 8.6% chance of being zero. The average error of this fit is 0.093 kcal/mol (root mean square error 0.117 kcal/mol), a considerable improvement over fit I.

Qualitatively, the two hypersurfaces are very similar; no plots are shown as they differ only slightly from Figures 2 to 5.

Table 2. Hypersurface equation II for n-butane.

```
BUTANE RIGID ROT, STO-3G, EXP GEOM: K KUCHITSU, BULL CHEM SOC JAP 32, 748 (1959)
THETA1 IS CENTRAL C-C ROT, THETA2 AND THETA3 ARE METHYL. (0,0,0)=(CIS,ECL,ECL)
```

```
FUNCTIONS FOR COORDINATE  1:

 1   1.0
 2   COS  3X
 3   EXP( -2.00000 ( 1X) ** 2)
 4   EXP( -2.75000 ( 1X) ** 2)
 5   EXP( -4.00000 (X -  15.00000) ** 2)
 6   EXP( -4.00000 (X +  15.00000) ** 2)
 7   EXP( -4.00000 (X -  30.00000) ** 2)
 8   EXP( -4.00000 (X +  30.00000) ** 2)
 9   EXP( -4.00000 (X -  45.00000) ** 2)
10   EXP( -4.00000 (X +  45.00000) ** 2)
11   EXP( -4.00000 (X -  60.00000) ** 2)
12   EXP( -4.00000 (X +  60.00000) ** 2)
13   EXP( -4.00000 (X -  90.00000) ** 2)
14   EXP( -4.00000 (X +  90.00000) ** 2)

FUNCTIONS FOR COORDINATE  2:

 1   1.0
 2   SIN  3X
 3   COS  3X
 4   COS  6X
 5   EXP( -1.00000 ( 3X) ** 2)
 6   SIN  3(X -  10.00000)
 7   SIN  3(X +  10.00000)
 8   SIN  3(X -  20.00000)
 9   SIN  3(X +  20.00000)
10   SIN  3(X -  40.00000)
11   SIN  3(X +  40.00000)
12   SIN  3(X -  50.00000)
13   SIN  3(X +  50.00000)

FUNCTIONS FOR COORDINATE  3:

 1   1.0
 2   SIN  3X
 3   COS  3X
 4   COS  6X
 5   EXP( -1.00000 ( 3X) ** 2)
 6   SIN  3(X -  10.00000)
 7   SIN  3(X +  10.00000)
 8   SIN  3(X -  20.00000)
 9   SIN  3(X +  20.00000)
10   SIN  3(X -  40.00000)
11   SIN  3(X +  40.00000)
12   SIN  3(X -  50.00000)
13   SIN  3(X +  50.00000)
```

TERMS IN SURFACE EQUATION:[a]

TERM	C	X1	X2	X3
1	5.745904	1	1	1
2	2.163391	2	1	1
3	13.605454	3	1	1
4	6.521184	4	3	3
5	2.014185	1	3	1
6	2.014185	1	1	3
7	.230440	2	3	1
8	.230440	2	1	3
9	6.945719	3	3	1
10	6.945719	3	1	3
11	-31.574354	3	5	5
12	36.558862	4	5	5
13	-8.284713	7	3	3
14	-8.284713	8	3	3
15	9.604671	9	3	3
16	9.604671	10	3	3
17	.074408	11	4	4
18	.074408	12	4	4
19	.085860	11	4	1
20	.085860	11	1	4
21	.085860	12	4	1
22	.085860	12	1	4
23	-.115500	7	3	4
24	-.115500	7	4	3
25	-.115500	8	3	4
26	-.115500	8	4	3
27	.216564	11	3	4
28	.216564	11	4	3
29	.216564	12	3	4
30	.216564	12	4	3
31	-5.204790	5	2	2
32	-5.204790	6	2	2
33	9.885180	7	6	6
34	9.885180	8	7	7
35	-8.790398	9	8	8
36	-8.790398	10	9	9
37	1.089013	11	8	8
38	1.089013	12	9	9
39	-1.566169	11	10	10
40	-1.566169	12	11	11
41	.153143	13	12	12
42	.153143	14	13	13
43	-2.929144	5	2	1
44	-2.929144	5	1	2
45	2.929144	6	2	1
46	2.929144	6	1	2
47	.207153	13	6	1
48	.207153	13	1	6
49	-.207153	14	7	1
50	-.207153	14	1	7
51	2.504177	7	10	1
52	2.504177	7	1	10
53	-2.504177	8	11	1
54	-2.504177	8	1	11

[a]These coefficients calculate the energy in kcal/mol relative to the ab initio energy of the (180°, 60°, 60°) conformer.

Stereochemical intuition may be used to predict the approximate structure of the various critical points since, for saturated hydrocarbons, staggered conformers are usually more stable than eclipsed ones. The critical structures, derived from this intuitive consideration, are shown in Figure 6. These geometries were used as initial guesses for the critical point searches on the fitted hypersurfaces I and II. The critical points from fit I were then used as starting points for the ab initio critical point optimizations. All optimizations were performed by minimizing S_g, equation (14), using the VA05AD method, as detailed in section 2.3. For the direct ab initio searches, only the torsional coordinates were allowed to vary, to correspond exactly with the analytic hypersurfaces.

The twelve unique critical points, as determined from fit I, fit II and by direct ab initio optimization, are listed in Table 3. Note that the location of half of the critical points may be determined by symmetry alone. It is readily apparent the fit I represents the ab initio data poorly: the root mean square

Figure 6a. Intuitive structures for the minima and first order saddle points of the n-butane hypersurface.

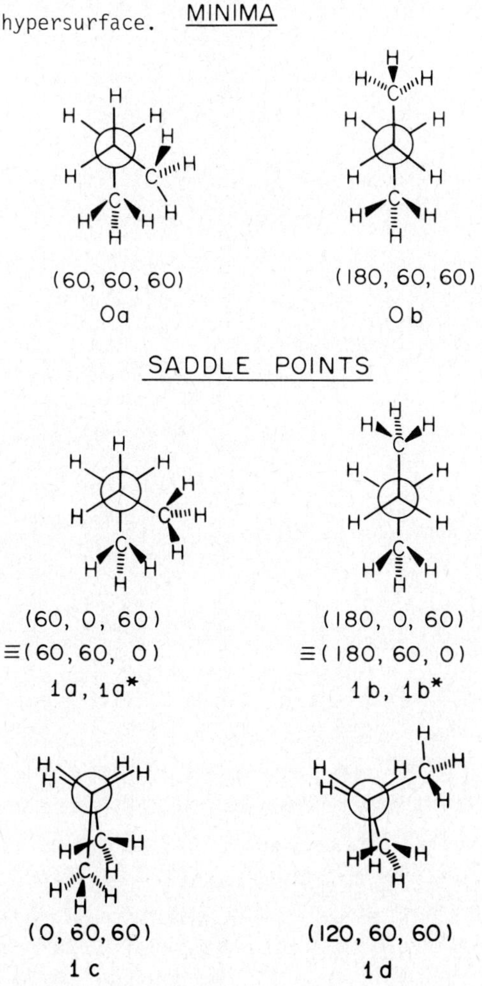

MINIMA

(60, 60, 60)
Oa

(180, 60, 60)
O b

This figure taken from ref. [95] with permission.

SADDLE POINTS

(60, 0, 60)
≡(60, 60, 0)
1a, 1a*

(180, 0, 60)
≡(180, 60, 0)
1b, 1b*

(0, 60, 60)
1c

(120, 60, 60)
1d

Figure 6b. Intuitive structures for the second order saddle points and maxima of the n-butane hypersurface.

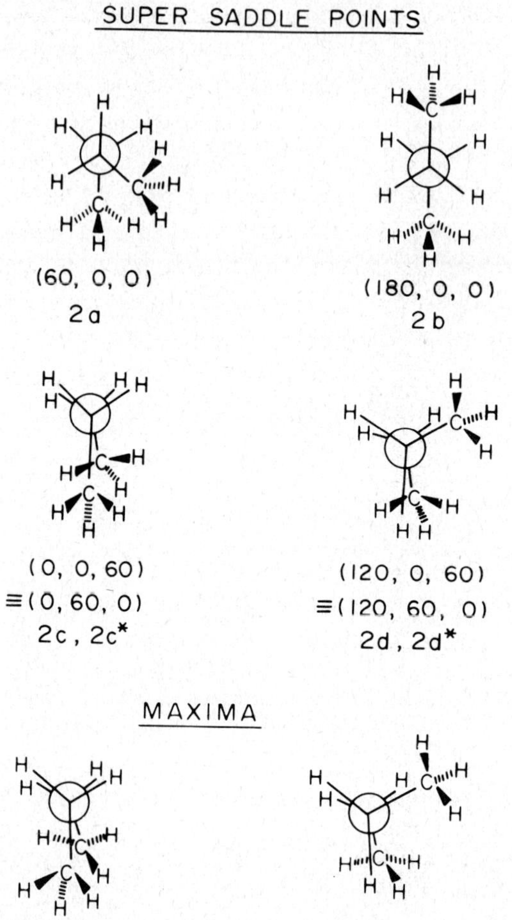

SUPER SADDLE POINTS

(60, 0, 0)
2a

(180, 0, 0)
2b

This figure taken from ref. [95] with permission.

(0, 0, 60)
≡(0, 60, 0)
2c, 2c*

(120, 0, 60)
≡(120, 60, 0)
2d, 2d*

MAXIMA

(0, 0, 0)
3a

(120, 0, 0)
3b

deviation of the twelve energies from the 'exact' ab initio energies is 0.39 kcal/mol, while the torsional angle deviation, for the non-symmetry-constrained critical points, is 4.3°. Fit II does substantially better for both the relative energies (rms error 0.06 kcal/mol) and angles (rms error 1.2°). Given the scatter in the experimental and previous ab initio determinations (Table 1 of ref. [95]), fit II is probably an adequate representation of the ab initio data, but by no means exact. Note that the average error of fit II is approximately ten times smaller than the A-G energy difference.

Table 3. n-Butane critical points from the fitted hypersurfaces and by direct optimization.

POINT	TYPE[a]	θ_1(deg)	θ_2(deg)	θ_3(deg)	ΔE(kcal/mol)[b]	
0a	Fit I	69.5	52.3	52.3	0.95	(1.27)
	Fit II	68.3	55.2	55.2	1.15	
	Opt	69.7	56.1	56.1	1.19	
0b*	Fit I	180.0	60.0	60.0	0.00	(0.00)
	Fit II	180.0	60.0	60.0	0.00	
	Opt	180.0	60.0	60.0	0.00	
1a	Fit I	74.8	- 9.7	60.7	3.46	(4.34)
	Fit II	70.9	- 3.6	61.5	4.40	
	Opt	71.2	- 2.7	60.6	4.44	
1b*	Fit I	180.0	0.0	60.0	3.63	(3.57)
	Fit II	180.0	0.0	60.0	3.57	
	Opt	180.0	0.0	60.0	3.57	
1c*	Fit I	0.0	60.0	60.0	9.32	(9.28)
	Fit II	0.0	60.0	60.0	9.31	
	Opt	0.0	60.0	60.0	9.28	
1d	Fit I	121.5	60.4	60.4	3.56	(3.56)
	Fit II	118.8	60.5	60.5	3.49	
	Opt	119.9	60.2	60.2	3.57	
2a	Fit I	58.1	- 6.5	- 6.5	8.06	(8.23)
	Fit II	63.4	- 2.6	- 2.6	7.90	
	Opt	66.2	- 0.4	- 0.4	7.75	
2b*	Fit I	180.0	0.0	0.0	7.25	(7.16)
	Fit II	180.0	0.0	0.0	7.13	
	Opt	180.0	0.0	0.0	7.16	
2c*	Fit I	0.0	0.0	60.0	18.14	(18.36)
	Fit II	0.0	0.0	60.0	18.37	
	Opt	0.0	0.0	60.0	18.36	
2d	Fit I	112.2	0.7	58.5	8.52	(7.81)
	Fit II	120.0	0.4	60.6	7.89	
	Opt	120.0	0.2	60.1	7.98	
3a*	Fit I	0.0	0.0	0.0	44.81	(45.00)
	Fit II	0.0	0.0	0.0	44.97	
	Opt	0.0	0.0	0.0	45.00	
3b	Fit I	125.5	- 0.2	- 0.2	12.98	(12.30)
	Fit II	120.4	0.6	0.6	12.34	
	Opt	119.8	0.2	0.2	12.41	

[a] Fit refers to results obtained from the fitted hypersurfaces, Opt to results obtained by direct optimization.

[b] Energies are measured relative to the energy of point 0b given by each method. The value in parentheses is the relative ab initio energy at the hypersurface coordinates from fit I.

[*] These points can be completely determined by symmetry.

Figure 7 shows the positions of the various critical points in a unit cell of the coordinate space. Note especially that there is a saddle point between each pair of adjacent minima and a super-saddle (second order saddle) point between each pair of adjacent maxima. Saddle points may be interconnected either through minima or super-saddle points, while in going from one super-saddle point to another, one must pass through either a saddle point or a maximum.

The results of Radom and Pople [44], who have also calculated ab initio the relevant parameters for rigid rotation of n-butane at the same experimental geometry [100], using the same basis set, are the most comparable to these calculations. As expected for rigid rotation, the energies of conformers where nonbonded atoms (here, the methyl group hydrogens) come into close proximity (as in the syn conformers 1c, 2c and 3a) are overestimated, as the geometry was not relaxed to reduce the interaction. This leads to overestimation of the gauche-syn-gauche barrier, as may be seen in Figure 4, which corresponds to curve D of Radom and Pople.

In the gauche conformation (0a) the methyl groups are rotated away from the "ideal" staggered angle (60°) to reduce the nonbonded hydrogen-hydrogen interaction. This lowers the gauche-anti energy difference by only 0.03 kcal/mol but reduces the gauche dihedral angle by 2.8° compared with that of Radom and Pople, who did not allow methyl group rotation. This methyl relaxation has virtually no effect on the anti-gauche barrier since the saddle point still occurs very near the "ideal" (120°, 60°, 60°) conformation (1d, see Table 6). The present results are in very close agreement with the empirical studies mentioned earlier, where the methyl groups were allowed to relax during the searches for the critical points.

The same effect is noticed for methyl group rotation in the G conformation; however, in this case, both the minimum (0a) and the saddle point (1a) structures are significantly shifted from the idealized values (see Table 6). The central C-C angle (θ_1) is predicted to increase by about 2° as the saddle point is approached, and the other methyl group angle (θ_3) to increase by about 8°.

Summarizing the results, the gauche dihedral angle is calculated to be 69.7°, with an A-G energy difference of 1.19 kcal/mol. Both of these values are slightly larger than the experimental results, as are most of the ab initio calculations. The A→G barrier (3.57 kcal/mol) and the methyl group rotation barrier (3.57 kcal/mol for the A conformer, 3.25 kcal/mol for G) are however in excellent agreement with the experiment. The G→G barrier, at 8.09 kcal/mol, is most certainly overestimated due to the rigid rotation employed in this study; the experimental results suggest a value near 6 kcal/mol.

Although it would be desirable to generate a complete hypersurface involving more variables (the C-C-C angles in particular), the three-coordinate

Figure 7. Unit cell for the n-butane coordinate space, showing the location and types of critical points: (●) minima, (■) first order saddle points, (◆) second order saddle points, (▲) maxima.

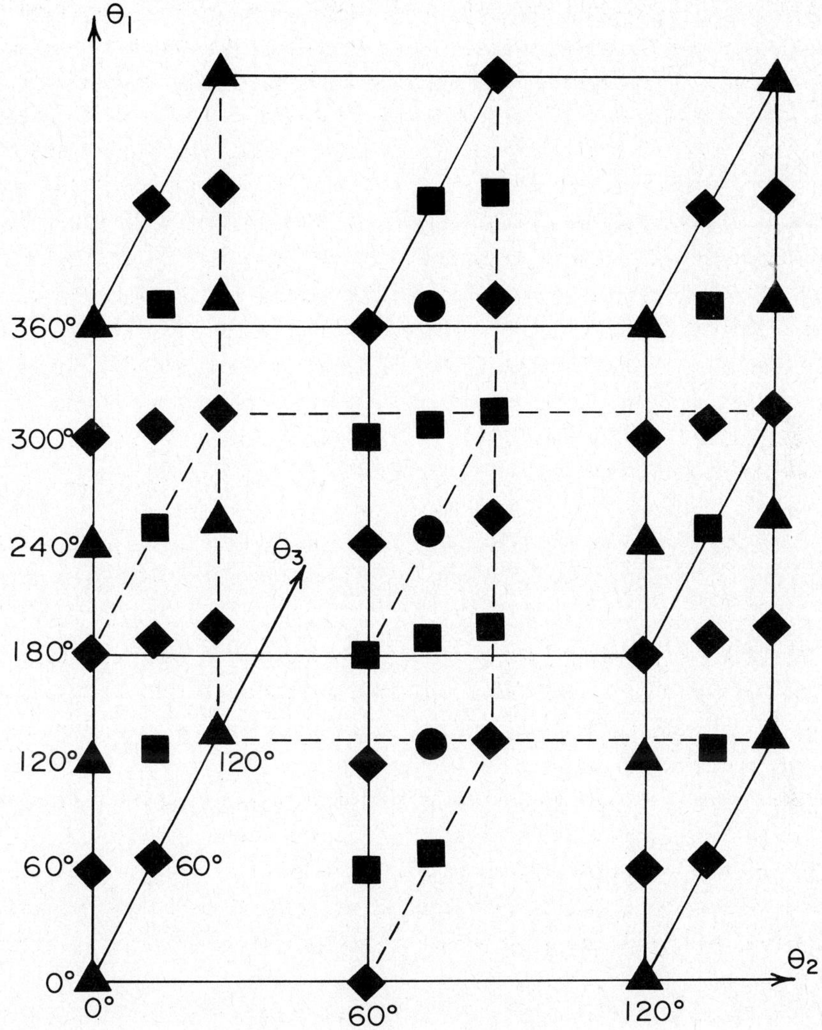

This figure taken from ref. [95] with permission.

hypersurfaces for rigid rotation, reported in this work, are already quite complex. The empirical force field results of Bartell [101] indicate that 82% of the stabilization energy on going from a rigid gauche (θ_1=60°) structure to the fully relaxed one came from the torsional modes and only 18% from bond length and bond angle adjustments. Recently, it has been demonstrated [102] that rigid rotation is adequate for the A→G portion of the θ_1 potential curve, and that only C-C-C bond angles need be optimized near the syn conformations to obtain the best results possible with a given basis set. A fit made to energies from such partially optimized geometries would likely be quantitative as the energy difference between the (0°, 0°, 0°) and (180°, 60°, 60°) conformers would be substantially reduced from the present 45 kcal/mol. Thus more terms could cater to the more subtle features of the hypersurface, rather than just to the large range of energy values.

3.2 Formic Acid

A topic of considerable interest is the interdependence of isomerization or tautomerization and conformational change. Although it is traditional to discuss these processes as independent phenomena, in effect they represent different cross sections of the same energy hypersurface. Perhaps the justification for the separation of these processes lies in the difference in barrier heights, but sometimes certain conformational transition states may also have relatively high energies. This situation occurs for methanol, for which the barrier to OH rotation has been calculated [103] to be 1.44 kcal/mol, while the barrier to in-plane inversion at oxygen was 32.5 kcal/mol. The inversion barrier for water was computed [103] to be 31.1 kcal/mol.

A molecule that possesses rotational, in plane inversion and tautomerization barriers is formic acid, as illustrated by the following Scheme, taken from ref. [116] with permission.

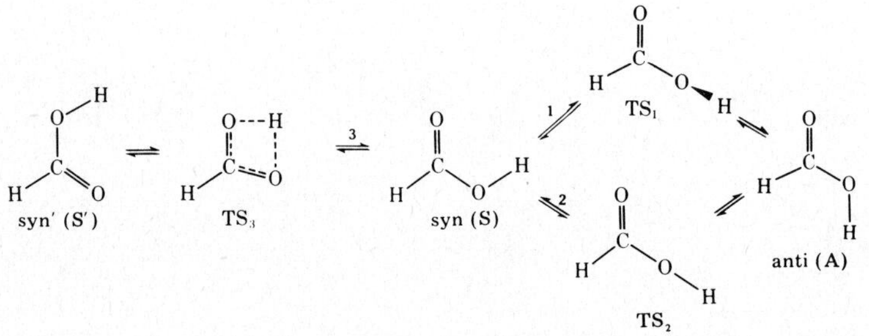

syn′ (S′) ⇌ TS_3 ⇌(3) syn (S) ⇌(1) TS_1 ⇌ anti (A); syn (S) ⇌(2) TS_2 ⇌ anti (A)

Table 4 collects previous experimental and theoretical data for the syn-anti energy difference (ΔE) and the barrier (ΔE_1) to OH rotation (path 1, through TS_1). The barrier to in-plane inversion (ΔE_2) may be estimated from water and methanol to be 30-35 kcal/mol, while TS_3 had been calculated [104] to lie

74.2 kcal/mol above the syn (S) conformation. This situation is illustrated in Figure 8, using the notation of the Scheme.

The geometrical parameters used to define the hypersurface are the O-H bond length (r), the C-O-H angle (α), and the O=C-O-H torsional angle (τ); all the critical points from the Scheme lie on this hypersurface.

The portion of the hypersurface involved in paths 1 and 2 can readily be described in terms of r, α, and τ, but a transformation of the hypersurface coordinates suitable for path 3, including the C_{2v} symmetry of TS_3, has not yet been found. Here, then, a mathematical fit of the conformational portion of the formic acid hypersurface is reported, and the geometries of TS_1 and TS_2, which cannot be obtained by direct energy optimization, determined.

The structures of syn, anti, and linear (which should approximate TS_2) formic acid were determined by minimal basis set ab initio calculations, by geometry optimization. Sixty-one more points were generated by varying the O-H bond length (r) between 0.91 and 1.10 Å, the C-O-H bond angle (α) between 90 and 210°, and the O=C-O-H torsional angle (τ) from 0 to 180° to systematically cover the entire hypersurface coordinate space. For each of these latter 61

Table 4. Syn-anti energy difference and barrier to OH rotation for formic acid.

ΔE(kcal/mol)[a]	ΔE_1(kcal/mol)[b]	Ref
Experimental		
--	17	105
2.0	10.9 (98°)	106
--	13.4	107
>4	--	108
4.0	13.8 (99.8°)	109
Theoretical		
8.1	--	110
9.46	14.20	111
8.1	13.0	112
6.3	12.2 (97°)	113
4.8	12.34	114

[a]Syn-anti difference

[b]Energy difference between TS_1 and S (barrier height for rotation). The torsional angle τ of TS_1 is indicated in parenthesis, if not assumed to be 90°.

Figure 8. Projected energy profile for the total conformational (processes S⇌A, S'⇌A') and tautomerization (process S⇌S') hypersurface of formic acid. The structures and pathways are labeled as in the Scheme.

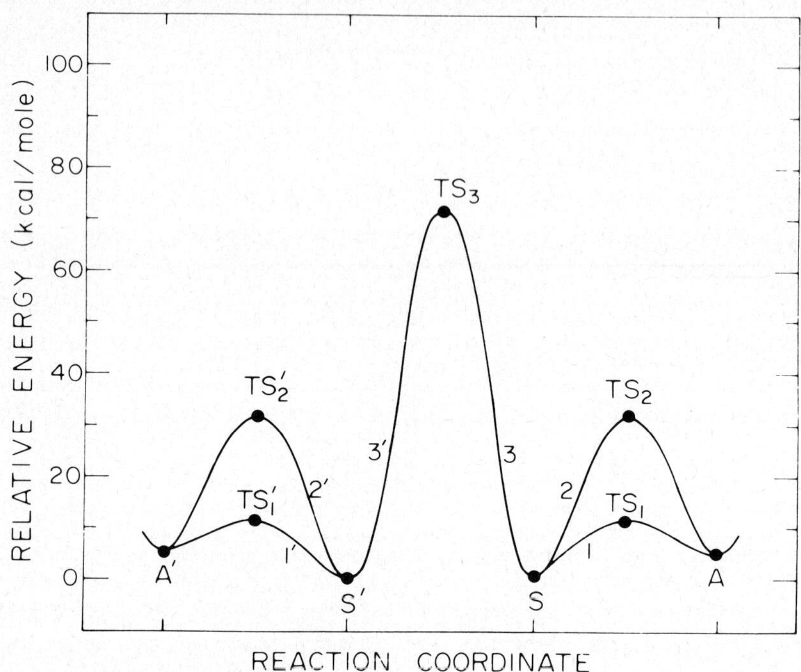

This figure taken from ref. [116] with permission.

points, the C=O and C-O bond lengths and the O=C-O angle were optimized. The C-H bond length and the O=C-H angle were fixed at 1.106 Å and 124°, respectively, based on the results of the preliminary geometry optimizations. Also, to maintain planarity at the carbon atom, the O-C-H angle was decreased by 1° for every 1° increase in the O=C-O angle during the optimization.

An analytic equation of the form

$$E = E(r,\alpha,\tau) = \sum_i C_i\, f_{1i}(r)\, f_{2i}(\alpha)\, f_{3i}(\tau) \qquad (65)$$

was determined by the stepwise regression technique outlined in section 2.2.

There are no symmetry restrictions on r or α, but the torsional coordinate τ must be symmetric about 0° due to the planarity of syn formic acid. In addition, for a linear C-O-H angle (α=180°), there is a line of degeneracy (constant energy) in the τ subspace. This was incorporated into the candidate terms by either having no τ dependence, or ensuring the function for α was zero for α=180°.

Polynomials up to third order in (r-0.99) were used for $f_1(r)$, and cos $n_3\tau$ (n_3 ranging from 0 to 4) was used for $f_3(\alpha)$. The α dependence included both sin $n_2\alpha$ and cos $n_2\alpha$ with n_2 = 0, 1, 2, 3 and 4, but with cos2α and cos4α replaced by the equivalent functions $\sin^2\alpha$ and $\sin^2 2\alpha$ respectively.

The first thiry-five candidate terms had no τ dependence:

$$(r-0.99)^{n_1} \sin n_2\alpha \tag{66}$$

$$(r-0.99)^{n_1} \cos n_2\alpha \tag{67}$$

The values of n_1 ranged from 0 to 3, but in (66) n_2 ranged from 1 to 4, and from 0 to 4 in (67). The term with n_1=n_2=0, a constant, was omitted. The cos2α and cos4α functions were replaced as mentioned previously.

The remaining candidate terms were also of two forms:

$$(r-0.99)^{n_1} \sin n_2\alpha \cos n_3\tau \tag{68}$$

$$(r-0.99)^{n_1} (\sin n_2\alpha)^2 \cos n_3\tau \tag{69}$$

In the former case, n_2 and n_3 ranged from 1 to 4; in the latter, n_2 was 1 or 2 while n_3 = 1, 2, 3 and 4. In both cases, n_1 ranged from 0 to 3. Note that the $(\sin n_2\alpha)^i$ function ensured the presence of the degenerate line on the hypersurface. There were sixty-four and thirty-two terms derived from (68) and (69) respectively, for a total of 131 candidate terms.

The equation selected is detailed in Table 5, in the same format as for n-butane. The average error of the fit was 0.014 kcal/mol (the root mean square error was 0.020 kcal/mol), and the largest deviation was 0.06 kcal/mol. Thus the hypersurface equation should accurately follow the true ab initio hypersurface, which was confirmed by examining plots of various cross sections of the fitted hypersurface (Figures 9 and 10). All terms were significant (i.e., different from zero) at the 1% confidence level.

Two notes should be made concerning the form of the equation: terms 1-13 have no τ dependence for all values of α, and terms 14-34 are identically zero when α is 180°. Also, since (r-0.99) is in the range 0.0-0.1, terms with $(r-0.99)^2$ and $(r-0.99)^3$ for $f_{1i}(r)$ are effectively multiplied by factors of the order 0.005 and 0.0005, respectively, accounting for the large coefficients for those terms.

Table 6 contains the coordinates and relative energies of the critical points of the analytic hypersurface. The hydroxyl rotation saddle point (TS_1) occurs very near τ=90° (perpendicular), and as r and α vary only slightly between the syn and anti minima, path 1 corresponds to essentially rigid rotation. The linear structure (TS_2) is actually a supersaddle point (order 2) and is predicted to lie quite close to the linear geometry. The geometries

Table 5. Formic acid analytic hypersurface equation.

FORMIC ACID (HCOOH) ROTATION HYPERSURFACE (STO-3G OPTIMIZED PTS)
E=E(R,A,T), R IS O-H DISTANCE, A IS C-O-H ANGLE, T IS OH TORSIONAL ANGLE

FUNCTIONS FOR COORDINATE 1:

1	1.0
2	(X - .99000) ** 1
3	(X - .99000) ** 2
4	(X - .99000) ** 3

FUNCTIONS FOR COORDINATE 2:

1	1.0
2	SIN 1X
3	SIN 2X
4	SIN 3X
5	SIN 4X
6	(SIN 1X) ** 2
7	(SIN 2X) ** 2
8	COS 1X
9	COS 3X

FUNCTIONS FOR COORDINATE 3:

1	1.0
2	COS 1X
3	COS 2X
4	COS 3X
5	COS 4X

TERMS IN SURFACE EQUATION:[a]

TERM	C	X1	X2	X3
1	142.351500	1	1	1
2	75.111592	2	1	1
3	730.793054	3	1	1
4	480.234801	4	1	1
5	-130.078881	1	6	1
6	-88.569668	2	6	1
7	-5.674825	1	7	1
8	-17.872444	2	7	1
9	-587.810981	4	7	1
10	77.939793	1	8	1
11	1349.906715	4	8	1
12	9.310406	1	9	1
13	13.217690	2	9	1
14	-3.359607	1	2	2
15	-.984099	2	2	2
16	-864.170567	4	2	2
17	-1.680256	1	3	2
18	-1.320314	1	4	2
19	-.175468	1	5	2
20	1.279823	2	5	2
21	6.101473	1	3	3
22	6.961888	1	4	3
23	1.937445	1	5	3
24	3.825868	1	6	3
25	3.058267	2	6	3
26	-3.846462	1	7	3
27	-3.260103	2	7	3
28	455.937948	4	4	4
29	.050697	1	5	4
30	-.712172	1	6	4
31	.238280	1	7	4
32	-.081897	1	6	5
33	.070640	1	7	5
34	-20.904187	3	7	5

[a]These coefficients calculate $E(r,\alpha,\tau)$ in kcal/mol relative to the energy of the (0.991 Å, 105.5°, 0°) syn conformer (-186.21780 hartrees).

and relative energies determined from the analytic equation compare favorably with those obtained directly from ab initio calculation, also presented in Table 6, along with some representative experimental results. The complete set of experimental geometry determinations for syn formic acid was given in the original hypersurface publication [116].

Figure 9. Cross-section of the formic acid hypersurface (α=105°).

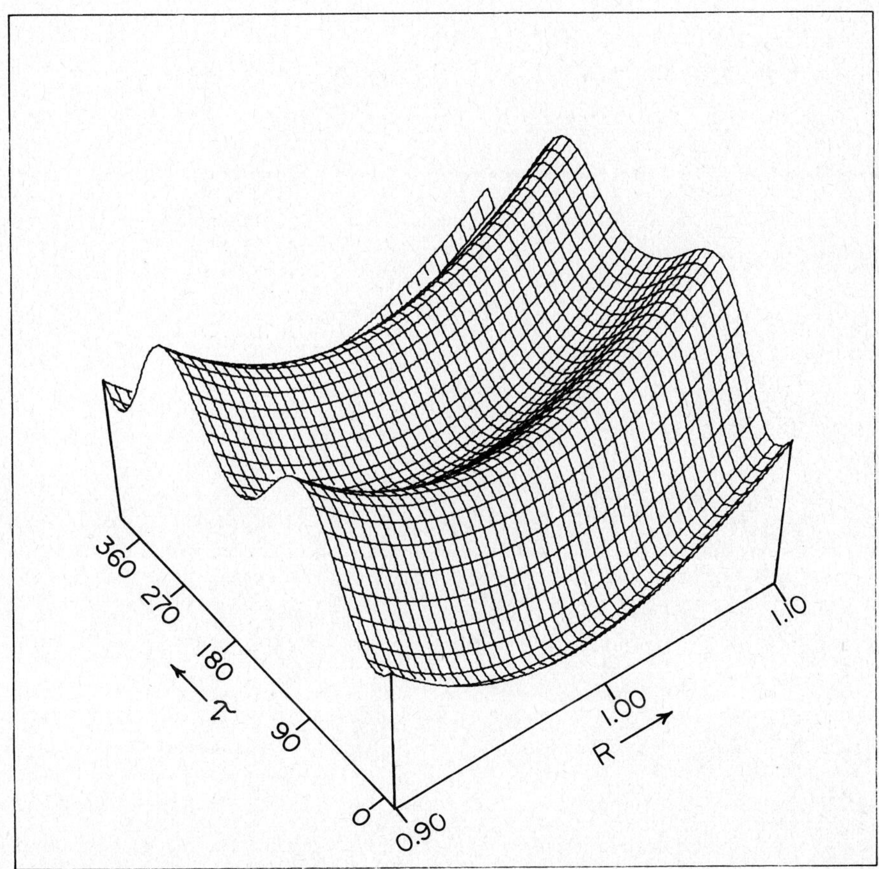

The unit cell, Figure 11, considers only the (α,τ) coordinate subspace, since the O-H bond (r) possesses a positive force constant over the entire hypersurface. Inspection reveals that the set of critical points from Figure 8 is not sufficient since only TS_2 occurs near the α=180° line of degeneracy. Noting that TS_2 is actually a maximum in the unit cell, a saddle point, labelled TS_4 in Figure 11, must occur between occurences of TS_2; this critical point was then located from the analytic hypersurface equation. However, it could not be directly optimized due to numerical difficulties, as one force constant, corresponding to the τ coordinate, was about four orders of magnitude smaller than the remaining force constants. This nearly degenerate eigenvalue

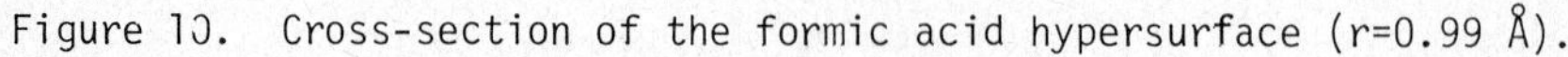

Figure 10. Cross-section of the formic acid hypersurface (r=0.99 Å).

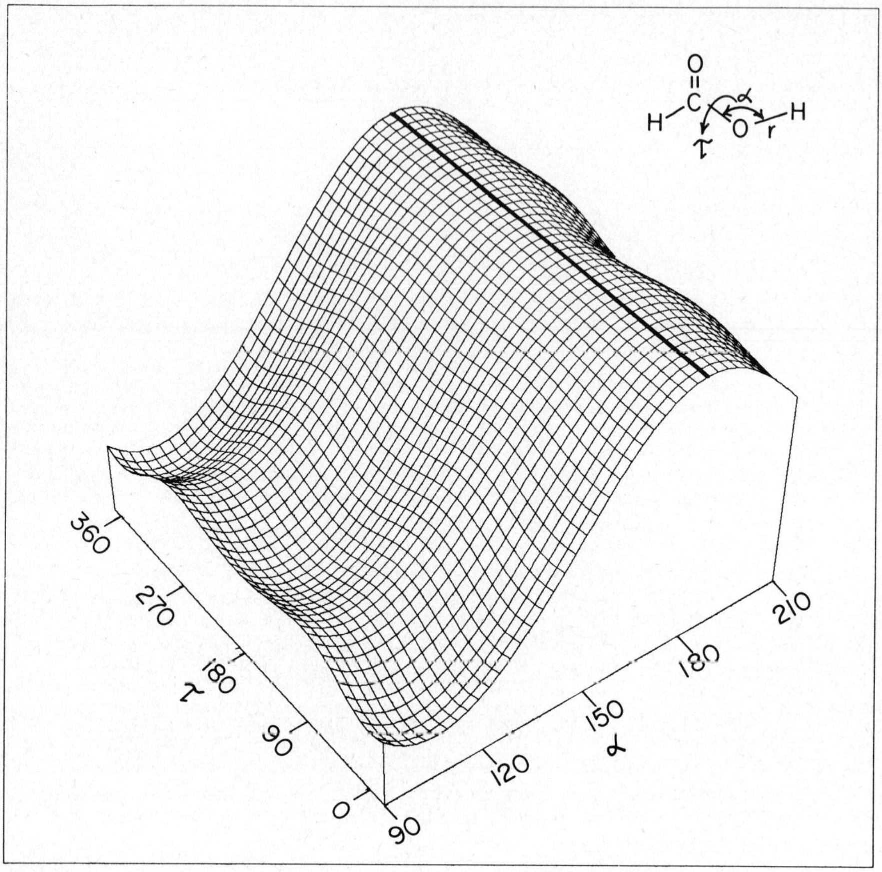

would not permit inversion of the force constants matrix, a necessary step in the VA05AD optimization technique. Therefore, TS_2 will be used as the transition state for linear inversion at O, since it lies very close to TS_4 in both energy and geometry (Table 6).

The ab initio totally optimized geometries of all conformational critical points are given in Table 7. The syn and anti geometries agree fairly well with the experimental gas phase structures, with the exception of the C-O bond length. All the trends in bond lengths and angles are reproduced. The syn geometry has also been previously optimized [117,118] with the STO-3G basis set, using derivative-free methods; there are only slight differences. Note that the only non-planar critical point is TS_1, where neither hydrogen atom lies in the OCO plane. The order of each critical point was confirmed by diagonalizing the Hessian matrix, calculated by gradient differences.

The calculated syn-anti energy difference (ΔE) was 4.4 kcal/mol, comparable to the experimental values of Table 4, and smaller than the previous ab initio

Table 6. Formic acid critical points from the fitted hypersurface and by direct optimization.

POINT	TYPE[a]	r(Å)	α(deg)	τ(deg)	ΔE(kcal/mol)[b]
Syn	Fit	0.990	104.7	0.0	0.00
	Opt	0.990	104.8	0.0	0.00
	Exp[c]	0.972	106.3	0.0	0.00
Anti	Fit	0.990	105.7	180.0	4.36
	Opt	0.988	105.8	180.0	4.44
	Exp[d]	0.956	109.7	180.0	4.0
TS1	Fit	0.993	104.8	89.7	9.55
	Opt	0.991	105.0	92.5	9.52
	Exp[d]	--	--	99.8	13.8
TS2	Fit	0.950	180.8	0.0	53.82
	Opt	0.950	181.1	0.0	53.89
TS4	Fit	0.950	180.0	109.2	53.80

[a]Fit refers to results from the fitted hypersurface, Opt to the direct optimization results, and Exp to the experimental results.

[b]Measured relative to the energy of the syn from for each method.

[c]Ref. 115.

[d]Ref. 109.

results. This could be due to the optimization of the anti geometry--all other ab initio studies had employed rigid rotation.

The computed barrier to rotation (through TS_1) is 3-5 kcal/mol lower than the previous ab initio and experimental values; the reason for this result is unclear, as minimal basis sets generally tend to overestimate barrier heights due to their lack of flexibility. In addition, the torsional angle at the critical point is closer to perpendicular than any of the other results in Table 4 where it was determined. The 54 kcal/mol barrier to in-plane inversion (through TS_2) is considerably higher than that of water or methanol.

Although the entire tautomerization hypersurface was not determined, the geometry of TS_3, a first order saddle point, was optimized utilizing the VA05AD routine, and is given in Table 8. Thus the calculated barrier to tautomerization is 56.8 kcal/mol, considerably lower than the 74 kcal/mol computed previously [104] using the same basis set. The present result was also obtained by Radom et al [118], who found that the barrier was increased by only 6 kcal/mol when a larger split-valence (4-31G) basis set was used. This relatively small change in relative energies at the more flexible basis set level lends credibility to the STO-3G results, as the TS_3 structure is considerably distorted, having an

Table 7. Geometries of the formic acid conformational critical points (bond lengths in Å, angles in degrees).

CONFORMER	C=O	C—O	C—H	O—C—O	O=C—H	O—H	C—O—H	O=C—O—H	ORDER
Syn (opt)	1.2142	1.3856	1.1036	123.63	126.00	0.9904	104.77	0.00	0
Syn (exp)[a]	1.202	1.343	1.097	124.9	124.1	0.972	106.3	0.0	0
Anti (opt)	1.2115	1.3925	1.1076	121.33	124.18	0.9880	105.79	180.00	0
Anti (exp)[b]	1.195	1.352	1.105	122.1	123.3	0.956	109.7	180.0	0
TS_1 (opt)[c]	1.2125	1.4100	1.1081	123.00	123.24	0.9914	104.99	92.46	1
TS_2 (opt)	1.2157	1.3146	1.1126	124.18	123.38	0.9496	181.15	0.00	2

[a] Ref. 115

[b] Ref. 109

[c] The dihedral angle between the H(C) and O about the C=O bond axis is 184.19°, the H(C) lying on the opposite side of the OCO plane to the H(O).

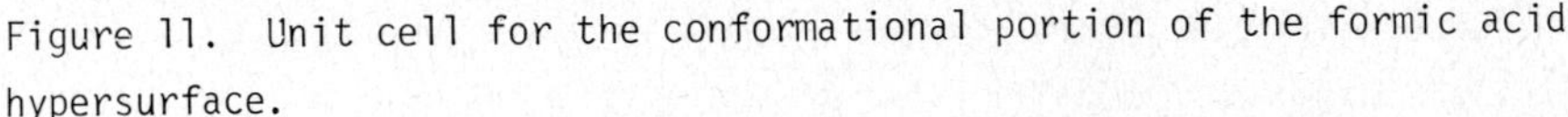

Figure 11. Unit cell for the conformational portion of the formic acid hypersurface.

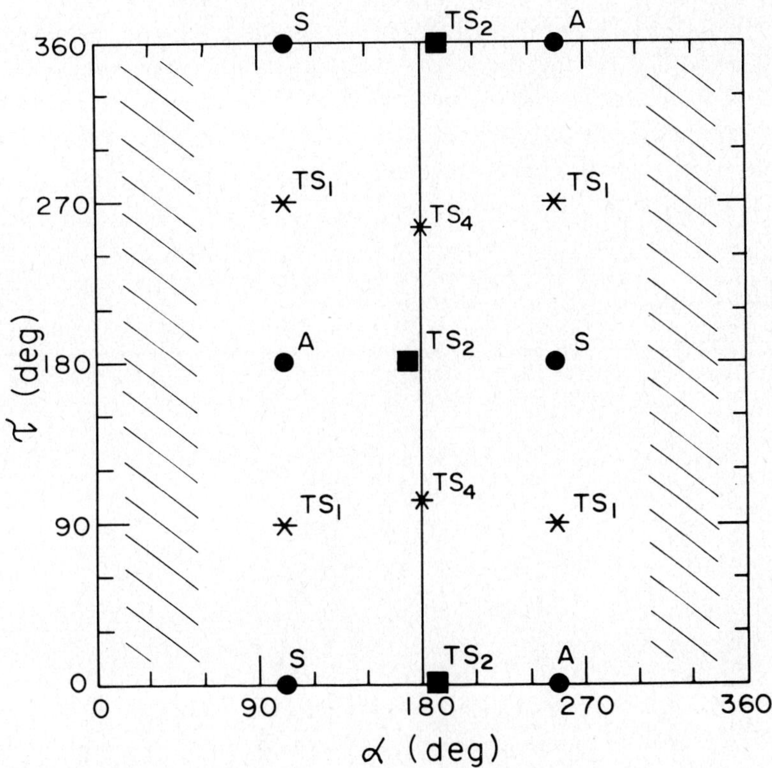

The shaded areas represent high energy regions of the PES.

Table 8. Optimized geometry for the transition state structure TS_3 for 1,3-H migration. H' is the migrating hydrogen, and the structure possesses C_{2v} symmetry.

C-O	1.2966 Å
C-H	1.0952 Å
C-H'	1.5260 Å
O-H'	1.2389 Å
O-C-O	102.56°
O-C-H	128.72°
C-O-H'	73.97°

OCO angle of 102.6°. Radom [118] rationalized the high barrier to H migration in terms of orbital interactions, and also noted that TS_3 contains four π electrons and is therefore antiaromatic.

Figure 12 collects all the calculated results, and is comparable with Figure 8, the estimated results. It perhaps should be emphasized that the conventially accepted notion that barriers along reactive coordinates of a hypersurface are always considerably higher than barriers along conformational coordinates of the same hypersurface (cf. Figure 12) need not necessarily be true. The present results (tautomerization barrier 57 kcal/mol, in-plane inversion 54 kcal/mol) with a difference of 3 kcal/mol in barrier heights would tend to support this conclusion.

Figure 12. Computed energy profile for the total conformational and tautomerization hypersurface.

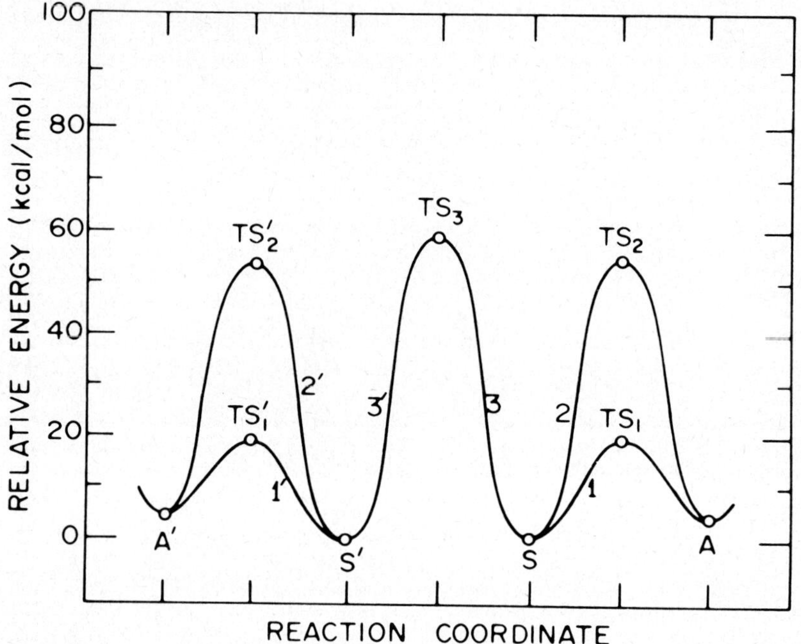

This figure taken from ref. [116] with permission.

3.3 Protonated Phosphoric Amide

In the course of a study on the sites of protonation of $H_2P(=O)NH_2$, a model phosphoric amide, it was found [119] that O protonation was favoured over N protonation by some 50 kcal/mol at the STO-3G level (all geometries were fully optimized by non-gradient methods). This clear-cut preference has recently been confirmed experimentally by an x-ray crystallographic study [120] on protonated phenyl-(t-butyl)-phosphinic amide.

There were two optimized conformers of the O-protonated phosphinamide: the syn- and anti-periplanar arrangements of the N lone pair and the POH group. The interconversion of these two stable species prompted the generation of a surface involving OH rotation (θ, 0° is syn to N) and inversion at N (γ, 180° corresponds to a planar NH_2 moiety).

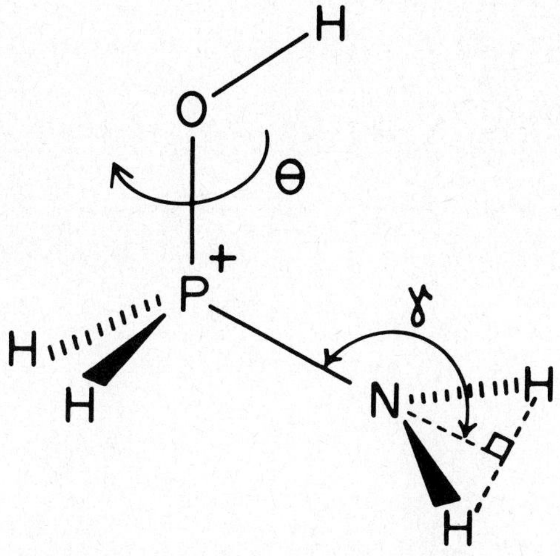

The syn structure corresponds to the point (0°,210°), and the anti form to the point (180°,134°) on the (θ,γ) surface.

Fifty-six points on the surface were computed for values of θ between 0 and 180°, and γ ranging between 105 and 240°. The remaining geometrical parameters were fixed at the average of the values optimized for the syn and anti structures [119], with the following changes. To facilitate the use of the γ coordinate, the NH bond length was replaced by the mean H-H distance (1.710 Å) and the mean distance from the center of the H-H bisector to the N atom (0.567 Å). This also removed the difficulty of changes in the PNH angle with inversion, which are now automatically incorporated.

The surface equation was generated in the usual manner, each candidate term t_i being a product

$$t_i = f_{1i}(\theta)\ f_{2i}(\gamma) \tag{70}$$

There is no symmetry in the γ coordinate, so sin $n_2\gamma$ and cos $n_2\gamma$ were used (n_2 ranging from 1 to 6 and 0 to 6 respectively) for that coordinate. The OH rotation must be symmetric about 0°, restricting f_1 to cos $n_1\theta$, with n_1 ranging from 0 to 6. All possible products of the f_1 and f_2 coordinate functions, except $n_1=n_2=0$ (simply a constant), were utilized in the stepwise regression analysis, for a total of ninety candidate terms.

The sixteen term equation, including a constant, detailed in Table 9, was chosen to represent the ab initio data. The root mean square deviation of the fitted and calculated values was 0.012 kcal/mol (an average error of 0.01 kcal/mol), indicating an excellent fit. All terms in the equation were significant at the 0.01% level.

Table 9. Surface equation for the O-protonated phosphinamide.

ROTATION - INVERSION SURFACE FOR PH2(OH)NH2+
THETA (X1) IS ROTATION (0 IS SYM), GAMMA (X2) IS INVERSION (180 IS PLANAR)

FUNCTIONS FOR COORDINATE 1:

1	1.0
2	COS 1X
3	COS 2X
4	COS 3X
5	COS 4X

FUNCTIONS FOR COORDINATE 2:

1	1.0
2	SIN 1X
3	SIN 2X
4	SIN 4X
5	COS 1X
6	COS 2X
7	COS 3X

TERMS IN SURFACE EQUATION:[a]

TERM	C	X1	X2
1	57.520713	1	1
2	.309613	4	1
3	-.029291	5	1
4	-1.688138	1	2
5	1.987440	2	2
6	.403340	3	2
7	.020569	5	2
8	-.543512	1	3
9	-.052191	4	3
10	-.043077	2	4
11	81.076942	1	5
12	.869186	2	5
13	.184661	3	5
14	29.010301	1	6
15	-.113197	2	6
16	2.423394	1	7

[a]These coefficients calculate the energy in kcal/mol relative to the optimized energy for (VIa) in Ref. 119.

The critical points were located on the fitted surface, and also by direct ab initio optimization varying only θ and γ, by minimizing the reduced gradient length S_g as in section 2.3.1. The results, presented in Table 10, demonstrate the quantitative agreement between the analytic equation and the direct optimization approaches.

Figure 13 shows the surface topology and the positions of the critical points, while Figure 14 is a pseudo-three-dimensional representation of the surface. Note that there are four unique minima, of which only two have been optimized previously [119]: m_1, which corresponds to the syn structure, and m_3,

Table 10. Critical points of the O-protonated phosphinamide surface.

CONFORMER	METHOD[a]	θ_{OH}(deg)	γ_{NH_2}(deg)	ORDER	E[b]	ΔE(kcal/mol)
m_1	Fit	0.00	217.19	0	0.01	0.00
	Opt	0.00	217.19	0	-0.189343	0.00
m_2	Fit	74.21	140.74	0	0.90	0.89
	Opt	74.21	140.74	0	-0.187886	0.91
m_3	Fit	180.00	137.37	0	0.21	0.20
	Opt	180.00	137.37	0	-0.189011	0.21
m_4	Fit	180.00	213.13	0	3.34	3.33
	Opt	180.00	213.70	0	-0.184071	3.31
s_1	Fit	0.00	147.34	1	2.02	2.01
	Opt	0.00	147.34	1	-0.186125	2.02
s_2	Fit	39.72	174.96	1	2.16	2.15
	Opt	39.72	174.96	1	-0.185898	2.16
s_3	Fit	113.50	138.28	1	1.12	1.11
	Opt	113.50	138.28	1	-0.187536	1.13
s_4	Fit	135.52	213.29	1	3.69	3.68
	Opt	135.52	213.86	1	-0.183503	3.66
s_5	Fit	180.00	190.43	1	3.70	3.69
	Opt	180.00	190.43	1	-0.183412	3.72

[a]Fit refers to the fitted equation, Opt to the direct ab initio optimization.
[b]Fitted energies are in kcal/mol relative to the ab initio energy of conformer VIa (ref. 119). Optimized energies are in hartrees; -467.0 should be added to each value.

which is the anti conformer. The small deviations of the locations of these minima from the fully optimized structures [119] are due to the use of an average geometry for the surface. The two new minima occur at the gauche conformation (m_2, θ=74°) and an anti structure with the NH bonds nearly eclipsing the PH bonds (m_4). The low barriers ($\lesssim$2.2 kcal/mol) to rotation-inversion, particularly along the path m_1-s_2-m_2-s_3-m_3, are indicative of considerable conformational flexibility in the O-protonated phosphoric amide.

This is probably the first case where the analytic equation has proved to be advantage: the plots and critical point searches immediately revealed the presence of the two additional minima, which were not considered in the ab initio geometry optimization [119]. In the previous applications, n-butane and formic acid, stereochemical intuition was used to locate, at least approximately, <u>all</u> the critical points on the hypersurfaces.

Figure 13. Topological map of the $PH_2(OH)NH_2^+$ rotation-inversion energy surface. Minima are shown as m_1 through m_4, while the saddle points are labelled s_1 through s_5. The low maxima are M_1 and M_2 while the shaded areas represent rapidly increasing energy as the inversion angle becomes too small or large (the planar PNH_2 moiety has γ = 180°).

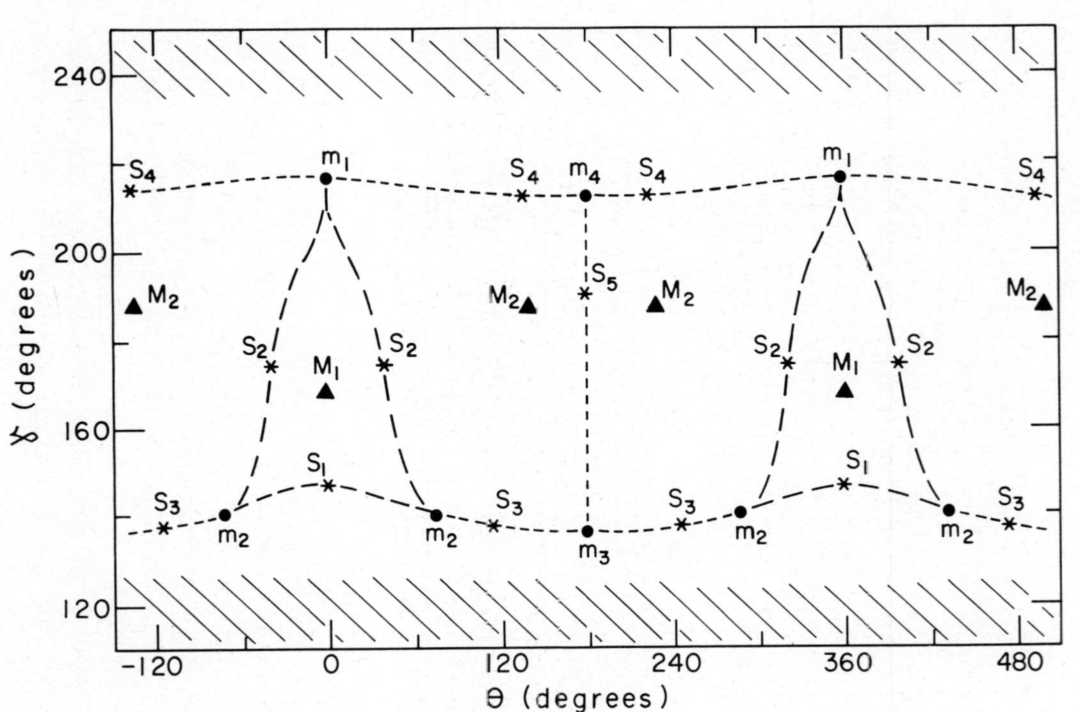

Figure 14. Pseudo-three-dimensional plot of the $PH_2(OH)NH_2^+$ rotation-inversion energy surface. The critical points are identified as in Figure 13. The surface has reflective symmetry in the lines θ = 0°, θ = 180° and θ = 360°.

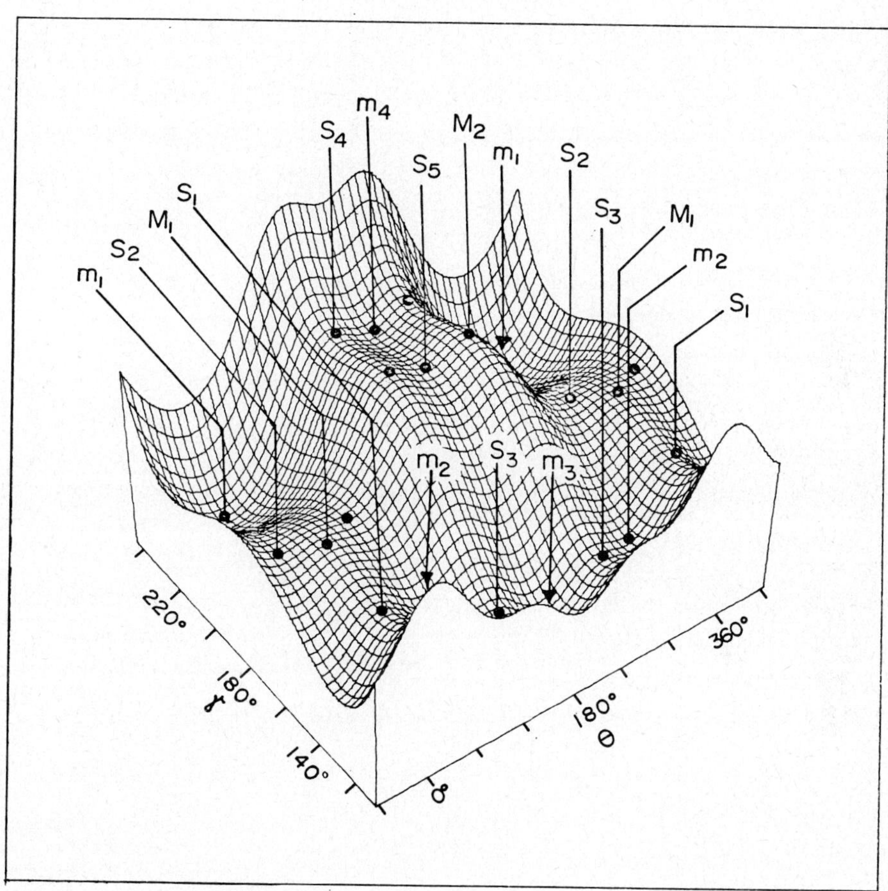

3.4 Ring-opened Triplet Methyloxirane Isomers

3.4.1 Introduction. The reaction of $O(^3P)$ atoms with olefins has been studied extensively in both the gas and liquid phases, mainly by Cvetanović and co-workers [121,122]. The $O(^3P)$ atoms have been shown to exhibit a distinct electrophilic character, attacking preferentially the less substituted of the unsaturated C atoms when the substituents are alkyl groups. When the largely suppressible fragmentation reactions are prevented, the final principal products are the epoxides and their carbonyl isomers. It has been suggested that the primary product of the reaction is a $\dot{C}$-C-$\dot{O}$ triplet biradical, synonymous with the (ring-opened) T_1 state of the epoxide, which can undergo intersystem crossing to the S_0 state and be stabilized as oxirane or isomerization via H atom or alkyl migration and be stabilized as a carbonyl compound.

The triplet states of oxiranes have also been implicated in the $Hg(^3P)$ sensitized decomposition of oxiranes [123-125].

In order to elucidate which of the low lying triplet states of an oxirane molecule can be populated in these reactions, a theoretical study of the lowest triplet energy manifold of the methyloxirane molecule and its isomers was carried out [126]. Methyloxirane was chosen instead of the simpler ethylene oxide to allow comparison of the two sites of attack (for O plus propylene), and the barriers to H$^•$ and $CH_3^•$ migration in the triplet intermediate(s).

The second phase of the study [127] was an investigation of the conformational behaviour of triplet methyloxirane itself. The initial study [126] had revealed the existence of three minima on the C_3H_6O (lowest triplet state) energy hypersurface corresponding to the three ring-opened species, of differing thermodynamic stabilities, as indicated in the following Scheme, reproduced from ref. [127] with permission.

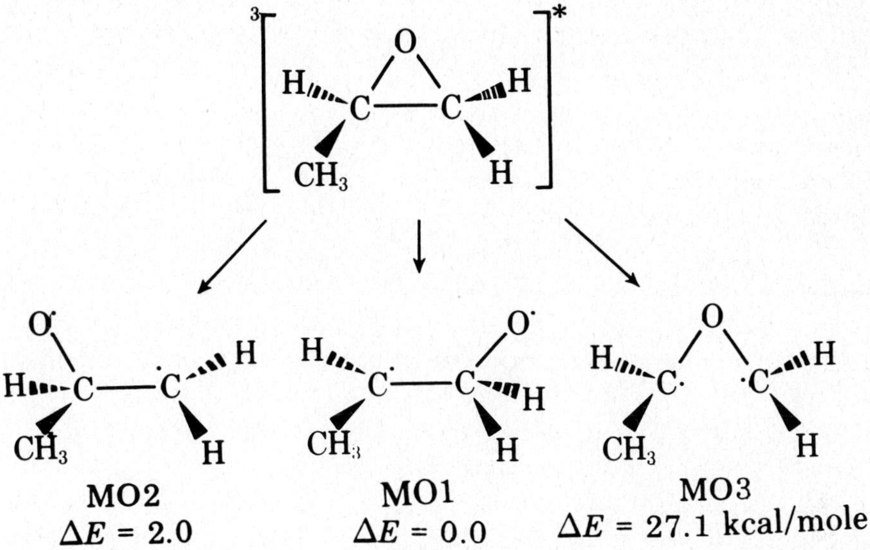

MO2 $\Delta E = 2.0$ MO1 $\Delta E = 0.0$ MO3 $\Delta E = 27.1$ kcal/mole

Scheme I

The conformational studies involved torsions about single bonds; in MO1 and MO2 this involved a methyl rotation (θ_1) and a rotation about the C-C bond (θ_2) which was part of the original three-membered ring. This led to the generation of an energy surface of the type $E=E(\theta_1,\theta_2)$. In MO3, in addition to the methyl rotation (θ_1), two non-equivalent C-O rotations (θ_2,θ_3) were considered, which led to an energy hypersurface $E=E(\theta_1,\theta_2,\theta_3)$. These energy surfaces represent rigid rotations at the geometries optimized previously [126] using fixed torsional angles.

The (0°,0°) and (0°,0°,0°) structures for the three isomers are shown below (reproduced from ref. [127] with permission).

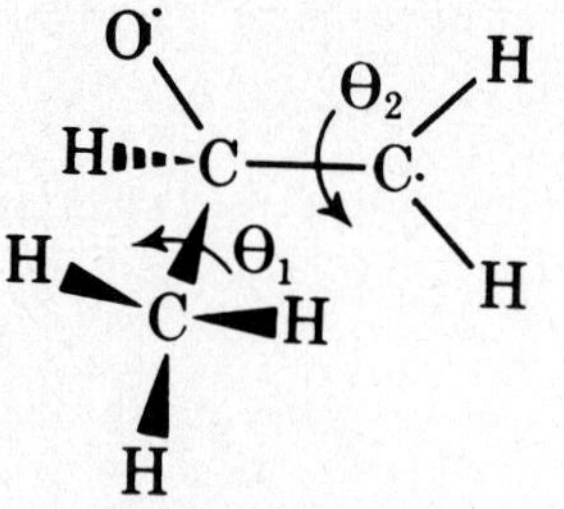

MO2(0°,0°) 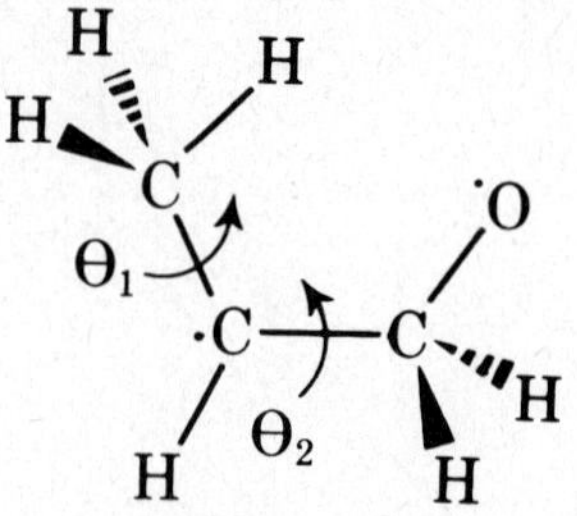 MO1(0°,0°)

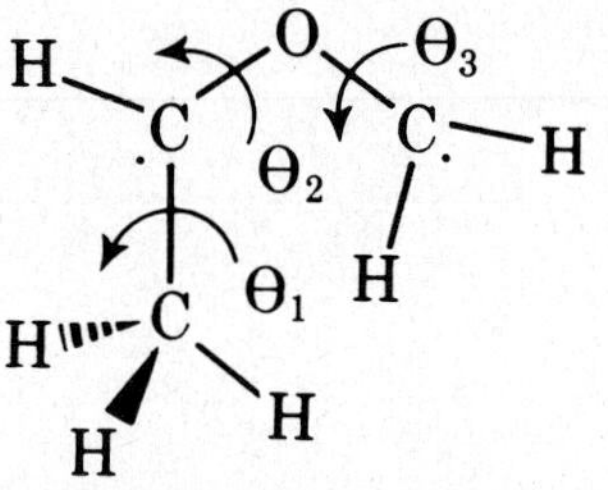

MO3(0°,0°,0°)

For MO2, θ_1=0° corresponds to a methyl hydrogen eclipsed with the oxygen atom. In all cases local C_{3V} symmetry was retained for the methyl group and carbon radical centers were held planar as before [126].

Each isomer will be discussed in turn. The final equation chosen always reduced the average error of the fit to 0.02 kcal/mol. During the direct ab initio optimizations, only the conformational coordinates were varied, to correspond exactly with the analytic (hyper)surfaces. The direct optimizations were started at the critical point geometries determined from the fitted equations, unless otherwise noted.

3.4.2 Methyloxirane 1. This surface possesses a center of inversion at (0,0), restricting candidate terms to

$$\cos n_1\theta_1 \cos n_2\theta_2 \tag{71}$$

$$\sin n_1\theta_1 \sin n_2\theta_2 \tag{72}$$

These terms were obtained in the same manner as for the n-butane hypersurface, by applying the inversion property to proposed terms. For example, the term $\sin n_1\theta_1$ is not invariant upon inversion in the (θ_1,θ_2) space, but the product (72) is acceptable.

The methyl rotation (θ_1) must have threefold periodicity, but the θ_2 coordinate possesses only onefold periodicity. There were fourteen candidate terms based on (71), with n_1 = 0, 3 or 6, and n_2 ranging from 0 to 4. The constant term, n_1=n_2=0, was omitted. Eight more candidate terms were derived from (72), with n_1 = 3 or 6, and n_2 ranging from 1 to 4.

The equation determined by the stepwise regression analysis, using eighteen unique ab initio points from a regular (30°,45°) grid in (θ_1,θ_2), is detailed in Table 11, and is plotted in Figure 15. The rms error of the ten term fit was 0.031 kcal/mol (average error 0.021 kcal/mol). The least significant term had a 4% probability of being zero.

Table 11. Parameters of the M01 analytic surface equation[a].

```
M01 T1 STO-3G CONFORMATIONAL SURFACE
THETA1 IS CH3 ROTATION, THETA2 IS CH(CH3). ROTATION
```

FUNCTIONS FOR COORDINATE 1:

1	1.0
2	COS 3X

FUNCTIONS FOR COORDINATE 2:

1	1.0
2	COS 1X
3	COS 2X
4	COS 3X
5	COS 4X

TERMS IN SURFACE EQUATION:

TERM	C	X1	X2
1	-485.619417	1	1
2	.224375	2	1
3	.362896	1	2
4	.054697	2	2
5	.538292	1	3
6	.086000	2	3
7	.392771	1	4
8	.149803	2	4
9	.085458	1	5
10	.040125	2	5

[a]The coefficients give the energy in milli-hartrees (189.0 hartrees must be subtracted to obtain the total energy).

The characteristics of the critical points located on the fitted surface and by direct optimizttion are summarized in Table 12. Five of the seven unique critical points occur at exactly the same coordinates on both the analytic fit and ab initio surfaces. The remaining stationary points show shifts averaging 4°, probably due to the very flat nature of the PES, leading to small force constants. Thus the energy changes very little in a relatively large neighbourhood of the critical points, and the precise location of the critical points becomes a very sensitive function of the actual PES used.

A single, unique conformational minimum was found on the PES. This minimum is interconnected by three distinctly different saddle points. A fourth saddle point (θ_1=0°, θ_2=180°) lies between two equivalent saddle points located at θ_1=±60° and θ_2=180°. Although the latter appear in Figure 15 as minima, in fact they are rather flat saddle points. The multitude of saddle points occur on this surface because of the existence of two non-equivalent maxima. The overall energy difference between the highest and lowest points of the surface is relatively small (<2 kcal/mol), which allows virtually free cis-trans isomerization of deuterium-labelled propylene oxide.

The minimum energy conformation corresponds to a gauche arrangement of the heavy atom skeleton. A methyl hydrogen eclipses the C-H bond of the -CH_2 radical

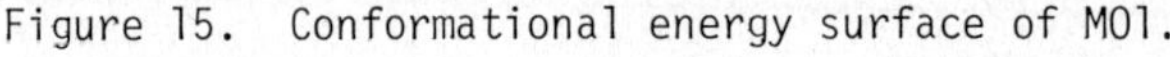

Figure 15. Conformational energy surface of MO1.

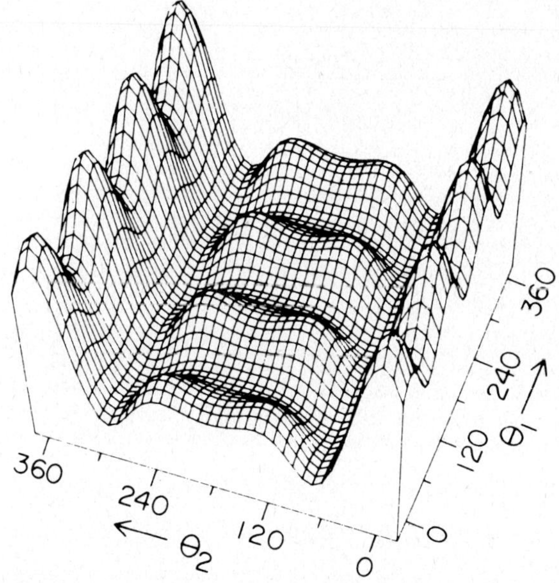

This figure taken from ref. [127] with permission.

site, although the calculated barrier to methyl group rotation is only 0.1 kcal/mol. The barrier to rotation about the other C-C bond (coordinate θ_2) was computed to be 1.0 kcal/mol via the syn conformation (θ_2=0°) and 0.3 kcal/mol via the anti conformer (θ_2=180°).

3.4.3 Methyloxirane 2. The MO2 surface possesses no symmetry; hence any surface terms with the correct periodicity may be utilized. The methyl rotation has threefold periodicity, while rotation about the C-$\dot{C}H_2$ bond is a twofold torsion. Each candidate term t_i was a product

Table 12. Characteristics of the critical points of the M01 conformational energy surface.

METHOD[a]	θ_1(deg)	θ_2(deg)	ORDER	ΔE(kcal/mol)[b]
FIT	60.0	76.7	0	0.00
OPT	63.6	72.8	0	0.00
FIT	0.0	68.9	1	0.08
OPT	2.6	63.1	1	0.13
FIT	0.0	180.0	1[c]	0.44
OPT	0.0	180.0	1[c]	0.48
FIT	60.0	0.0	1[c]	0.95
OPT	60.0	0.0	1[c]	0.97
FIT	60.0	180.0	1[c]	0.26
OPT	60.0	180.0	1[c]	0.27
FIT	0.0	0.0	2[c]	1.64
OPT	0.0	0.0	2[c]	1.68
FIT	0.0	127.5	2	0.56
OPT	0.0	127.5	2	0.48

[a]FIT refers to results determined from the fitted surface, OPT to the direct optimization results.

[b]Measured relative to the absolute minimum as determined by each method.

[c]These critical points may also be located by symmetry considerations.

$$t_i = f_{1i}(\theta_1)\, f_{2i}(\theta_2) \tag{73}$$

in the usual manner. The functions f_1 were sin $n_1\theta_1$ (n_1 = 3 and 6) and cos $n_1\theta_1$ (n_1 = 0, 3 and 6). The functions f_2 were sin $n_2\theta_2$ (n_2 = 2 and 4) and cos $n_2\theta_2$ (n_2 = 0, 2 and 4). All possible cross-products, except the constant term from n_1=n_2=0, were included in the stepwise regression analysis, for a total of twenty-four candidate terms.

The seven term equation chosen, detailed in Table 13 and plotted in Figure 16, gave an average fitting error of 0.020 kcal/mol (rms error 0.024 kcal/mol) to twenty-four ab initio points, chosen in a regular 30° grid in both coordinates. All terms were significant to the 0.5% level, at least.

An analysis of the fitted surface and direct optmization revealed the existence of only a single minimum. In addition, two non-equivalent saddle points and a single maximum were located. The characteristics of these critical points are summarized in Table 4. There is quantitative agreement on the critical point locations, but there are 0.2-0.3 kcal/mol shifts in the energies of the two θ_2=144.9° stationary points.

Table 13. Parameters of the M02 analytical surface equation[a].

M02 T1 STO-3G CONFORMATIONAL SURFACE
THETA1 IS CH3 ROTATION, THETA2 IS CH2. ROTATION

FUNCTIONS FOR COORDINATE 1:

1	1.0
2	SIN 3X
3	COS 3X

FUNCTIONS FOR COORDINATE 2:

1	1.0
2	SIN 2X
3	COS 2X
4	COS 4X

TERMS IN SURFACE EQUATION:

TERM	C	X1	X2
1	-480.501375	1	1
2	-.037500	2	1
3	2.351917	3	1
4	-.608816	1	2
5	.313250	1	3
6	.049667	2	3
7	-.068750	1	4

[a]The coefficients give the energy in milli-hartrees (189.0 hartrees must be subtracted to obtain the total energy).

Figure 16. Conformational energy surface of M02.

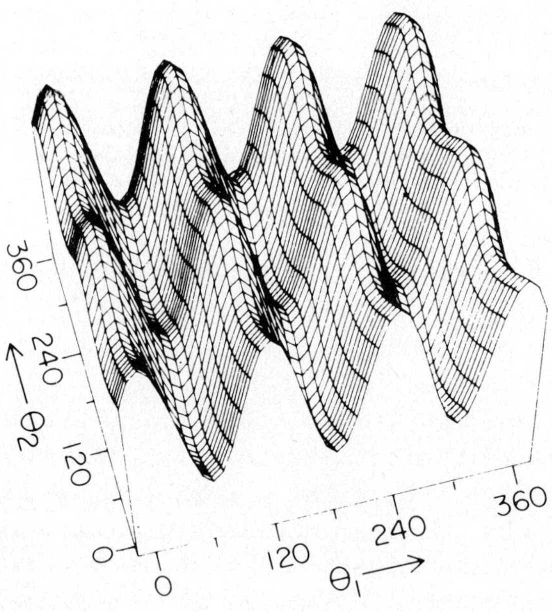

This figure taken from ref. [127] with permission.

The minimum energy structure features a staggered methyl group (θ_1=59°) and the $-\dot{C}H_2$ radical site bisects the C-O and C-C bonds (θ_2=64°). The barrier to methyl group rotation was calculated to be 3.0 kcal/mol, which is similar to the ground state barriers for other saturated compounds (c.f. n-butane, section 3.1) with the exception of methanol (1.12 kcal/mol) [113]. The $-\dot{C}H_2$ radical rotation barrier (0.6 kcal/mol) indicates essentially free rotation of that group. Note that both conformational changes follow least motion paths--the coordinate not being rotated changes by less than 1°.

The low barrier to θ_2 rotation in both MO1 and MO2 suggests that cis-trans isomerization, as might be observed in deuterium-substituted propylene oxide, is energetically quite accessible in the T_1 state. This is consistent with the experimental finding that both cis- and trans-epoxides result from the addition of $O(^3P)$ to 2-butenes [128].

Table 14. MO2 critical points from the fitted surface and by direct optimization.

METHOD[a]	θ_1(deg)	θ_2(deg)	ORDER	ΔE(kcal/mol)[b]
FIT	59.4	64.3	0	0.00
OPT	59.4	64.3	0	0.00
FIT	119.4	64.2	1	2.95
OPT	119.4	64.2	1	2.95
FIT	59.8	144.9	1	0.87
OPT	59.8	144.9	1	0.64
FIT	119.8	144.9	2	3.82
OPT	119.8	144.9	2	3.65

[a]FIT refers to results determined from the fitted equation, OPT to the direct optimization results.

[b]Measured relative to the energy of the absolute minimum as determined by each method.

3.4.4 Methyloxirane 3. The three rotors in MO3 have periodicities three, one and two respectively, and the hypersurface possesses a center of inversion. For θ_2 = 0 or 180°, the molecule may have higher symmetry: if θ_1 is 0 or 60°, then θ_3 must be symmetric, while if θ_3 is 0 or 90°, θ_1 must be symmetric. These special requirements must be incorporated into the hypersurface equation terms.

The first group of forty-four candidate terms was derived from

$$\cos n_1\theta_1 \cos n_2\theta_2 \cos n_3\theta_3 \qquad (74)$$

where n_1 was 0, 3 or 6, n_2 ranged from 0 to 4, and n_3 was 0, 2 or 4. The term n_1=n_2=n_3=0, a constant, was omitted as usual. Since cosine is a symmetric function, the special symmetries are ensured.

Although a single sine function is antisymmetric under inversion, a product of two sine functions leaves the term invariant. The next nine candidate terms arose from

$$\sin 3\theta_1 \sin n_2\theta_2 \cos n_3\theta_3 \tag{75}$$

where $n_2 = 1$, 2 or 3, and $n_3 = 0$, 2 or 4. Five more candidate terms had the form

$$\sin 3\theta_1 \cos n_2\theta_2 \sin 2\theta_3 \tag{76}$$

where n_2 ranged from 0 to 4. The next nine candidate terms were obtained from

$$\cos n_1\theta_1 \sin n_2\theta_2 \sin 2\theta_3 \tag{77}$$

with $n_1 = 0$, 3 or 6, and n_2 ranging from 1 to 3. The ranges of n_1 are limited by the uniform grid used for the ab initio data points (30° for θ_1, 45° for θ_2 and θ_3). The special symmetry properties are ensured because the sin $n_i\theta_i$ forces all terms of the type (75), (76) and (77) to zero at such points.

The remaining candidate terms involved repeating Gaussians, the first being

$$\exp(-6(3\theta_1)^2) \exp(-6(\theta_2)^2) \exp(-6(2\theta_3)^2) \tag{78}$$

Nine more terms arosc from

$$\cos n_1\theta_1 \exp(-6(\theta_2)^2) \cos n_3\theta_3 \tag{79}$$

for $n_1 = 0$, 3 or 6, and $n_3 = 0$, 2 or 4. In addition, the term

$$\sin 3\theta_1 \exp(-6(\theta_2)^2) \sin 2\theta_3 \tag{80}$$

was included.

The candidate terms requried a sum of two components, as for n-butane, when the θ_2 repeating Gaussian was moved from the origin:

$$f_1(\theta_1) \exp(B(\theta_2-A)^2) f_3(\theta_3) + f_1(-\theta_1) \exp(B(\theta_2+A)^2) f_3(-\theta_3) \tag{81}$$

The values of (A,B) were (50°,-2.5) and (40°,-4.0); the functions f_1 were 1, $\sin 3\theta_1$, $\cos 3\theta_1$ and $\cos 6\theta_1$, and the functions f_3 were 1, $\sin 2\theta_3$, $\cos 2\theta_3$ and $\cos 4\theta_3$. Thus (81) resulted in thirty-two candidate terms, for a grand total of 11C terms in the stepwise regression analysis.

The analysis, utilizing sixty-eight unique ab initio grid points, resulted in an equation containing twenty-nine candidate terms (plus a constant) which gave an average error of 0.021 kcal/mol. Subsequently, the values of A and B in (81) were optimized by the VA05AD routine to reduce the average error slightly to 0.020 kcal/mol (rms error 0.027 kcal/mol). The optimized equation is given in Table 15; the remaining eleven terms listed there resulted from the extra

Table 15. Parameters of the MO3 analytical hypersurface equation.[a]

MO3 TRIPLET HYPERSURFACE - RING-OPENED METHYLOXIRANE - STO-3G CALCULATIONS
THETA1 IS METHYL ROT, THETA2 IS CCOC ROT, THETA3 IS CH2. ROT

FUNCTIONS FOR COORDINATE 1:

```
 1   1.0
 2   SIN  3X
 3   SIN -3X
 4   COS  3X
 5   COS  6X
 6   EXP( -6.00000 ( 3X) ** 2)
```

FUNCTIONS FOR COORDINATE 2:

```
 1   1.0
 2   SIN  1X
 3   SIN  2X
 4   SIN  3X
 5   COS  1X
 6   COS  2X
 7   COS  3X
 8   COS  4X
 9   EXP( -6.00000 X ** 2)
10   EXP( -1.54700 (X -  39.07900) ** 2)
11   EXP( -1.54700 (X +  39.07900) ** 2)
12   EXP( -4.33260 (X -  38.76300) ** 2)
13   EXP( -4.33260 (X +  38.76300) ** 2)
```

FUNCTIONS FOR COORDINATE 3:

```
 1   1.0
 2   SIN  2X
 3   SIN -2X
 4   COS  2X
 5   COS  4X
 6   EXP( -6.00000 ( 2X) ** 2)
```

TERMS IN SURFACE EQUATION:

TERM	C	X1	X2	X3
1	-444.808310	1	1	1
2	.070230	4	1	1
3	.199050	4	5	1
4	-.031352	1	8	1
5	-.557267	1	1	4
6	.275283	1	6	4
7	.124945	1	1	5
8	.074236	2	3	1
9	-.686729	1	2	2
10	.470625	1	3	2
11	-.562729	1	4	2
12	1.563996	6	9	6
13	6.597595	1	9	1
14	3.315548	4	9	1
15	5.653404	1	9	4
16	2.786433	4	9	4
17	.787595	1	9	5
18	.532647	4	9	5
19	-1.556965	2	9	2
20	2.548004	1	10	1
21	2.548004	1	11	1
22	-.199121	2	10	2
23	-.199121	3	11	3
24	-1.131589	2	12	1
25	-1.131589	3	13	1
26	1.140145	2	12	2
27	1.140145	3	13	3
28	-.187143	4	12	2
29	-.187143	4	13	3
30	.813418	1	12	4
31	.813418	1	13	4
32	-.397673	2	12	4
33	-.397673	3	13	4
34	-.533825	4	12	4
35	-.533825	4	13	4
36	-.104588	1	12	5
37	-.104588	1	13	5
38	.076220	2	12	5
39	.076220	3	13	5
40	-.171712	4	12	5
41	-.171712	4	13	5

[a]The coefficients give the energy in milli-hartrees (189.0 hartrees must be subtracted to obtain the total energy).

components in some candidate terms due to the inversion centers. All terms were significant to the 0.5% level, at least.

Various cross-sections of the analytic hypersurface are depicted in Figure 17. Analysis of the hypersurface equation for critical points revealed the existence of a single, unique minimum and two unique maxima. In addition, six first order and seven second order saddle points were located, as indicated in Table 16.

Table 17 lists the critical points located by direct optimization on the ab initio conformational hypersurface. The eight symmetry constrained critical points, at θ_2 = 0 and 180°, are retained, as is the minimum, the first order saddle point on the θ_2=140° plane, and two saddle points from the θ_2=110° plane. The remaining critical points have shifted considerably, but most are little changed in energy. Some difficulty was encountered in locating several of the critical points due to the extreme flatness of the hypersurface away from the θ_2=0° plane. To obtain reasonable estimates of the Hessian by gradient differences, steps of length 0.1 radians (5.73°) were necessary--the step length usually used was 0.01-0.02 radians. Even so, the force constants were generally on the order of 0.005 mdyne-Å/radian^2 (torsional force constants are usually on the order of 0.1 mdyne-Å/radian^2). These small force constants lead to uncertainty in some of the critical point coordinates--typically, differences of up to 1° were found upon optimizing towards the same stationary point from alternate starting points, even with a required gradient length of 0.0001 mdyne-Å/radian, five times smaller than the normal termination criterion. Note that the ratio of the smallest energy difference between adjacent critical points of the analytic equation to the fitting error is about three in the case of MO3 (the ratio is approximately four and forty for MO1 and MO2 respectively).

The topology of the critical points (obtained from the analytic equation) characteristic of a unit cell is shown in Figure 18. This unit cell is analogous in principle to the unit cell described earlier for the n-butane hypersurface, but with one marked distinction: the non-regular distribution of first and second order saddle points around the minimum of the MO3 hypersurface. In n-butane the minima were surrounded by a regular grid of first order saddle points which, in turn, were surrounded by minima and second order saddle points, etc. This regularity does not prevail in the present MO3 hypersurface because a string of critical points, such as minimum → first order saddle point → minimum, which is characteristic of a "reaction coordinate", does not lie along a least motion path, whereas they do for n-butane. For example, the saddle points for the rotation of θ_3 (i.e., the torsion of the $-\dot{C}H_2$ radical end) are found on the θ_2=180° plane (cf. Figure 18) at (0°, 180°, 90°) or on the θ_2=110° plane at (27°, 113°, 107°). Both paths involve rotations about all three bonds along the reaction coordinate. This topological difference probably occurs

Table 16. Characteristics of the critical points of the analytic conformational energy hypersurface for MO3.

θ_1(deg)	θ_2(deg)	θ_3(deg)	ORDER	ΔE(kcal/mol)[a]
0.0	0.0	0.0	3[b]	15.97
60.0	0.0	0.0	2[b]	6.57
0.0	0.0	90.0	2[b]	4.65
60.0	0.0	90.0	1[b]	2.86
55.4	51.9	60.0	1	2.16
50.0	66.4	87.1	2	2.30
26.5	113.4	107.3	1	1.82
90.5	110.5	104.5	2	1.87
13.4	142.1	37.7	0	0.00
73.8	140.9	37.5	1	0.14
15.0	135.6	119.7	2	1.88
74.3	138.0	120.4	3	2.01
0.0	180.0	0.0	1[b]	0.84
60.0	180.0	0.0	2[b]	1.00
0.0	180.0	90.0	1[b]	1.19
60.0	180.0	90.0	2[b]	1.35

[a]Energies are relative to the energy of the minimum (13.4°, 142.1°, 37.7°).
[b]These critical points may also be located by the symmetry constraints.

because the molecule has little symmetry in most conformations.

Starting from the minimum, the computed ab initio barrier to methyl rotation is 0.1 kcal/mol through the saddle point (73°, 140°, 37°) from Table 17. Rotation about the internal C-O bond (coordinate θ_2) through the anti conformation (0°, 180°, 0°) has a computed barrier of 0.8 kcal/mol, but through the lowest syn structure (60°, 0°, 90°), it is 2.8 kcal/mol. Rotation of the terminal $-\dot{C}H_2$ radical group has a calculated barrier of 1.2 or 2.0 kcal/mol via the processes already discussed (the latter saddle point occurs at (28°, 111°, 111°) on the ab initio hypersurface). Although methyl group rotation occurs along least motion paths, the remaining two C-O bond torsions are highly coupled processes in all regions of the hypersurface.

It is evident in Table 17 that all critical points, except those on the θ_2=0 plane, lie within 2.1 kcal/mol of one another, an indication of the considerable conformational flexibility of the MO3 triplet state. This is

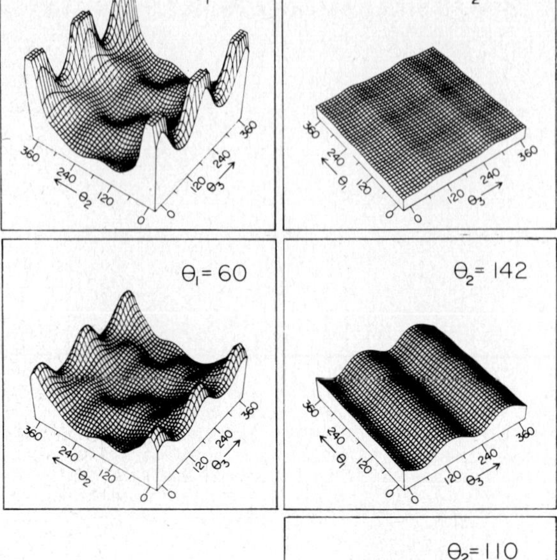

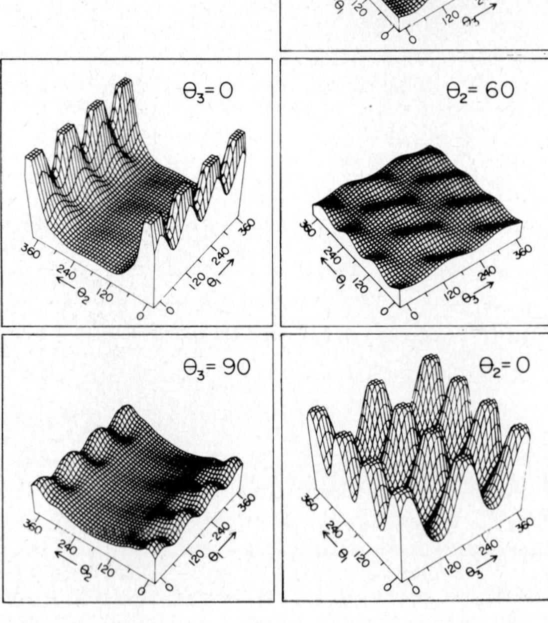

Figure 17. Various cross-sections of the M03 fitted hypersurface, all plotted with the same scale factor to show the large energy variations. All cuts are parallel to one face of the unit cell (Figure 18).

This figure taken from ref. [127] with permission.

Table 17. Critical points of the MO3 hypersurface by direct optimization.

θ_1 (deg)	θ_2 (deg)	θ_3 (deg)	ORDER	ΔE (kcal/mol)[a]
0.0	0.0	0.0	3[b]	15.96
60.0	0.0	0.0	2[b]	6.55
0.0	0.0	90.0	2[b]	4.63
60.0	0.0	90.0	1[b]	2.83
35.2	78.9	165.3	1[c]	1.03
112.1	87.2	176.1	2[c]	1.10
37.5	94.7	97.8	1	2.00
29.2	108.1	108.8	2	2.00
90.8	108.4	107.7	3	2.10
13.7	139.9	37.7	0	0.00
74.7	139.5	36.9	1	0.13
0.0	180.0	0.0	1[b]	0.83
60.0	180.0	0.0	2[b]	0.95
0.0	180.0	90.0	1[b]	1.17
60.0	180.0	90.0	2[b]	1.36

[a]Energies are relative to the direct optimized energy of the minimum (13.7°, 139.9°, 37.7°).

[b]These points can also be located from the molecular symmetry constraints.

[c]These points were located by inspection of critical points already located for θ_2 ranging from 75 to 110°.

consistent with the possibility of a "rapid equilibrium of intermediates" during the cis-trans isomerization of irradiated 2,3-diphenyloxirane [129], which isomerizes by opening the C-C bond of the oxirane, as do most aryl-substituted epoxides. The energy calculated for the (0°, 0°, 0°) conformation is probably too high because the geometry was not reoptimized to reduce the nonbonded interaction of the methyl and the $\dot{C}H_2$ groups.

The main results of this study of the three isomeric ring-opened methyl-oxiranes may be summarized as follows:

(1) The conformational energy surfaces are generally fairly flat and lead to low rotational barriers, which indicate considerable conformational flexibility and the possibility of facile cis-trans isomerization via triplet state oxiranes.

(2) The energies of the two C-O ring-opened triplets are similar and the global minima on each surface are quite close to the structures optimized at fixed torsional angles [126]. The most stable C-C ring-opened structure (MO3)

Figure 18. Critical point topological map for the unit cell of the M03 analytic hypersurface: (•) minima, (*) first-order saddle points, (+) second-order saddle points, and (▲) maxima.

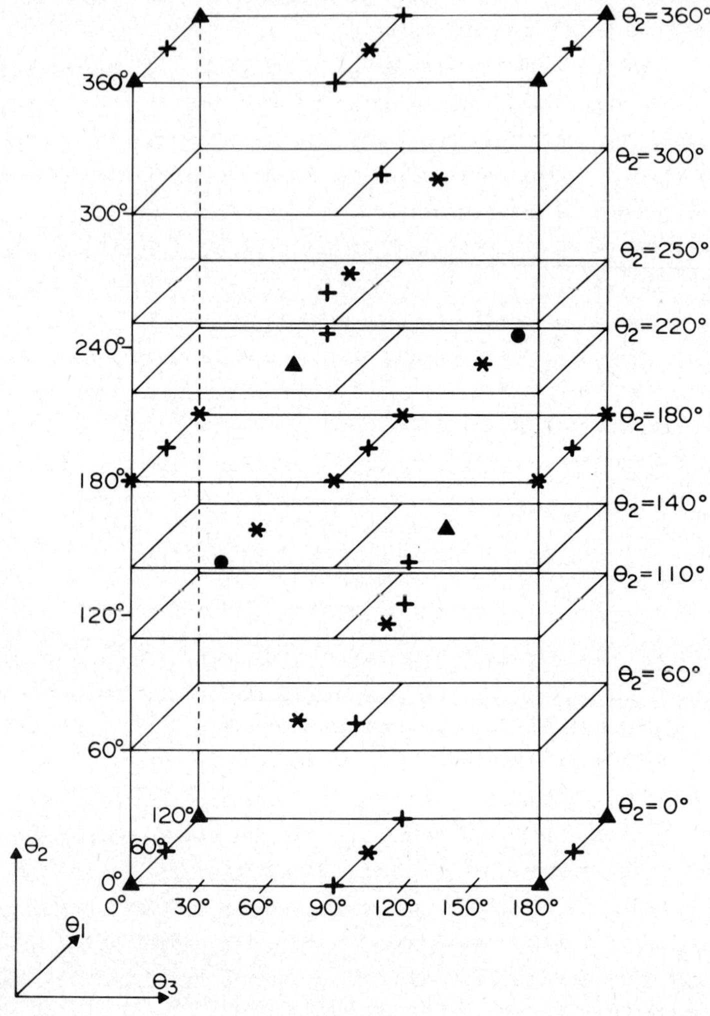

This figure taken from ref. [127] with permission

is 2.0 kcal/mol lower in energy than the optimized point [126] but still lies above M01 and M02. Thus among the three species, M01, the more favoured product of the $O(^3P)$ + propylene reaction [122,128], is lowest in energy. It is now predicted that M02 will lie 1.8 kcal/mol and M03 25.1 kcal/mol above M01 as a result of the direct torsional angle optimizations. That M01 is more stable than M02 provides an explanation for the predominance of propionaldehyde over acetone in the end products.

(3) The barrier for methyl rotation is reduced from ca. 3 kcal/mol if the α-carbon is saturated to 0.1 kcal/mol if the α-carbon is a radical. Barriers for rotation about $\dot{C}$HR-X bonds, where X is a saturated C or O atom but not a methyl group, are in the 0.6-1.2 kcal/mol range.

(4) Despite the difficulties caused by the flatness of the surfaces and hypersurface, they were still very valuable in that they provided good initial guesses for the direct optimizations. For M03, locating several of the critical points of the analytic equation involved dozens of fruitless searches and a few detailed contour plots before a new critical point was found--such exhaustive searches are very expensive directly on the ab initio hypersurface, where most of the critical points were rapidly located by simply starting at each of the analytic equation stationary points. The analytic equations have again proved advantageous in cases, like the present ones, where stereochemical intuition is of little use. Possibly as a result of more theoretical studies, some stereochemical knowledge of excited state conformational preferences will be built up. To this end, conformational hypersurfaces for triplet state propionaldehyde and acetone have recently been determined and analyzed by the techniques described in this work [130].

4. CONCLUSION

The generally good agreement between the fitted equation and direct optimization results in terms of critical point types, locations and relative energies indicates that analytic equations can represent (ab initio) energy data with reasonable accuracy. However, it is apparent that automatic selection of the terms included in the equation is essential due to the magnitude of the problem. The stepwise regression technique utilized has proved very satisfactory and efficient for determining surface and hypersurface equations.

With the advent of direct optimization on the ab initio hypersurface via force methods, there is probably little advantage to fitting an analytic equation when stereochemical considerations may be used to locate the critical points, at least approximately.

However, there are at least three occasions when the fitted equations may be useful, the first being when contour or pseudo-three-dimensional plots of the potential energy are desired. The fit need not be highly accurate for this purpose, provided the critical points are actually located by direct optimization.

The utility of the technique increases with the size of the molecular system under study and level of calculation employed: if only a few internal coordinates are varied, as in a purely conformational study with rigid or partially relaxed rotation, a large portion of the gradient calculation is wasted on the other less-interesting internal coordinates. Provided the ratio of the smallest energy difference between adjacent critical points (e.g. the smallest barrier), to the fitting error is sufficiently large, the analytic equation results should adequately reflect the ab initio results, and there would be little need to confirm or improve them by direct optimization. The minimum suitable ratio, based on the previous sections, is near ten.

The analytic equations have proved their value for the cases where classical stereochemistry cannot readily be applied: $H_2P(OH)NH_2^+$ and the triplet methyloxirane isomers. In the former case, several new critical points were immediately revealed that might well have been missed in a direct optimization study. In all the applications except M03, there was a direct one-to-one correspondance of critical points between the fitted euqations and direct optimization, and the former provided excellent starting points for the direct searches. Even for M03, most of the directly optimized critical points were located from the analytic equation critical points.

Due to the extreme flatness af the M03 hypersurface, the exact locations of the critical points is probably not crucial. However, the fitted results might be improved by choosing a finer grid of ab initio points; this applies to all examples studied, but is especially relevant to the M03 case. The average error of the fit would also have to be reduced, if possible, by including more terms in the equation.

ACKNOWLEDGEMENTS

We thank Dr. H. Bernhard Schlegel for providing a copy of his FORCE program, and for many discussions, and Dr. G.H. Goodnight of the Statistical Analysis System Institute in Raleigh, North Carolina, for permission to adopt ideas embodied in his stepwise regression program. M.R.P. thanks the Natural Sciences and Engineering Council (NSERC) of Canada for the award of a Postgraduate Scholarship, and the Imperial Oil Company of Canada for the award of a Graduate Research Fellowship. The continuous financial support of NSERC of Canada is gratefully acknowledged. We thank Dr. R.A. Poirier for his assistance in the preparation of the manuscript.

REFERENCES

1 GAUSSIAN 70: W.J. Hehre, W.A. Lathan, R. Ditchfield, M.D. Newton and J.A. Pople, Program 236, Quantum Chemistry Program Exchange, Dept. of Chemistry, Indiana University, Bloomington, Indiana, 47405.
2 M. Dupuis, J. Rys and H.F. King, J. Chem. Phys., 65 (1976) 111.
3 Several techniques were reviewed in Modern Theoretical Chemistry, Vol 3, H.F. Schaeffer III (Ed.), Plenum Press, New York, 1977.
4 P. Pulay, Mol. Phys. 17 (1969) 197.
5 M. Born and J.R. Oppenheimer, Ann. Phys., 84 (1927) 457.
6 A. Rauk, S. Wolfe and I.G. Csizmadia, Can. J. Chem., 47 (1969) 113.
7 R.E. Kari and I.G. Csizmadia, J. Chem. Phys., 50 (1969) 1443.

8 S. Wolfe, A. Rauk and I.G. Csizmadia, J. Am. Chem. Soc., 91 (1969) 1567.
9 S. Wolfe, L.M. Tel and I.G. Csizmadia, Can. J. Chem., 51 (1973) 2423.
10 P.G. Mezey, A.J. Kresge and I.G. Csizmadia, Can. J. Chem., 54 (1976) 2526.
11 W. Gerhartz, R.D. Poshusta and J. Michl, J. Am. Chem. Soc., 98 (1976) 6427.
12 C.W. Bauschlicher, Jr., H.F. Schaefer III and C.F. Bender, J. Am. Chem. Soc., 98 (1976) 1653.
13 R.D. Levine and R.B. Bernstein, "Molecular Reaction Dynamics", Oxford University Press, New York, 1974.
14 M.H. Lien, A.C. Hopkinson, M.R. Peterson, K. Yates and I.G. Csizmadia, Prog. Theoret. Org. Chem., 2 (1977) 162.
15 P. Coppens in International Review of Science, Physical Chemistry Series Two, Vol 11, J.M. Robertson (Ed.), Butterworths, London, 1975, p. 21.
16 P. Schuster, H. Lischka and A. Beyer, Prog. Theoret. Org. Chem., 2 (1977) 89.
17 S.C. Nyburg and W. Wong-Ng, Proc. R. Soc. Lond. A, 367 (1979) 29.
18 D.C. Finster, W.C. Hutton and R.N. Grimes, J. Am. Chem. Soc., 102 (1980) 400.
19 I.G. Csizmadia, J.C. Polanyi, A.C. Roach and W.H. Wong, Can. J. Chem., 47 (1969) 4097.
20 W. Kutzelnigg, V. Staemmler and C. Hoheisel, Chem. Phys., 1 (1973) 27.
21 C.F. Giese and W.R. Gentry, Phys. Rev. A, 10 (1974) 2156.
22 N. Sathyamurthy and L.M. Raff, J. Chem. Phys., 66 (1977) 2191.
23 G.D. Billing and A. Hunding, Chem. Phys. Lett., 44 (1976) 30.
24 M.H. Alexander and A.E. DePristo, J. Chem. Phys., 65 (1976) 5009.
25 D.G. Truhlar and C.J. Horowitz, J. Chem. Phys., 68 (1978) 2466.
26 J.S. Wright and S.K. Gray, J. Chem. Phys., 69 (1978) 67.
27 R.L. Vance and G.A. Gallup, J. Chem. Phys., 69 (1978) 736.
28 R. Schinke and W.A. Lester, Jr., J. Chem. Phys., 70 (1979) 4839.
29 N. Agmon and R.D. Levine, J. Chem. Phys., 71 (1979) 3034.
30 W.C. Ermler and C.W. Kern, J. Chem. Phys., 55 (1971) 4851.
31 T.H. Dunning, Jr., R.M. Pitzer and S. Aung, J. Chem. Phys., 57 (1972) 5044.
32 K.S. Sorbie and J.N. Murrell, Mol. Phys., 29 (1975) 1387.
33 P. Hennig, W.P. Kraemer and G.H.F. Diercksen, Theoret. Chim. Acta, 47 (1978) 233.
34 J. Almlof, Chem. Phys. Lett., 17 (1972) 49.
35 W.L. Hase, G. Mrowka, R.J. Brudzynski and C.S. Sloane, J. Chem. Phys., 69 (1978) 3548.
36 F. Bernardi, I.G. Csizmadia, H.B. Schlegel and S. Wolfe, Can. J. Chem., 53 (1975) 1144.
37 F. Bernardi, I.G. Csizmadia, H.B. Schlegel, M. Tiecco, M.-H. Whangbo and S. Wolfe, Gazz. Chim. Ital., 104 (1974) 1101.
38 S. Wolfe, H.B. Schlegel, I.G. Csizmadia and F. Bernardi, J. Am. Chem. Soc., 97 (1975) 2020.
39 S.C. Nyburg, G. Theodorakopoulos and I.G. Csizmadia, Theoret. Chim. Acta, 45 (1977) 21.
40 V. Lucchini and G. Modena, Prog. Theoret. Org. Chem., 2 (1977) 268.
41 A. Kapur, R.P. Steer and P.G. Mezey, J. Chem. Phys., 69 (1978) 968.
42 P.G. Mezey, Prog. Theoret. Org. Chem. 2 (1977) 127.
43 J.R. Hoyland, J. Chem. Phys., 49 (1968) 1908.
44 L. Radom and J.A. Pople, J. Am. Chem. Soc., 92 (1970) 4786.
45 L.A. Curtiss, S.R. Langhoff and G.D. Carney, J. Chem. Phys., 71 (1979) 5016.
46 J.N. Murrell, Structure and Bonding, 32 (1977) 93.
47 a) J.A. Pople and M. Gordon, J. Am. Chem. Soc., 89 (1967) 4253;
b) W.J. Hehre, R. Ditchfield, R.F. Stewart and J.A. Pople, J. Chem. Phys., 52 (1970) 2769;
c) L. Radom, W.J. Hehre and J.A. Pople, J. Am. Chem. Soc., 93 (1971) 289;
d) J.D. Dill, P.v.R. Schleyer and J.A. Pople, J. Am. Chem. Soc., 97 (1975) 3402;
e) J.D. Dill, P.v.R. Schleyer, J.S. Binkley and J.A. Pople, J. Am. Chem. Soc., 99 (1977) 6159.

48 M.R. Peterson and I.G. Csizmadia, unpublished results.
49 M.J.S. Dewar and G.P. Ford, J. Am. Chem. Soc., 101 (1979) 5558.
50 T.A. Halgren, D.A. Kleier, J.H. Hall, Jr., L.D. Brown and W.N. Lipscomb, J. Am. Chem. Soc., 100 (1978) 6595.
51 P.W. Payne and L.C. Allen, in Modern Theoretical Chemistry, Vol 4, H.F. Schaefer III (Ed.), Plenum Press, New York, 1977, p. 29.
52 a) W.J. Hehre, R.F. Stewart and J.A. Pople, J. Chem. Phys., 51 (1969) 2567;
b) W.J. Hehre, R. Ditchfield, R.F. Stewart and J.A. Pople, J. Chem. Phys., 52 (1970) 2769.
53 J.A. Pople, in Modern Theoretical Chemistry, Vol 4, H.F. Schaefer III (Ed.), Plenum Press, New York, 1977, p. 1.
54 A. Veillard, in "Internal Rotation in Molecules", W.J. Orville-Thomas (Ed.), J. Wiley and Sons, New York, 1974, p. 385.
55 C.C.J. Roothaan, Rev. Mod. Phys., 23 (1951) 69.
56 I.G. Csizmadia, Prog. Theoret. Org. Chem., 1 (1976) 77-305.
57 R. McWeeny and B.T. Sutcliffe, "Methods of Molecular Quantum Mechanics", Academic Press, New York, 1969, Ch. 5.
58 J.A. Pople and R.K. Nesbet, J. Chem. Phys., 22 (1954) 571.
59 a) C.C.J. Roothaan, Rev. Mod. Phys., 32 (1960) 179;
b) W.J. Hunt, T.H. Dunning and W.A. Goddard III, Chem. Phys. Lett., 3 (1969) 606;
c) J.S. Binkley, J.A. Pople and P.A. Dobosh, Mol. Phys., 28 (1974) 1423;
d) F.W. Bobrowicz and W.A. Goddard III, in Modern Theoretical Chemistry, Vol 3, H.F. Schaefer III (Ed.), Plenum Press, New York, 1977, p. 79.
60 J.A. Pople and W.J. Hehre, J. Comp. Phys., 27 (1978) 161.
61 N.R. Draper and H. Smith, "Applied Regression Analysis", J. Wiley and Sons, New York, 1966.
62 R.J. Abraham and R. Stolevik, Chem. Phys. Lett., 58 (1978) 622.
63 J.H. Goodnight, Procedure STEPWISE, in A.J. Barr, J.H. Goodnight, J.P. Sall and J.T. Helwig, SAS 76, Statistical Analysis System Institute Inc., P.O. Box 10066, Raleigh, N.C. 27605.
64 G.M. Furnival and R.W. Wilson, Jr., Technometrics, 16 (1974) 499.
65 J.N. Murrell and K.J. Laidler, Tran. Far. Soc., 64 (1968) 371.
66 J.W. McIver, Jr., Accts. Chem. Res., 7 (1974) 72.
67 R.E. Stanton and J.W. McIver, Jr., J. Am. Chem. Soc., 97 (1975) 3632.
68 E.B. Wilson, Jr., J.C. Decius and P.C. Cross, "Molecular Vibrations", McGraw-Hill, New York, 1955.
69 D.F. McIntosh and M.R. Peterson, "General Vibrational Analysis Program", Program 342, Quantum Chemistry Program Exchange, Dept. of Chemistry, Indiana University, Bloomington, Indiana 47405.
70 K. Ishida, K. Morokuma and A. Komornicki, J. Chem. Phys., 66 (1977) 2153.
71 J.W. McIver, Jr. and A. Komornicki, J. Am. Chem. Soc., 94 (1972) 2625.
72 T.A. Halgren and W.N. Lipscomb, Chem. Phys. Letts., 49 (1977) 225.
73 D. Poppinger, Chem. Phys. Letts., 35 (1975) 550.
74 A. Komornicki, K. Ishida, K. Morokuma, R. Ditchfield and M. Conrad, Chem. Phys. Lett., 45 (1977) 595.
75 T.A. Halgren, I.M. Pepperberg and W.N. Lipscomb, J. Am. Chem. Soc., 97 (1975) 1248.
76 M.R. Peterson and I.G. Csizmadia, Prog. Theoret. Org. Chem., 2 (1977) 117.
77 P.G. Mezey, M.R. Peterson and I.G. Csizmadia, Can. J. Chem., 55 (1977) 2941.
78 K. Muller and L.D. Brown, Theoret. Chim. Acta, 53 (1979) 75.
79 H.B. Schlegel, private communication.
80 R. Fletcher, in "Numerical Methods for Unconstrained Optimization", W. Murray (Ed.), Academic Press, London, 1972, Ch. 8.
81 M.J.D. Powell, in "Numerical Methods for Nonlinear Algebraic Equations", P. Rabinowitz (Ed.), Gordon and Breach, London, 1970, Ch. 6 and 7.
82 M.J.D. Powell, Subroutine VA05AD, AERE Harwell Subroutine Library, Harwell, Didcot, Berkshire, U.K., 1971.
83 H.B. Schlegel, S. Wolfe and F. Bernardi, J. Chem. Phys., 63,(1975) 3632.

84 P. Pulay, in Modern Theoretical Chemistry, Vol 4, H.F. Schaefer III (Ed.), Plenum Press, New York, 1977, p. 153.
85 W. Meyer and P. Pulay, Proceedings of the Second Seminar on Computational Problems in Quantum Chemistry, Strassburg, France, 1972, p. 44.
86 J.D. Goddard, N.C. Handy and H.F. Schaefer III, J. Chem. Phys., 71 (1979) 1525.
87 A. Tachibana, K. Yamashita, T. Yamabe and K. Fukui, Chem. Phys. Lett., 59 (1978) 255.
88 H.B. Schlegel, K. Mislow, F. Bernardi and A. Bottoni, Theoret. Chim. Acta, 44 (1977) 245.
89 H.L. Sellers, V.J. Klimkowski and L. Schafer, Chem. Phys. Lett., 58 (1978) 541.
90 P. Pulay, G. Fogarasi, F.P. Pang and J.E. Boggs, J. Am. Chem. Soc., 101 (1979) 2550.
91 H.B. Schlegel, private communication. The program was described in ref. 61 and in H.B. Schlegel, Ph.D. Thesis, Queen's University, Kingston, 1975.
92 D.F. McIntosh, K.H. Michaelian and M.R. Peterson, Can. J. Chem., 56 (1977) 1289.
93 M.R. Peterson and R.A. Poirier, MONSTERGAUSS, a program for ab initio calculations, unpublished.
94 GAUSSIAN 76: J.S. Binkley, R.A. Whitehead, P.C. Hariharan, R. Seeger, J.A. Pople, W.J. Hehre and M.D. Newton, Program 368, Quantum Chemistry Program Exchange, Dept. of Chemistry, Indiana University, Bloomington, Indiana 47405.
95 M.R. Peterson and I.G. Csizmadia, J. Am. Chem. Soc., 100 (1978) 6911.
96 J.B. Hendrickson, J. Am. Chem. Soc., 89 (1967) 7036.
97 A.W. Burgess, L.L. Shipman, R.A. Nemenoff and H.A. Scheraga, J. Am. Chem. Soc., 98 (1976) 23.
98 S. Fitzwater and L.S. Bartell, J. Am. Chem. Soc., 98 (1976) 5107.
99 L.S. Bartell, private communication. The force field used was MUB', described in ref. 98.
100 K. Kuchitsu, Bull. Chem. Soc. Jpn., 32 (1959) 748.
101 E.J. Jacob, H.B. Thompson and L.S. Bartell, J. Chem. Phys., 47 (1967) 3736.
102 S. Scheiner, J. Am. Chem. Soc., 102 (1980) 3723.
103 L.M. Tel, S. Wolfe and I.G. Csizmadia, J. Chem. Phys., 59 (1973) 4047.
104 J. Koller, D. Hadzi and A. Azman, J. Mol. Struct., 17 (1973) 157.
105 R.G. Lerner, B.P. Bailey and J.P. Friend, J. Chem. Phys., 26 (1957) 680.
106 T. Miyazawa and K.S. Pitzer, J. Chem. Phys., 30 (1959) 1076.
107 D.L. Bernitt, K.O. Hartman and I.C. Hisatsune, J. Chem. Phys., 42 (1965) 3553.
108 D.R. Lide, Jr., Trans. Am. Crystallogr. Assoc., 2 (1966) 106.
109 E. Bjarnov and W.H. Hocking, Z. Naturforsch., 33a (1978) 610.
110 P. Ros, J. Chem. Phys., 49 (1968) 4802.
111 A.C. Hopkinson, K. Yates and I.G. Csizmadia, J. Chem. Phys., 52 (1970) 1784.
112 M.E. Schwartz, E.F. Hayes and S. Rothenberg, J. Chem. Phys., 52 (1970) 2011.
113 L. Radom, W.A. Lathan, W.J. Hehre and J.A. Pople, Aust. J. Chem., 25 (1972) 1601.
114 M. Perricaudet and A. Pullman, Int. J. Pept. Protein Res., 5 (1973) 99.
115 G.H. Kwei and R.F. Curl, Jr., J. Chem. Phys., 32 (1960) 1592.
116 M.R. Peterson and I.G. Csizmadia, J. Am. Chem. Soc., 101 (1979) 1076.
117 J.E. Del Bene, G.T. Worth, F.T. Marchese and M.E. Conrad, Theor. Chim. Acta, 36 (1975) 195.
118 W.J. Bouma, M.A. Vincent and L. Radom, Int. J. Quantum Chem., 14 (1978) 767.
119 T.A. Modro, W.G. Liauw, M.R. Peterson and I.G. Csizmadia, J. Chem. Soc. Perkin II (1979) 1432.
120 W. Wong-Ng, S.C. Nyburg and T.A. Modro, Chem. Comm. (1980) 195.
121 R. J. Cvetanović, Can. J. Chem., 36 (1958) 623.
122 H. Hirokame and R.J. Cvetanović, J. Am. Chem. Soc., 96 (1974) 3738.
123 C.M. Neeley, Diss. Abstr. Int. B, 30 (1970) 3064.

124 G.R. DeMaré and O.P. Strausz in "Proceedings of a Symposium on Chemical Kinetics in honour of Academician N.N. Semenov on the occasion of his 80th birthday," Moscow, April 12-14, 1976, to be published.
125 G.R. DeMaré, J. Photochem., 7 (1977) 101.
126 O.P. Strausz, R.K. Gosavi, G.R. DeMaré, M.R. Peterson and I.G. Csizmadia, Chem. Phys. Letts., 70 (1980) 31.
127 G.R. DeMaré, M.R. Peterson, I.G. Csizmadia and O.P. Strausz, J. Comp. Chem., 1 (1980) 141.
128 R.J. Cvetanović, J. Chem. Phys., 25 (1956) 376.
129 A. Albini and D.R. Arnold, Can. J. Chem., 56 (1978) 2985.
130 M.R. Peterson, G.R. DeMaré, I.G. Csizmadia and O.P. Strausz, J. Mol. Struct. (Theochem), in press.

Molecular Structure and Conformation: Recent Advances,
I.G. Csizmadia (Ed.), *Progress in Theoretical Organic Chemistry*, Volume 3

QUANTUM CHEMICAL STUDIES ON THE MECHANISM OF ENZYME ACTION

G. NÁRAY-SZABÓ and T. BLEHA

1 INTRODUCTION

The wide application of sophisticated experimental techniques, like X-ray diffraction and different branches of spectroscopy, in the study of enzymatic processes lead to a deeper understanding of finer details. An especially important result is the precise description of active-site regions where biochemical reactions actually take place. Detailed information on three-dimensional structures of crystalline enzymes including their active sites can be gained from the Protein Data Bank [1] which provides atomic coordinates for several highly complicated biological macromolecules. Processing these data with effective computer programs [2] allows the scientist to get a pictorial and precise representation of the active-site geometry. This facility gave enormous impetus on research work in the field in the last decade.

Developing from a completely different area of research quantum chemistry became an essential tool in understanding chemical reaction mechanisms [3]. Since high-speed computers with large storage capacity are presently available at a reasonable cost, semiempirical, or even minimal basis set *ab initio*, molecular orbital calculations became routine work for systems containing not more than 20-30 first-row atoms. Simultaneously, efficient methods were developed to take environmental effects, like solvation or intraproteic fields, into consideration when discussing elementary processes. As a counterpart, empirical atom-pair potential calculations were also refined offering an alternative to molecular orbital studies.

Simultaneous progress, both in experimental and theoretical, techniques renders possible a thorough understanding of enzymatic processes. Experimental results, be they most precise, are often problematic to interpret since pieces of informations, corresponding to different regions (active site, protein environment, hydration shell, etc.), get mixed. Theory may help in clearing details and it gives a sound basis on which experimental results can be discussed.

In the past few years several scientists performed molecular orbital or atom-pair potential calculations to study enzymatic processes in detail. Most of these studies model the active site by an adequate set of small molecules. Other parts of the reacting system are neglected completely or they are considered as a continuous or structured medium perturbing the active site. The number of papers, where the three-dimensional structure of the enzyme is taken explicitely into consideration, is continuously growing therefore it seems to be timely to summarize work on this field. In this paper we start with computational methods afterwards applications to serine proteinases, papain, lysozyme, carboxypeptidase and other enzymes are discussed. A separate section is devoted to possible theoretical explanations of the enormous rate enhancement found in enzymatic processes relative is the analogous chemical reactions.

2 COMPUTATIONAL METHODS

Theoretical treatment of a chemical phenomenon needs an adequate model on which calculations can be done. To construct this model a considerable amount of experimental information should be available. This requirement is mandatory for all but the simplest systems.

Since enzymatic processes are especially complicated, to construct an appropriate model, much more experimental data are needed than for a common chemical reaction. Consequently, computational results have to be discussed with great care while making clear their relation to experimental facts. Theory should, above all, explain phenomena. In some cases it is also able to complete observations where current laboratory techniques fail. This "theoretical observation" means a kind of interpolation between diverse experimental findings. The final justification should come, however, from experiments which may be designed on the basis of theoretical predictions.

In the past decade several simplified models were proposed which, more or less, reflect essential features of enzymatic processes. In the following we discuss these from the methodological point of view.

2.1 Models for the enzymatic reaction

Complex rate processes, like an enzymatic reaction, are usually interpreted in terms of the transition-state theory [4]. The typical change of the standard Gibbs free energy, ΔG^{o}, along the reaction path is presented in Fig.1. An important feature of the enzymatic process is the formation of the enzyme-substrate (or Michaelis)complex in the first stage of the reaction. This is stabilized by weak non-covalent interactions (hydrogen bonds, van der Waals forces, etc.) between atoms of the enzyme and the substrate. Its formation is indicated by a saturation kinetics which is characteristic for

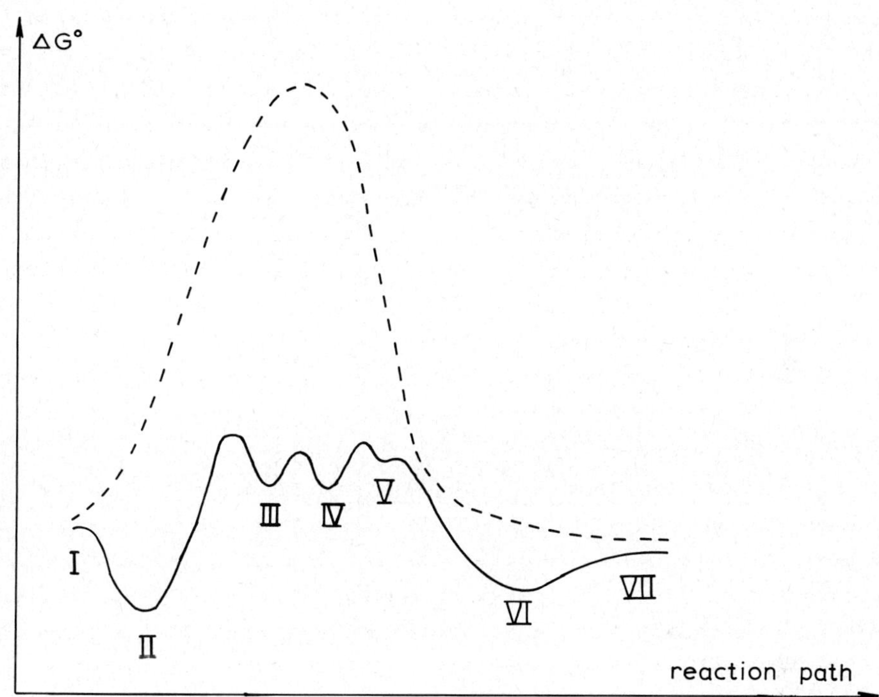

Fig. 1. Typical change of the standard Gibbs free energy along the reaction path. Full line: schematic reaction path for serine proteinases (see Sec. 3.1.); dashed line: analogous reaction in the gas phase.

enzymatic reactions. In most cases the final reaction products are not formed from the Michaelis complex in a single elementary step. For example, cleavage of peptide bonds by serine proteinases occurs via several, relatively stable, intermediates (see Sec. 3.1. and ref. 5). This results in a further acceleration of the rate process. The existence of each intermediate is more or less indicated by kinetic measurements, X-ray diffraction and other structural studies on enzymatic systems.

Enzyme proteins are macromolecules containing typically 200 or more amino-acid residues i.e. about 3,500 atoms and 15,000 electrons. In most cases only relatively few atoms are directly involved in the catalytic process, other parts of the enzyme seem to have a secondary importance. These atoms, forming protein backbone and side chains, compose the active site of the enzyme [6]. This is illustrated in Fig. 2 where the schematic view of α-chymotrypsin, a

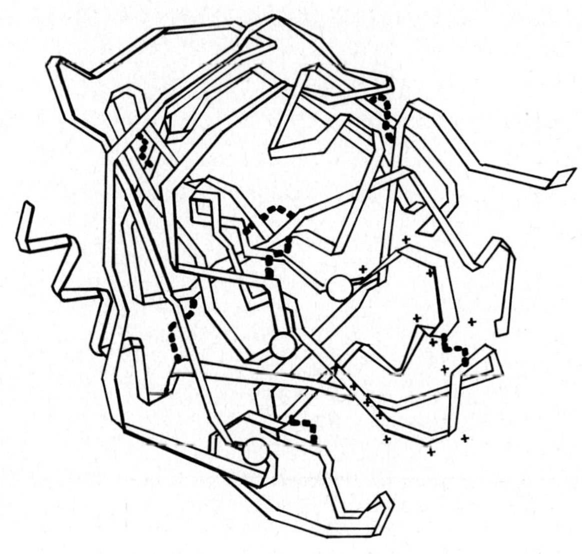

Fig. 2. Schematic view of α-chymotrypsin. Residues of the catalytic triad are indicated by open circles. Crosses show other residues of the active site: Ser-189, Ser-190, Cys-191, Met-192, Gly-193, Asp-194, Val-213, Ser-214, Trp-215, Gly-216, Ser-217, Ser-218, Thr-219, Cys-220. Disulfide bridges are marked with dotted lines. After D.M. Blow, Acc. Chem. Res. 9 (1976) 785.

serine proteinase (see Sec. 3.1), is depicted. Three-dimensional structure of the protein backbone is stabilized by disulfide bridges and hydrogen bonds. Due to the folding of the polypeptide chain side chains, which are separated in the amino-acid sequence of the protein, may get close to each other and they act as a whole. For example, in α-chymotrypsin the hydroxyl of Ser-195, the imidazole of His-57 and the carboxylate of Asp-102 are grouped into a catalytic triad. This is the central machinery of the active site. Other side chains are involved in the stabilization of the Michaelis-complex or of other intermedicate states. The active site is extensively hydrated and positive or negative counter ions may be located in its vicinity, also.

In application of computational methods to the description of enzymatic activity the proper design of active-site models is of primary importance. First, the three-dimensional structure is needed which is determined by X-ray diffraction. Further, since the catalytic region may contain several dozens of atoms, a reasonable truncation has to be done. Most advanced current methods [7-10] classify the atoms of the active site into two separate groups. The first includes the central machinery which is directly involved in the catalytic process by participating in bond formation or cleavage. In our example, α-chymotrypsin, this is the catalytic triad. It is treated by an, <u>ab initio</u> or semiempirical, molecular orbital method replacing Ser by ethanol or methanol , His by imidazole and Asp by acetate or formate [11]. Other parts of the catalytic region and of the enzyme are considered as the environment which is treated by molecular mechanics [7,8,10,12] or by some quantum chemical method [9,13-15]. Early studies are focused exclusively on the central machinery while the effect of the secondary catalytic groups, which are responsible for positioning and orienting the substrate during reaction, enter only indirectly.

Hydration can be mimicked either by a structured [15] or by a continuum model [10,12] at various levels of sophistication (for a review see [16]).Counter ions are best represented by point charges using simple electrostatics [15]. The central problem in modelling hydration and counter-ion effects is to establish the proper number and position of water molecules and counter ions around the active site. Experimental methods yield only limited information therefore hydration shells should be constructed on a theoretical basis. The simplest way is to use chemical evidence: atoms bearing lone pairs (N, O or S) are combined with the O-H bond of water while acidic X-H bonds (X = N, O, S or eventually C) interact with the oxygen lone pair. Though the positions of strongly bound counter ions may be determined by X-ray diffraction (see <u>e.g.</u>[5])it is unclear whether these are also present in the biological system where enzymes, in fact, exert their activity.

Enzymatic processes are often compared with simple analogous chemical reactions in solution to clear the role of the protein environment. Molecular orbital calculations on a simple model of the central machinery refer to the gas phase therefore erroneous conclusions may be drawn from them. The best way to avoid misinterpretations is to mimick environmental effects both for the enzymatic process and for the simple reference reaction. While doing so the consistency of both models has to be ensured i.e. the same level of sophistication has to be followed. For example, if only the first hydration shell is considered in the enzymatic reaction, this has to be done for the analogous chemical process, too.

2.2 Molecular orbital treatment of the central machinery

In the previous section we called "central machinery" of an enzyme the ensemble of those side chain and backbone atoms which are involved in bond formation or breaking during reaction. This ensemble may be modelled by a combination of simple molecules, like water, formic acid or imidazole, but even the smallest model contains typically 15-20 atoms or 60-80 electrons. Therefore early and several present studies on these systems are based on semiempirical molecular orbital calculations. Recently *ab initio* computations, though

TABLE I

Examples for molecular orbital calculations on protein central machinery models.

Method	Basis set	Protein	Model[a]	Ref.
ab initio	double zeta	papain	MeSH...Im...HCHO	[17]
ab initio	4-31G [18]	serine proteinases	MeOH...Im...HCOOH	[19]
ab initio	STO-4G + 3d and 4p on Zn	carboxy-peptidase	$Zn(OH)(NH_3)_2^+$	[20]
ab initio	STO-3G [21]	glyoxylase I	MeCOCH(OH)SMe	[22]
PRDDO [23]	minimal	serine proteinases	H_2NCHO...MeOH...Im ...HCOOH	[24]
Hartree-Fock-Slater [25]	—	high potential iron protein	$Fe_4S_4(SH)_4$	[26]
MINDO/3 [27]	minimal	cytochrome P450	CH_4 or C_2H_4 plus O	[28]
CNDO/2 [29]	minimal	α-chymotrypsin	MeOH...5-Me-Im ...MeCOOH	[30]
EHT [31]	minimal	cytochrome P450	ferrous-ferric porphyrine...MeSH	[32]

[a] Me and Im stand for methyl and imidazole.

costly, are also available. We have summarized some typical examples for application of both *ab initio* and semiempirical methods in Table I. To give more information on the field not only enzymes but other proteins are also included.

Since enzymatic processes are, in general, described within the frame of transition-state theory [4], calculations are done to gain some information on characteristic points (minima or cols) of the energy hypersurface. This involves at least a partial optimization of the geometry of the reacting species. Geometry optimization is best made, using some version of the force method (for a review see [33]). This leads, however, to a serious multiplication of the numerical work therefore selection of the appropriate molecular orbital method and of the geometry parameters to be optimized is an important problem. Certainly, the decision has to be made individually for each system under study. However,general directives can be gained from comparative studies on simple chemical reactions. In the following we discuss some of them.

First of all, we would emphasize that model and method have to correspond to about the same level of sophistication. For example, if environmental effects are not considered no far-reaching conclusions for the enzymatic process as a whole can be drawn even if the most sophisticated *ab initio* calculation is done on the gas-phase model of the central machinery. A central problem is whether the estimate of a given physical property, calculated by one or another molecular orbital method, is accurate enough to substitute or explain experimental facts. General trends may be extracted from recent reviews [34-37]. The most important molecular properties which are needed in theoretical description of an enzymatic process are (1) differences between energies of intermediate states (*i.e.* local minima of the hypersurface); (2) activation energies determining rates of elementary reaction steps; (3) charge distribution for the estimation of environmental effects and reactivity and (4) molecular geometry.

Determination of relative energies of reaction intermediates and products is a central problem of computational quantum chemistry. For a recent review see [34]. If atomic environment and number of electron pairs is conserved during reaction (*e.g.* protonation) correlation energy effects are small. Double zeta plus polarization type basis sets reproduce the Hartree-Fock limit within 5-10 kJ/mol while the error with a 4-31G basis set lies within $\pm$80 kJ/mol [34].

For example, experimental protonation energy of the formate anion is -1430 kJ/mol [38] while -1488 or -1499 kJ/mol is obtained with a 4-31G basis set with slightly different geometries [24,39]. However, this relatively good agreement is not achieved for energy differences of proton transfer reactions which are of primary importance in enzymatic processes. Reliable results are obtained only with a double zeta plus polarization type basis set. Unfortunately, such calculations are too costly, if possible at all, for models of the central machinery containing at least 50-60 electrons. If a 4-31G basis set is used the interaction energies are wrongly estimated as a consequence of the incorrect charge distribution [34]. To obtain qualitatively correct proton transfer energies reasonable corrections, based on experimental data, can be made [13, 24, 39]. In case of the protonated imidazole-formate anion diad, a part of the central machinery in serine proteinases, protonation energies were calculated by the _ab initio_ molecular orbital method with a 4-31G basis set while the interaction energy between imidazole and formic acid was estimated from STO-3G minimal basis set results corrected with an empirical factor [39].

If the number or the neighbourhood of electron pairs changes during a chemical process (_e.g._ $HCN + 3H_2 \rightarrow CH_4 + NH_3$ or $H_2O \rightarrow OH + + H$) correlation effects may play a decisive role. Since models of the central machinery are, in general, too large, _ab initio_ calculation of their correlation energy is impracticable at present. For such reactions some version of the MINDO methods can be used (MINDO/3 [27], MNDO [40] or MNDOC [41]). Though the verification of these approximations is unclear and their adequacy is often questioned they describe a considerable number of chemical processes with success and they are applied to enzymes, too [28, 42].

Despite the deficiencies in their predicting power _ab initio_ minimal basis set and semiempirical (CNDO/2 or Extended Hückel) molecular orbital calculations were also performed on models of the central machinery [11, 30, 32, 43-56]. These studies should be, and are often in fact, of comparative nature with the closest possible reference to experiment. If computational results are discussed with sufficient care essential and reliable informations can be gained. One example is the work of Kitayama and Fukutome [30] who discuss factors which determine the basicity of histidine in serine proteinases. They state that the pK_a of His-57 in α-chymotrypsin is raised by the basic assistance of the neighbouring

Asp-102 and lowered by the effect of the distorted hydrogen bond of the ligand water to His-57. These effects cancel and the apparent pK_a is about 7 being equal to that measured in water.

Quantitative estimation of activation energies need consideration of correlation effects and a large basis set. This is practicable only for models with not more than 15-20 electrons. In some cases the semiempirical MINDO methods can be applied for large systems with success [27, 40, 41]. Due to the extreme complexity of enzymatic processes, elementary reaction steps (i.e. transfer of the system from one minimum of the potential surface via a col, to the other one) can be studied only exceptionally. As a rule, experimental reaction rates refer to the entirety of several elementary steps therefore comparison with theoretical activation energies is highly problematic. Therefore, we feel, that, at the present state of art, calculation of the activation energies is not a central problem in the theoretical study of enzymatic reactions.

Charge distributions can be best obtained from ab initio STO-3G basis set [34] or semiempirical CNDO/2 calculation [29]. The 4-31G basis, though larger than the minimal STO-3G one, yields incorrect results. This has to be borne in mind when such calculations are performed. Molecular geometries are satisfactorily calculated with the ab initio molecular orbital method using a STO-3G basis set [57] though distances between hydrogen-bonded atoms are underestimated [34]. The CNDO/2 method yields also relatively good results even for heterocyclic molecules [58]. Here again too small hydrogen bond distances are obtained.

As it is seen from the above overview different properties have to be calculated with different methods to achieve about the same reliability. Further, to spare computer time, semiempirical methods have to be applied, if possible. To reach the highest efficiency in the theoretical study of enzyme action it is advisable to combine several methods while comparing and discussing results carefully.

2.3 Molecular mechanics calculations

Molecular mechanics (or empirical force field) methods for reviews see [59-61]) are based on the observation that bond properties are more or less transferable within certain classes of molecules. Regions of potential surfaces around their local minima can be approximated as a sum of simple analytical expressions. The constants of these expressions are fitted to reproduce a great number of experimental quantities, like equilibrium bond lengths and angles, stretching and bending vibrational frequencies, rotational barriers and others. These parameters are than used to calculate properties of molecules differing from those included into the original fitting set. This procedure is a kind of interpolation and, if applied carefully, may yield very accurate results at a low cost even for large systems involved is enzymatic processes.

It is possible to write the total potential energy of a molecule approximately as a sum of bonded interactions, intermediate, or 1-4, interactions between atoms separated by two atoms and nonbonded interactions between all other atoms [61]:

$$E = E_b + E_i + E_{nb} \tag{1}$$

Bonded interactions are approximated by a sum of stretching, bending and cross terms:

$$E_b = \frac{1}{2}\Sigma k_r (r - r_o)^2 + \frac{1}{2}\Sigma k_\vartheta(\vartheta - \vartheta_o)^2 + \text{cross terms} \tag{2}$$

r, r_o, ϑ and ϑ_o denote actual and equilibrium bond lengths and angles, respectively, k_r and k_ϑ stand for quadratic force constants. Summation runs over all stretching and bending degrees of freedom. Cross terms, if considered, include stretch-stretch, bend-bend or stretch-bend contributions.

1-4 interactions are described by torsional potentials which can be expressed as a periodic function of the torsional angle, φ :

$$E_i = \frac{1}{2}\Sigma V_\varphi^n (1 - \cos n\varphi) \tag{3}$$

V_φ^n is the n-fold rotational barrier (*e.g.* n = 3 for ethane) and summation runs over all torsional degrees of freedom.

The nonbonded term of Eq. (1) can be described as an atom-atom interaction potential neglecting three-body, long-range interactions:

$$E_{nb} = \epsilon^{-1}\Sigma q_i q_j r_{ij}^{-1} + \Sigma V_{Hb}(d_{ij}) + \Sigma C_{ij}[(r_{ij}^*)^n - \frac{n}{6}(r_{ij}^*)^6] \tag{4}$$

The first term corresponds to electrostatic interactions between atoms i and j bearing net charges of q_i and q_j, ϵ denotes the dielectric constant. Atomic net charges are taken either from quantum chemical calculations on model molecules [62-64] or from fitting to experimental properties of hydrocarbon, amide or amino-acid crystals [61]. The ε dielectric constant is chosen somewhat arbitrarily its value may range between 1 and 4 from author to author [7, 8, 65-67]. The second term is the hydrogen-bonding potential which may have different forms [68-69]. Finally, van der Waals interactions are accounted for in the last term. $r^*_{ij} = r^o_{ij}/r_{ij}$, C_{ij} and r^o_{ij} are adjustable parameters differing for various types of atom pairs. n is equal to 12 in most cases but it can be chosen as an adjustable parameter, too. Sometimes the repulsive part of the van der Waals potential, $(r^*_{ij})^n$, is represented by an exponential term, $A\exp(-\alpha r_{ij})$.

There exist several sets of well-balanced parameters suitable for calculations of the structure of enzymes and of enzyme-substrate

TABLE II
Molecular mechanics calculations for enzymatic systems.

Enzyme	Scope	Ref.
α-chymotrypsin	substrate specificity	[70]
papain	effects of environment on the course of catalytic action	[7, 8]
lysozyme	preferred conformations of inhibitors and substrates bound to the active site	[65,69, 71,72]
lysozyme	role of electrostatics and strain in stabilization of reaction intermediate	[10]
lysozyme	hydration	[73,74]

complexes [66, 68, 70, 75]. An overview on applications is given in Table II. Molecular mechanics studies can help in answering such questions as fitting the substrate or inhibitor into the enzyme cleft or conformational changes in substrate or in enzyme side chains and backbone to ensure maximum interaction. The conformational strain of the substrate or enzyme, due to complexation, can be quantified and data on substrate binding energies can be obtained. Molecular mechanics methods can be combined with quantum chemical calculations to consider effects of the protein environment on enzymatic processes (see Sec. 2.4.).

The main advantage of molecular mechanics calculations lies in their simplicity and efficiency. Due to the simple analytical expressions complete geometry optimization of the system under study is possible [61]. Further, if the parameter set is carefully fitted, an impressive accuracy of 2 kJ/mol can be achieved [70]. On the other hand, no single version became widely popular up to now, each research group uses its own. Therefore results, coming from different laboratories, are difficult to compare and the untrained reader has, in general, no firm basis to judge the reliability of the results.

2.4 Simulation of the protein environment

It is known that chemical reactions are more or less influenced by the medium in which they take place. Several theoretical methods have been proposed to consider environmental effects explicitely (for a review see [16]). These effects are more pronounced in water since structured hydration shells, creating anisotropic electrostatic field, are formed around the reacting molecules which may alter reaction or activation enthalpies considerably (see *e.g.* [76-83]). Electrostatic field anistropy, due to protein environment around the central machinery may be even larger than that of hydration shells.

The importance of protein field originating from dipoles in the vicinity of polar enzyme-substrate complexes was stressed intuititively by Stern as early as in 1935 [84]. Explicite statements on the role of electrostatics in enzymatic processes are given by Yomosa [85]. Intraproteic fields were first estimated quantitatively by Johannin and Kellersohn [86]. They used a simple point-charge model which included all atoms of the protein peptide groups. Similar calculations were performed for the α-helix existing in many enzymes [87, 88]. Hayes and Kollman considered also side-chain atoms into their model of carboxypeptidase A and emphasized the importance of charged groups neighbouring the central machinery in the enzymatic process [89, 90]. They used point charges as obtained from *ab initio* STO-3G or 4-31G calculations. A further example is carbonic anhydrase where intraproteic field effects were shown to influence enzyme reactivity considerably [91]. In a recent work the electrostatic potential around the catalytic triad of subtilisin BPN′ was calculated [15] (see Fig. 3). Based on the idea of Scrocco and Tomasi [92] the potential is approximated as a sum of transferable bond contributions [93, 94]. According to comparative studies on small model molecules this simplified method estimates true electrostatic potentials around these molecules satisfactorily resulting in a qualitatively correct picture. A computer program has been written [95] which calculates electrostatic isopotential maps for molecules containing up to 600 first-row atoms. It can be used with considerable ease as it needs only 20,000 words storage capacity and very limited computer time.

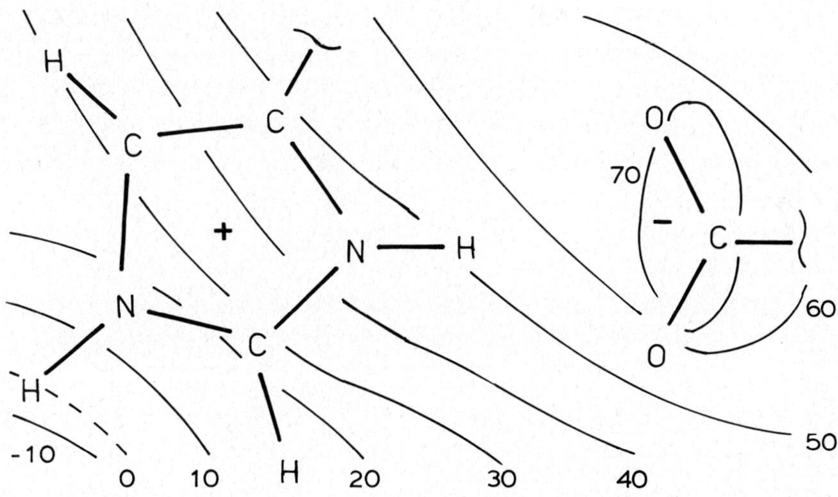

Fig. 3. Isopotential map of the intraproteic electrostatic field around the His-64...Asp-32 diad in subtilisin [96] (in kJ/mol). Note the direction of the gradient which stabilizes the imidazolium...aspartate ion pair.

More realistic description of environmental effects on enzymatic processes may be achieved through decomposition of the whole reacting system into catalytic site and environment and applying different computational methods for both parts. The total energy change of the reaction is partitioned as follows:

$$\Delta \underline{E} = \Delta \underline{E}_{\underline{c}} + \Delta \underline{E}_{\underline{c}/\underline{env}} + \Delta \underline{E}_{\underline{env}} \quad (5)$$

$\Delta \underline{E}_{\underline{c}}$ corresponds to the gas-phase model of the central machinery and it can be calculated with a semiempirical or ab initio molecular orbital method (see Sec. 2.2). $\Delta \underline{E}_{\underline{c}/\underline{env}}$ stands for the interaction between central machinery and environment, $\Delta \underline{E}_{env}$ denotes the energy change of the environment (protein core, hydration shell, counter ions, etc.). While simple gas-phase models retain only $\Delta \underline{E}_c$ in Eq. (5) further terms can also be taken into consideration at various levels of sophistication.

The interaction energy between catalytic site and environment may be approximated by the following expression [92]:

$$\Delta \underline{E}_{c/env} = \epsilon^{-1} \Sigma[\underline{V}_f(\underline{r}_m)\underline{q}_{fm} - \underline{V}_i(\underline{r}_m)\underline{q}_{im}] \tag{6}$$

Here $\underline{V}(\underline{r}_m)$ denotes the electrostatic potential of the environment at the site of the $\underline{m}$-th atomic centre bearing a net charge of $\underline{q}_m$. The subscripts $\underline{f}$ and $\underline{i}$ refer to the final and initial states. The dielectric constant ϵ is equal to unity if the first hydration shell is considered since in this case there is no dielectric medium between the water molecules and the hydrated system. For the protein backbone ϵ may range from 1.36 [10] to 4.0 [7, 8] according to different authors. Summation includes all atoms of the gas-phase model. This is cut out of the whole enzyme while $\Delta \underline{E}_c$ is calculated for a molecular system where free valences of the above model are saturated by hydrogen atoms. For example, in case of a serine side chain, the model of Eq. (6) includes $-CH_2-OH$ while molecular orbital calculations are performed on methanol. The $\underline{V}$ electrostatic potential is calculated from simple net atomic charges but it can be constructed of bond fragments [93] , too.

A more sophisticated way, followed by some authors [13, 17, 90, 97, 98], is to include an environmental electrostatic potential term into the Hamiltonian of the gas-phase model system. This approximation allows to consider the polarization of the charge distribution of the central machinery, due to the environment, explicitely. The reverse polarization of the protein core by the catalytic groups, charge transfer and exchange are not considered by this method. Hayes and Kollman [90] emphasized that not all fractional charges of the protein core atoms are equally important. All charged functional groups like NH_3^+ of lysine or COO^- of aspartate have to be included but other atoms outside a sphere of 0.5 nm radius around the central machinery can be neglected. Van Duijnen et $\underline{al}$. considered the dipole field of the α-helix, located near the catalytic triad of papain, only [17, 98]. They state that the dipole effect increases as the length of the helix gets as large as 1.0 nm. Longer helices do not produce stronger fields.

Tapia and coworkers included not only the electrostatic potential but also a reaction field term into their model Hamiltonian [9, 16]. This means that the polarization of the protein core by the cage (*i.e.* by the gas-phase model) is also considered. They claim that the polarization terms may be important (*e.g.* they range between 100-300 kJ/mol) for the proton-relay system of liver alcohol dehydrogenase [9].

Warshel and Levitt [10] considered van der Waals terms besides electrostatic effects and polarization too, when calculating $\underline{E}_{c/env}$ of Eq. (5) for the cleavage of hexasaccharides by lysozyme. These do not influence the reaction equilibrium considerably. Here again charge-charge interaction and polarization play a decisive role. For $\Delta\underline{E}_{env}$ they used the empirical method proposed by Levitt [99]. They studied strain effects in the ground state and conformation of the transition state. A similar way was followed by Clementi and coworkers for the reaction catalyzed by papain [7, 8]. At present calculation of $\Delta\underline{E}_{env}$ is possible at the molecular mechanics level, only.

We can conclude from the above short overview that several methods, relatively simple or sophisticated, are available at present to treat enzymatic reactions in detail. Though these are, in some cases [7, 8, 10] very complicated most of them can be used at a reasonable cost while ensuring sufficient accuracy to clarify most important terms which determine course, and products of a catalytic process. It is highly probable that further improvements and rationalizations will be made in the near future which may enable experienced workers to carry out large-scale calculations on any enzyme if its three-dimensional structure is available.

2.5 Limitations of theoretical calculations in elucidation of enzymatic action

The central problem in modelling enzymatic processes, just like simple chemical reactions, is to find out the sequence of elementary steps leading from the initial state to the final product. This may be very simple in some cases. For example, most proton transfers, like $H_2O...HO^+H_2 \rightarrow H_2O^+H...OH_2$, are one-step reactions which can be described by a double-well potential curve. One of the minima corresponds to the initial the other one to the final state. They are transforming into each other through a hypothetical transition "state" which corresponds to the maximum on the potential curve. Once the gas-phase model is constructed solvent effects can be modelled relatively simply [76]. Most real chemical reactions, however, follow through several elementary steps and often concurrent reaction paths exist. Furthermore, the solvent itself can act as a reagent. The situation gets extremely complicated for enzymatic processes where the number of atoms of the reacting system is very large.

Complete theoretical description of an enzymatic reaction would mean that only the amino-acid sequence of the given protein and the nature of the reacting species have to be known. In an ideal,though artificial, case secondary, tertiary and quaternary structure, geometry of the enzyme-substrate complex, hydration sites, positioning of counter ions, protonation equilibria, etc. should be obtained by minimizing the total Gibbs free energy of the system. This consists of an energy and an entropy part. Both have to be estimated reliably. Even if energy and entropy could be calculated with chemical accuracy (i.e. with a maximal error of 2 kJ/mol) one problem remains which is not solvable with mere theoretical computations. Several stable states of the whole enzyme can be located by geometry optimization. These correspond to local minima of the complete energy hypersurface. The real enzyme, which exists in nature, corresponds to only one of these. Information on the true state can be obtained only by experiment.

Maybe the most important point in the theoretical study of an enzymatic process is the construction of a realistic model. As we have pointed out above, this is partly at hand if three-dimensional structure of the enzyme is available. X-ray diffraction studies often locate positions of strongly bound water molecules around the enzyme therefore construction of the first hydration shell of the central machinery is sometimes also straightforward. However, even further hydration shells may influence the enzymatic process. For these no direct experimental data are available therefore their construction is problematic. One of the principal question is whether a given water molecule of the hydration shell is bound firmly enough to produce an anisotropic environment in the vicinity of the central machinery. If the interaction is equal to or weaker than between molecules of bulk water statistical fluctuation do not allow the formation of structured hydration shells. In this case bulk water has to be considered as a continuous dielectric medium which exerts much weaker effect on the chemical process [10]. A logical way to decide on the adequate modelling of the structured hydration shells is proposed by Pullman et *al.* [101]. These contain those water molecules which are bound to the enzyme with higher energy than to their own hydration sphere in the enzyme-free solvent.

Since experimentalists compare enzymatic processes almost exclusively to analogous, but simpler, chemical reactions which take place in aqueous medium, theoretical models have to count for these, too. Reliable comparisons can be made only if models for the enzymatic and for the simple chemical process correspond to each other. For example, if only first hydration shell was considered for the former this has to be done for the latter, too.

Once the adequate model is constructed possible errors in the applied computational procedures have to be discussed. In the following the adequacy of Eq. (5) and of the approximations for its terms are discussed. Let us first treat ΔE_c. Reliability of standard quantum chemical methods, applied to the calculation of this term, was shortly discussed in the preceding section. We amend textbooks [29, 34, 36, 37] and reviews [3, 102] for further reading. Despite spectacular developments of computational procedures in the last decade applicability of quantum chemical methods to chemical reactivity remains a delicate problem. Estimation of ΔE_c may be the weakest point of the whole theoretical study of an enzymatic process.

Even if the energy hypersurface of the reaction, or at least its local minima, are calculated with a sufficient accuracy, approximations, inherent in the transition-state theory, may influence final results. Relativistic effects and the breakdown of the Born-Oppenheimer approximation do not have an important role in the description of the gas-phase model of the central machinery. On the other hand, differences in zero-point energies of the final and initial state may affect the magnitude of the reaction enthalpy [34]. As it is known, zero-point energies are derived from vibrational frequencies of the reacting system. Normal modes of a simple molecule with covalent bonds can be estimated relatively precisely by an *ab initio* method even with a minimal basis set [33, 34]. However, central machineries of enzymes often contain a group of molecules which is held together by hydrogen bonds or by other weak intermolecular forces. It is questionable whether vibrational frequencies, corresponding to these weak bonds, are also reproduced correctly by a minimal basis set calculation.

Kinetic isotope effects, often studied for protonation steps of enzymatic reactions [103] can be understood only in terms of quantum mechanical tunnelling [104]. To estimate relative rates of reactions precise knowledge of their minimum energy path is needed. As we have pointed out in Sec. 2.2 this is possible only in exceptional cases of very simple chemical reactions but, in general, not for enzymes. Proton transfer potential curves may be estimated empirically and in this case tunnelling can be described theoretically. This has been done for the proton transfer step leading to the formation of the acyl-enzyme in α-chymotrypsin [105].

Equilibrium constants of chemical reactions are also influenced by entropy changes. These can be expressed for gas-phase reactions by summing electronic, translational, rotational and vibrational terms [34]. The calculation of the entropy change for such reactions is relatively straightforward and reliable estimates, can be given [106]. In general, electronic and translational terms do not play an important role. Rotational entropy has to be considered only if size and shape of the reacting system change essentially during reaction (e.g. association). If only covalent bonds are broken or formed the vibrational entropy term is negligible. Just the opposite holds for weak links, like hydrogen bonds, where changes in the vibrational state of the bond may affect the entropy considerably.

Let us treat now limitations of calculation of the $\Delta E_{c/env}$ term of Eq. (5). First of all, the partitioning of the energy is an approximation in itself which may lead to errors in ΔE. It seems that the weakest point is the "cut out" of the active functional groups from their environment. Since $\Delta E_{c/env}$ is estimated by supposing that the interacting atoms lie relatively far from each other no exchange and charge transfer between active site and environment are considered. The error can be reduced if the gas-phase model is cut out of the environment in such a way that "insulators", like $-CH_2-$ bonds, are positioned between them. This would mean that an aspartate side chain should be modelled as a $-CH_2COO^-$ moiety rather than a $-COO^-$ functional group. According to chemical evidence the inductive effect (i.e. charge transfer and polarization) propagating through the methylene group is relatively small.

The calculation of the electrostatic potential, used in $\Delta E_{c/env}$, is based on more or less rough approximations if enzymes are considered. Either a model dipole field is only estimated [17,86-88] or point charges, positioned at atomic centres, are used [7, 10, 13]. Neither of these methods treat lone pairs producing an anisotropic, instead of spheric, potential around atoms bearing them correctly. Lone pairs are treated adequately in the bond-increment approximation [93] which, in turn, overestimates the value of the electrostatic potential [94]. Polarization terms are approximated with atomic polarizabilities taken from experiment [107]. Since these are not directly measurable quantities certain ambiguities in their derivation are possible. It seems that van der Waals interactions do not play an important role in enzymatic processes where strongly polar intermediate states are formed [10, 70].

Atoms, lying far from the central machinery, do not interact *in vacuo* with the latter but in a space which is partly filled with polarizable groups. This means that the electrostatic potential and field are partly shielded, the ϵ dielectric constant differs from unity. The simplest solution to this problem is to use some theoretical value which may be somewhat arbitrary. It varies from author to author indicating the incertainty in its choice. Van Duijnen and coworkers considered the dependence of the dielectric constant on the interatomic distance with a simple expression [17]. Improper choice of ϵ may lead to erroneous estimation of the electrostatic interaction energy part of $\Delta \underline{E}_{c/env}$.

Complete modelling of an enzymatic process should include theoretical description of the conformational changes in the environment produced by the reaction which takes place at the catalytic site. The change in conformation causes alteration in the electrostatic potential of the protein which affects $\Delta \underline{E}_{c/env}$ in Eq. (5). Presently available papers, which report on such calculations [7, 8, 10, 65, 70], utilize empirical force-field models (see Sec. 2.3). These are able to reproduce gross conformational characteristics and, consequently, significant changes in the electrostatic potential.

Though the calculation of the entropy change for the gas-phase model of the central machinery is straightforward and its contribution to $\Delta \underline{E}_c$ is relatively small, entropy effects may become important in case of $\Delta \underline{E}_{env}$ and especially of $\Delta \underline{E}_{c/env}$. In solution translational entropy should be expressed in terms of aperiodic vibrations [108] and only approximate treatment of the problem is available at present [109]. The importance of the entropy term in $\Delta \underline{E}_{env}$ is seen if hydration and dehydration of intermediate states are investigated. If one of these states is highly polar while the other is nonpolar the structure of their hydration shells should be different. The polar intermediate state produces a hydration shell which is structured in a larger region around it than in case of a nonpolar state. The structured hydration shell has a lower entropy since formation of such ordered configurations is of lower probability. Accordingly, standard Gibbs free energy of polar intermediate states gets larger due to hydration.

We have to point out that the partition of the total energy, as in Eq. (5), is not necessarily correct. Some authors claim [110-112] that collective effects, involving all atoms of the reacting enzymatic system may also be important in the catalytic process. The Born-Oppenheimer approximation which is supposed to be valid for enzymatic reactions [3a may break down and coupling between electronic or translational and vibrational terms gets to an important role. This hypothesis needs further experimental and theoretical confirmation.

At last we call the attention to the dynamic aspects of reactions which are catalyzed by enzymes. Precise handling of a macroscopic process involves statistical methods, i.e. an average over all possible states, weighted by the relative energy content, has to be considered [16]. This is best done by the Monte Carlo technique which is conceivable also for enzymatic systems but at a very high cost [73, 74]. According to a recent hypothesis fluctuations in the positions of enzyme atoms may change their chemical reactivity [113]. This proposition gained some support from experiments which indicate that the enzyme catalytic rate varies with solvent viscosity.

3. APPLICATIONS

With the growing amount of information on steric arrangement of active sites in enzymes a profound understanding of the nature of the catalytic process becomes possible. In the following we discuss primarily those studies which make use of the complete three-dimensional structure of the protein. This is possible only for a limited number of systems. In other cases theoretical considerations are rather speculative and they are often based on the properties of the substrate molecule alone [114,115] A few examples will be treated in the last section.

3.1 Serine proteinases

These enzymes catalyze the hydrolysis of peptide, ester or amide bonds using a uniquely reactive serine side chain. Two families, which are sequentially homologous and closely similar in geometry, are especially well studied. Members of the trypsin family occur mostly in vertebrates while subtilisin acts in bacilli. The reaction sequence of hydrolysis is shown schematically in Fig. 4 [5]. The active serine oxygen attacks at the carbon atom of the amide or ester carbonyl group and via a loosely bound Michaelis complex (II) a tetrahedral intermediate (III) is formed. During this process the proton is transferred from the serine to the neighbouring imidazole side chain. After this step the tetrahedral intermediate

$$\underset{\text{I}}{R-\overset{O}{\overset{\|}{C}}-X} + E\text{-}OH \rightleftharpoons \underset{\text{II}}{R-\overset{O}{\overset{\|}{C}}-X : E\text{-}OH} \rightleftharpoons \underset{\text{III}}{R-\overset{O^-}{\overset{|}{C}}(-X\cdots H^+)-O-E}$$

$$\text{III} \longrightarrow \text{IV} + HX$$

$$\underset{\text{VII}}{R-\overset{O}{\overset{\|}{C}}-Y} + E\text{-}OH \rightleftharpoons \underset{\text{VI}}{R-\overset{O}{\overset{\|}{C}}-Y : E\text{-}OH} \rightleftharpoons \underset{\text{V}}{R-\overset{O^-}{\overset{|}{C}}(-Y\cdots H^+)-O-E} \xleftarrow{HY} \underset{\text{IV}}{R-C(=O)-O-E}$$

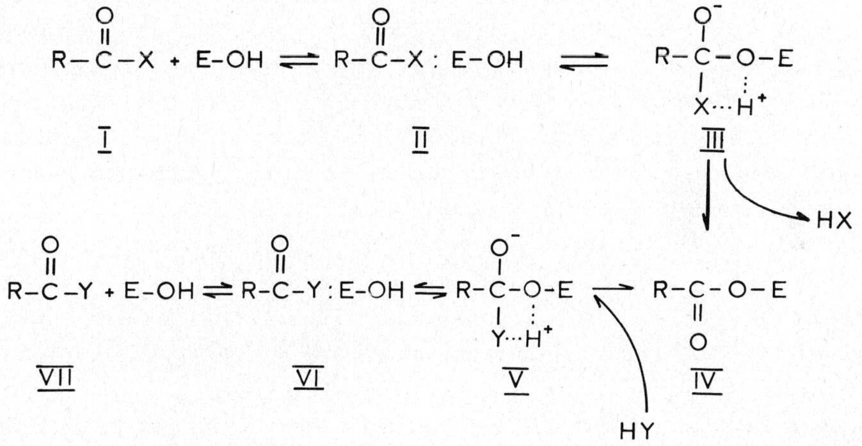

Fig. 4. Schematic sequence of hydrolytic reactions catalyzed by serine proteinases. X = R'NH or R'O, Y = OH. II: Michaelis complex, III: tetrahedral intermediate, IV: acyl-enzyme.

breaks down to give an acyl-enzyme (IV). The acyl-enzyme is hydrolyzed via the reverse route.

According to X-ray diffraction studies [116] a negatively charged aspartate residue is located near the catalytic site. The active serine, the histidine which serves as a proton acceptor during formation of the tetrahedral intermediate and as a proton donor during its break-down to the acyl-enzyme, and the aspartate form a catalytic triad (see Fig. 5). This triad is present in all serine proteinases in a similar steric arrangement. Due to the hypothesis of Blow and coworkers [116] a double proton transfer occurs as the substrate approaches the serine oxygen. During the catalytic process the aspartate should become protonated and its negative charge should be relayed to the serine side chain. This "charge relay" system would account for the enhanced reaction rate by increasing the nucleophilicity of the serine oxygen. Though this hypothesis was questioned by Polgár and Bender [117] just after its publication it was accepted by most authors. Only recent nuclear magnetic resonance [118, 119] and neutron diffraction [120] experiments have shown that the histidine side chain is the ultimate proton acceptor during catalysis. Accordingly, an ion pair is formed between the

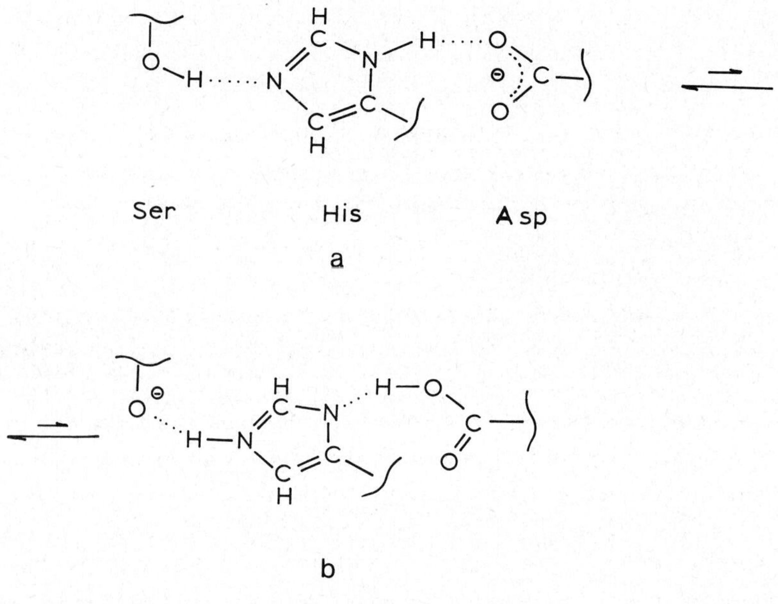

Fig. 5. Schematic representation of the catalytic triad in serine proteinases. (a) unperturbed form, (b) hypothetical activated form.

aspartate and the protonated histidine (Fig. 6).

The majority of quantum chemical studies on serine proteinases discusses the role of the buried aspartate side chain in catalysis [11, 13, 15, 19, 24, 30, 39, 45-49, 54, 55, 121]. All authors agree in the distinguished importance of aspartate in the enzymatic process. According to CNDO/2 [30, 45, 48], PRDDO [24] and *ab initio* calculations with STO-3G [13] and with 4-31G [19] basis sets the aspartate side chain is ionized in the unperturbed state, *i.e.* if

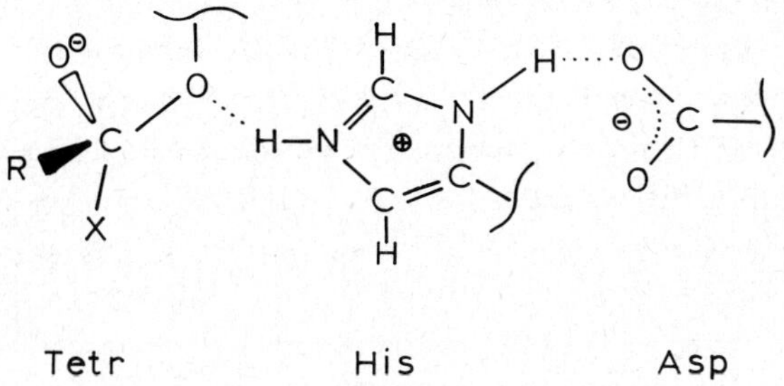

Fig. 6. Schematic representation of the catalytic triad after formation of the tetrahedral intermediate (Tetr) . X = NHR′ or OR′.

no substrate is present. In agreement with chemical evidence [117] calculations prove that, if the gas-phase model of the catalytic triad is treated, the serine oxygen cannot be ionized by the negatively charged aspartate alone. Proton transfer can be triggered by the presence of the substrate only [14, 54, 93].

Hayes and Kollman [13] have found that the intrinsic energy of the Ser$^-$...His$^+$...Asp$^-$ form of the catalytic triad in α-chymotrypsin is only +21 kJ/mol relative to the Ser...His...Asp$^-$ form. The energy difference becomes +111 kJ/mol if the ion-pair and neutral forms of the Ser... His couple are considered alone. These figures call the attention to the enormous stabilization effect of the Asp$^-$ side chain on the ion pair. Such an effect was indicated by recent pK measurements on α-chymotrypsin, too [122]. Hayes and Kollman have found that protein environment, other than the Asp$^-$ side chain, has also a considerable influence on the formation of the ion pair. This is hindered by the environment in α-chymotrypsin and facilitated in case of subtilisin. This would mean that intraproteic fields are essentially different in these enzymes.

Early calculations [24, 30, 45, 48] agree with early experiments [123] in predicting the buried aspartate residue to be the ultimate proton acceptor if the catalytic triad is protonated at higher pH values. This is disproved in the light of recent experiments [118-120]. The above agreement is fortuitous and it is due to errors in the model and in the applied quantum chemical method. Two basic types of error were made in the calculations mentioned above.

First, all minimum basis set methods overestimate the protonation energy of those anionic species, like aspartate, where the negative charge is highly concentrated [24]. Thus the energy curve of the proton transfer between imidazole and formic acid, a model for the His...Asp diad in the gas phase, is predicted by *ab initio* STO-3G basis set calculations to have only one single minimum near the oxygen atom of formate [39]. Due to chemical intuition, at least a shoulder should be present in the vicinity of imidazole indicating the possible existence of an ion pair. 4-31G basis set calculations on the $H_3N...HF$ system [124] confirm this presupposition.

Secondly, the intraproteic field (see Fig. 3) strongly influences the protonation equilibrium in serine proteinase [13-15]. This adds to the effect of hydration and of counter ion. While the stable form of the His...Asp couple is neutral in the gas phase, due to the decisive role of environment, the ion-pair form dominates in subtilisin [39]. Using the bond-increment method to calculate the electrostatic potential of the environment [93] we have found that the intraproteic field in α-chymotrypsin [96] is similar to that of subtilisin [14, 15], both stabilize the $His^+...Asp^-$ ion pair. This means that we are in disagreement with Hayes and Kollman [13] who have found electrostatic potentials to be essentially different. Further studies are needed to clear this problem.

Formation of the tetrahedral intermediate seems to have a considerable stabilizing effect on the ion-pair form of the His...Asp system. We have modelled it by a counter ion [15, 39] and our calculations have shown that its stabilizing effect is 50 kJ/mol. This result is supported by the calculations of Nakagawa et *al.* [19] who followed the proton transfer in the $Ser^-...His...Asp$ system. Using a 4-31G basis set in their *ab initio* calculations they have found that with the assistance of the Ser^- side chain, which may also serve as a model of the tetrahedral intermediate, the ion pair is stabilized by 55 kJ/mol relative to the neutral form even in the gas phase. The excellent agreement with our figure, obtained from a different model with another approximation, is certainly fortuitous. However, sign and order of magnitude seem to be correctly estimated.

Though it is clear that the buried aspartate side chain does not act as a proton acceptor in the catalytic process, its vital importance may not be overlooked. Amidon [48] has pointed out that the net charge on the serine oxygen changes by -57 millielectrons with respect to the free state if it is bound to the His...Asp$^-$ diad. Histidine alone shifts this charge by only -43 millielectrons. This means that, due to the polarizing effect of aspartate, the nucleophilicity of the serine oxygen increases. Apart from this through-bond mechanism a through-space activation is also observed [96]. The protein environment of Ser-221 in subtilisin is modelled by 33 neighbouring amino-acid residues which are located within a sphere of 1 nm radius around the centre of mass of the imidazole of His-64 [15]. These residues produce a strong negative electrostatic potential around the oxygen atom which originates primarily from Asp-32. Fig. 7 depicts the electrostatic isopotential map at the vicinity of the serine side chain. Its minimum value is -699 kJ/mol for the

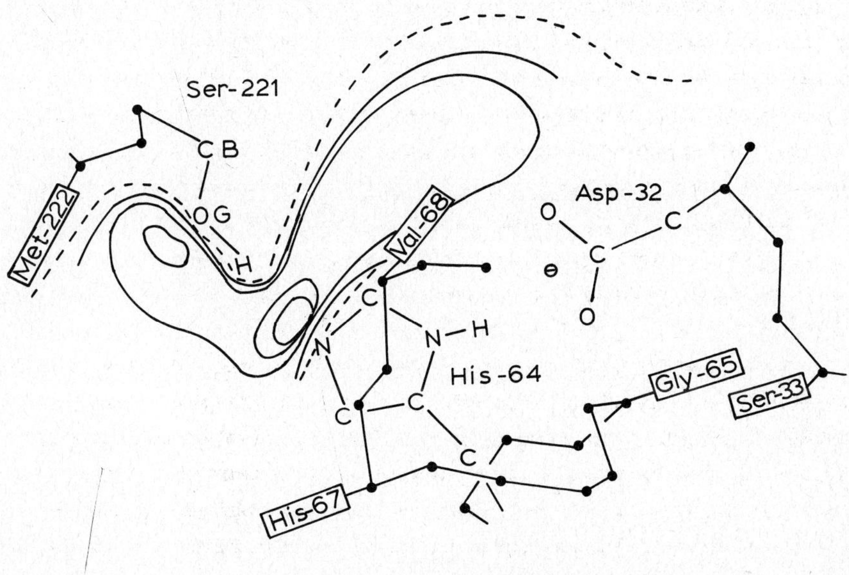

Fig. 7. Electrostatic isopotential map around the active serine side chain in subtilisin [96]. Dashed line indicates zero potential. Full lines correspond to potential values of 200, 400, 600 and 800 kJ/mol, respectively. Dots represent backbone atoms.

enzyme model. This can be considered as a measure of nucleophilicity [92]. The lower the potential the higher is the probability that a nucleophilic attack by the oxygen atom takes place. The minimum value of the potential around the oxygen atom of free ethanol is only -343 kJ/mol indicating a much lower reactivity.

Kitayama and Fukutome [30] discussed the acid-base property of the catalytic triad in α-chymotrypsin. It is known that the active site of this enzyme contains an ionizable group with pK_a of about 7. This group is identified as the imidazole side chain of His-57 in the catalytic triad. The observed pK_a is about equal to that measured for imidazole in aqueous solution which is somewhat astonishing. The above authors performed CNDO/2 calculations to study this problem.

First they have studied the assistance of Asp-102 to the pK_a of His-57 and they found that the negatively charged side chain raises the pK_a of this residue by about 4 pK units. This is qualitatively understood in terms of the same through-space effect which was discussed above for subtilisin (see Fig. 7). The minimal value of the potential around $N^{\epsilon 2}$ changes from -636 [94] to -919 kJ/mol as free imidazole is placed into the protein environment and it forms the side chain of His-64. Lowering of the minimal potential around the nitrogen atom corresponds to raising of the pK_a [92].

Simultaneously with the pK_a-raising effect of the buried aspartate side chain another factor lowers the basicity. This is attributed to the distorted structure of imidazolium cation formed after

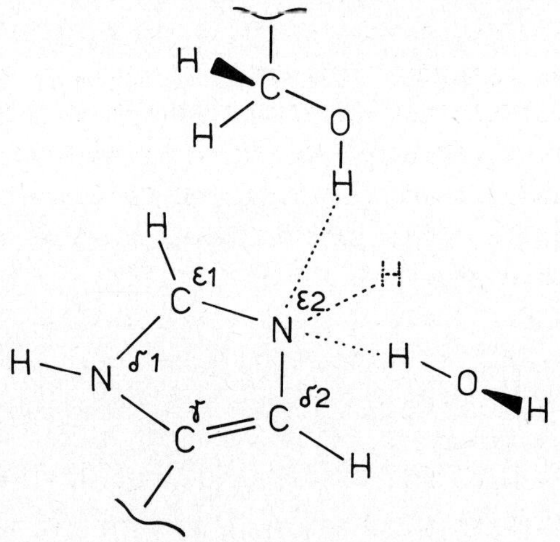

Fig. 8. Steric arrangement of side chains of His-57 and Ser-195 and of a strongly bound water molecule in α-chymotrypsin. Ideal position of the proton at $N^{\epsilon 2}$ is indicated by dashed lines.

protonation of neutral imidazole in the enzyme (see Fig. 8). As it is known, protonation in aqueous solution occurs via a proton transfer from one of the solvent water molecules to the proton acceptor atom. This is $N^{\epsilon 2}$ of His-57 in α-chymotrypsin. According to X-ray diffraction studies [125] a strongly bound water molecule is situated in the vicinity of $N^{\epsilon 2}$ but, due to the proximity of the serine hydroxyl group, in a distorted, instead of ideal, orientation. Protonation may occur only with assistance of this water molecule. The proton, attached to $N^{\epsilon 2}$, will thus be in a distorted position, the energy content of the imidazolium cation raises. This results in a lowering of pK_a by about 4 units since extraction of the proton from the cation gets easier.

Raising and lowering effects nearly cancel each other leading to a pK_a which is about equal to that measured in aqueous solution. Since proton transfer from the serine hydroxyl to the imidazole does not involve such a strong distortion through-space assistance of the aspartate ion acts alone. Proton abstraction from Ser-195 is much better than from the fixed water therefore the catalytic triad may work as a rate accelerating machinery in the state of nucleophilic attack of Ser-195 to the substrate.

The exciting problem of substrate specificity was discussed by DeTar [70]who used a molecular mechanics method for the α-chymotrypsin tetrahedral intermediate. The positioning of the substrate in the enzyme cleft is shown in Fig. 9. DeTar calculated the binding energy difference between D and L forms of a real substrate, Ac-Trp-NH_2. He found a numerical value of 29 kJ/mol. This means that the L form reacts about 10^5 times faster than the D form in nice agreement with experiment. Only 8-10 kJ/mol of this amount corresponds to van der Waals interactions, the main difference is due to distortions of H-bonds between the enzyme and the tetrahedral intermediate. While the L form allows optimal contacts the corresponding D form does not.

Umeyama and Nakagawa [55] studied the relative stability of bifurcated and linear hydrogen-bonded structures of the neutral Asp...His couple in the gas phase (see Fig. 10). Their *ab initio* calculations with a STO-3G basis set predicted the linear rather than the bifurcated structure to be more stable. This is in agreement with the recent X-ray diffraction measurements of Matthews et *al*. [126]. The stabilization energy is 75 kJ/mol if both structures are compared at their equilibrium N...O distances.

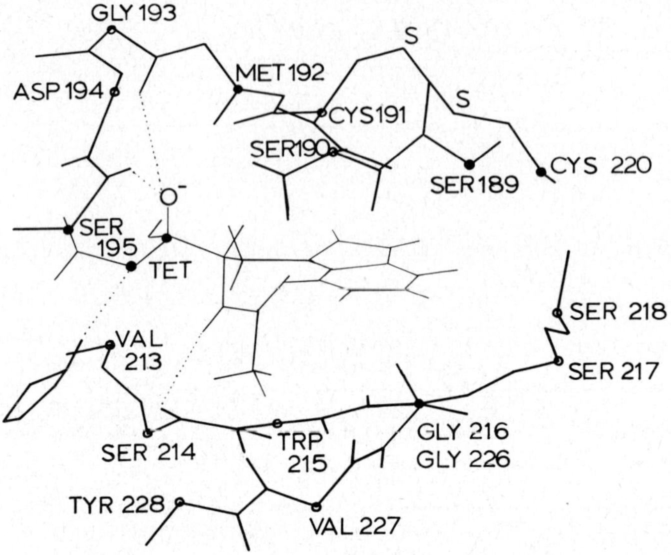

Fig. 9. View of Ac-Trp-NH_2 as tetrahedral intermediate (TET) at the active site of α-chymotrypsin. The α-carbon atoms of the residues are dotted. Dotted lines show the critical hydrogen bonds. After DeTar [70].

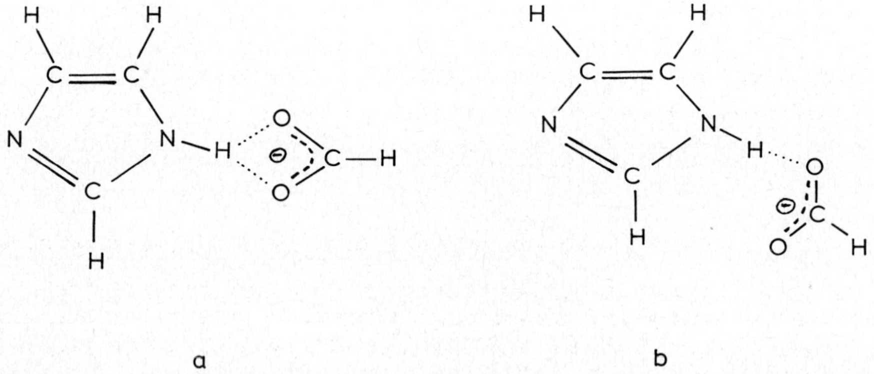

Fig. 10. Possible hydrogen-bonded structures of the imidazole... ...formate couple. (a) bifurcated, (b) linear.

3.2 Papain

Papain is a plant enzyme which requires a thiol group for activity. It belongs to the family of cysteine proteinases which, like serine proteinases, catalyze the hydrolysis of peptide bonds. Besides, acyl transfer reactions are also considerably accelerated by this enzyme. The steric arrangement of amino-acid residues in the main chain of the papain molecule is seen in Fig. 11. The active site is formed by the following segments of the polypeptide

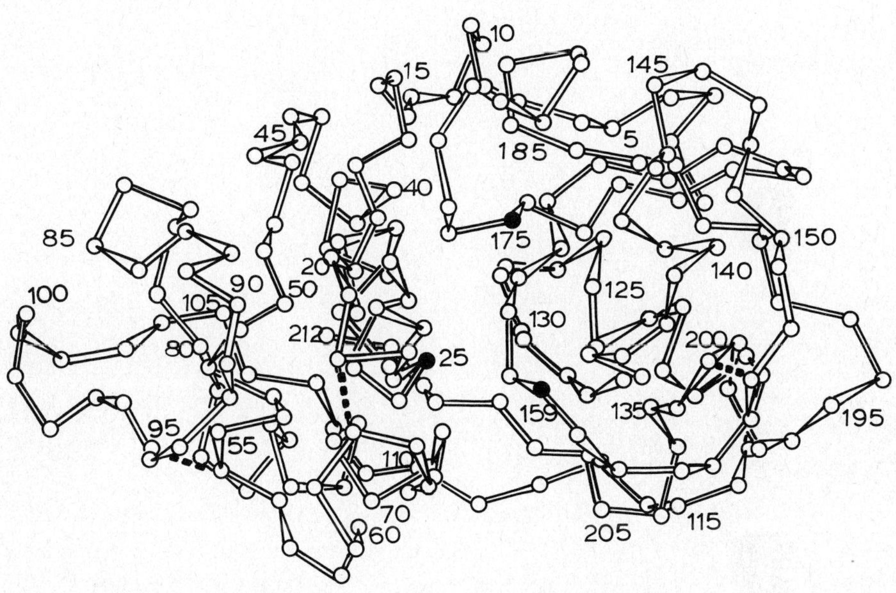

Fig. 11. The folding of the main chain in a papain molecule. Only positions of α-carbon atoms are indicated. Residues, belonging to the central machinery, are indicated by full circles. Dashed lines stand for disulfide bridges [128].

chain: Asn-18 to Trp-26, Tyr-61 to Trp-69, Val-133 to Glu-142, Val-157 to Ala-160, Asn-175 to Trp-177 and Thr-204 to Phe-207. The central machinery consists of the triad of a thiol group of Cys-25, an imidazole ring of His-159 and the side chain of Asn-175. These groups are fixed by a hydrogen-bond network like in the catalytic triad of serine proteinases (see Fig. 12) [128].

The reaction mechanism is outlined in Fig. 13 and it is similar to that in serine proteinases [129]. In the active state of the enzyme the side chains of Cys-25 and His-159 form an ion pair,

$-CH_2S^-...HIm^+CH_2-$ with the sulfur atom lying approximately coplanar with the imidazolium ring of His-159. The formation of the ion pair should be favoured by a hydrogen bond between the imidazole ring and the side chain of Asn-175. When a peptide substrate binds to the enzyme to form the Michaelis complex the carbonyl bond of the scissile peptide link forms a hydrogen bond with the amino group

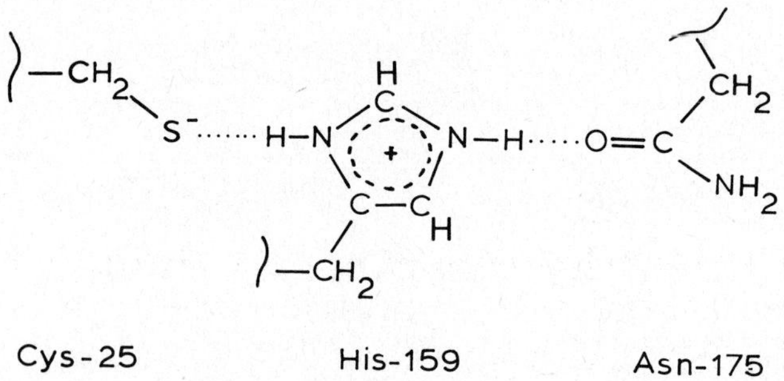

Fig. 12. The catalytic triad in papain.

of Glu-19. In the tetrahedral intermediate the NH bond of the scissile peptide link points to the carbonyl oxygen of Asp-158 while the nitrogen lone pair is oriented towards the imidazolium ring of His-159 which is rotated by about 30° from its initial position. Then the ring donates its proton to the NH of the tetrahedral intermediate and the acyl enzyme is formed. Deacylation occurs in a similar manner but inversely: a water molecule donates a proton to the imidazole ring, forms a second tetrahedral intermediate then the CS bond breaks to yield the products and the active enzyme.

Most theoretical calculations on papain [7, 8, 17, 130, 133] treat the problem of formation of the $ImH^+...S^-$ ion pair. This was an unclear point in experimental studies, some authors stated that the thiol group reacts in its non-dissociated form [131]. On the other hand, presence of mercaptide anion and histidine cation in the catalytically active state was supported by others [132]. In order to clarify this problem Broer and coworkers made _ab initio_

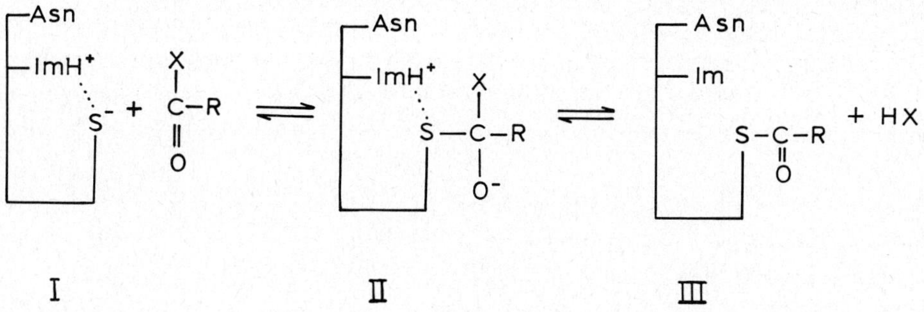

Fig. 13. Schematic reaction mechanism of peptide hydrolysis catalysed by papain.

molecular orbital calculations with a double-zeta type basis set and they found two energy minima for a proton moving along the S...N line. One corresponds to the ion pair and one to the neutral form the latter lying about 80 kJ/mol lower. This result was confirmed by Bolis et *al*. [7, 8] (see Table III). It is seen that the residues Asn-175 and Ala-160, which form H-bonds with the diad, stabilize the ion-pair form. However, their stabilization effect is insufficient alone, the presence of the α-helix near the active site, composed of residues Ser-24 to Gly-43, is also needed. The importance of dipole fields of α-helices near active sites of enzymes was claimed by others, too [86-88]. Hayes and Kollman [13] corrected the minimal basis set error using experimental proton affinities of imidazole and CH_3S^-. With this correction they have found that the effect of Asn-175 alone is sufficient to stabilize the His...Cys diad in the ion-pair form. However, as they themselves point out, their procedure may overcorrect minimal basis set errors thus, probably, their values are too low (see Table III).

Bolis et *al*. [7, 8] studied the structure of the Michaelis complex between papain and a model substrate, N-methylacetamide. Using a combination of *ab initio* and molecular mechanics methods they have estimated the interaction energy to be -47 kJ/mol. The main part, -31 kJ/mol, arises from the interaction between N-methyl-

TABLE III

Proton transfer energies in the Im...HSMe diad as calculated for different models (in kJ/mol). Im = imidazole, Me = methyl.

Model	E (ion pair) -	E(neutral)
	Refs.[7, 8][a]	Ref.[13][b]
diad in the gas phase	75	428 (21)
diad + Asn-175[c]	56	402 (-17)
diad + Asn-175 + Ala-160	22	--
diad + Asn-175 + Ala-160 + α-helix[d]	-41	--

[a] With a basis set of 7s/3p functions for C, N and O atoms, 3s functions for the H atom and 9s/6p functions for the S atom.

[b] With a minimal STO-3G basis set. Values, corrected for minimal basis set error, are given in parentheses

[c] Van Duijnen et al. [17] obtained 63 kJ/mol. They replaced Asn-175 by formaldehyde.

[d] With a simple point-charge representation of the α-helix [17].

acetamide and the thiolate side chain of Cys-25. However, the stabilizing effect of the neighbouring His-159, Ala-160 and Glu-19 residues is also considerable. The same authors studied the tetrahedral intermediate, too. They have found that it is less stable than the Michaelis complex which contradicts to the mechanism proposed by Drenth et al. [128]. They analyze this problem and they conclude that the result is probably an artefact due to incomplete geometry optimization and other simplifications. Further, they state that if the Asn-175 residue is replaced by a negatively charged aspartate the stabilization of the $ImH^+...S^-$ ion pair increases by 76 kJ/mol. This pronounced effect is understood in terms of the through-space electrostatic mechanism outlined in Sec. 3.1. On the other hand, it seems that the $-CONH_2$ side chain of Asn-175 exerts its stabilizing effect on the ion pair by stretching the $N^{\delta 1}$-H bond of imidazole of His-159 by about 20 pm. This makes the proton transfer from S to $N^{\epsilon 2}$ easier [47].

3.3 Lysozyme

Among those enzymes which catalyze the hydrolysis of glycosidic bonds the reaction mechanism is best understood for lysozyme. The natural substrate for this enzyme is mucopolysaccharides from bacterial cell wall. These are polymers of N-acetyl-β-D-glucosamine and β-N-acetylmuramic acid (see Fig. 14). The schematic three-dimensional structure of the enzyme with part of a bound substrate is shown in Fig. 15 [134]. The commonly accepted reaction mechanism, can be summarized as follows [135].

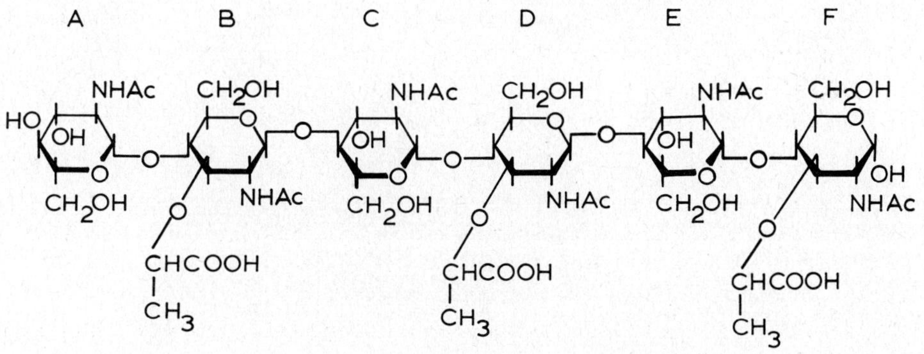

Fig. 14. Schematic representation of a polysaccharide substrate for lysozyme. The polysaccharide is composed of alternating residues of N-acetyl-β-D-glucosamine (rings A, C and E) and β-N-acetylmuramic acid (rings B, D and F).

Six saccharide units of the polymer substrate are bound to the active site of the enzyme by several hydrogen bonds (see Fig. 15). Each unit is bound to a subsite lettered by A, ..., F. Cleavage of the substrate occurs between D and E sites with the assistance of Glu-35 and Asp-52 (see Fig. 16). Upon binding of the substrate to the active site the sugar ring in subsite D is distorted from a chair towards a half-chair or sofa conformation weakening the bonds between atoms C1 of D and O4 of E (see Fig. 16). This bond is further weakened as Glu-35 donates a proton to atom O4. Cleavage of the C1-O4 bond leads to the formation of the transition state. In

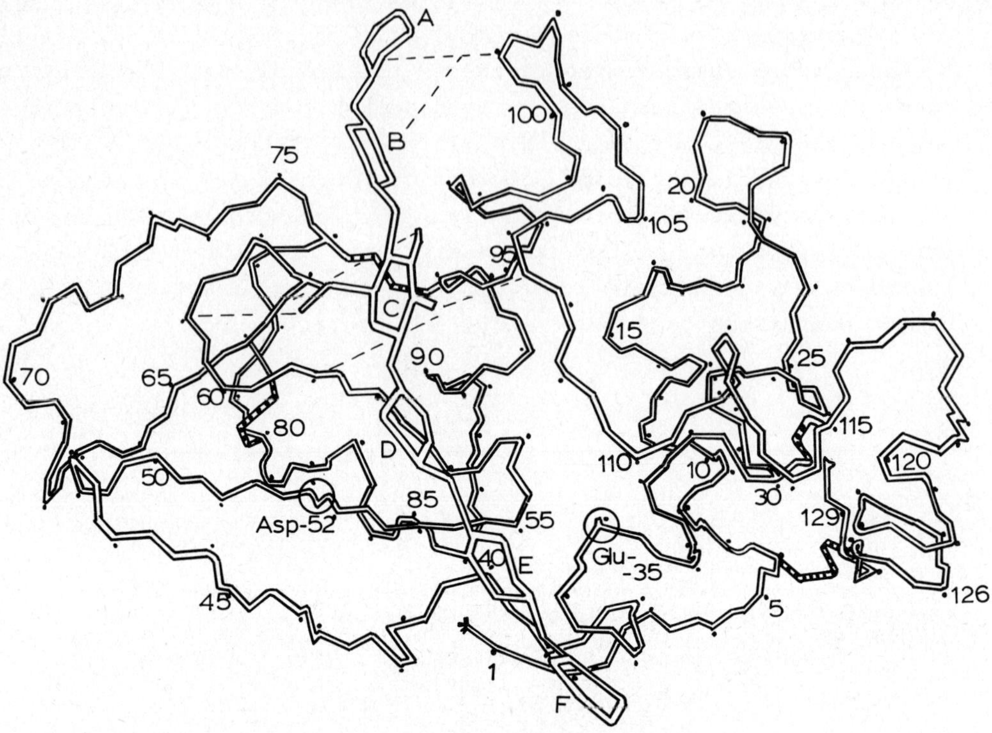

Fig. 15. Schematic view of the main chain in the lysozyme molecule. Substrate saccharide units are denoted by letters A, ..., F. Important hydrogen bonds are indicated by dashed lines. Heavy dotted lines denote disulfide bridges.

this, C1 and O5 atoms of the ring in subsite D share a positive charge: an oxycarbonium ion is formed. This is stabilized by the negative Asp-52 side chain. The C2C1O5C5 torsion angle becomes planar corresponding to the half-chair conformation. If the bond between the two saccharides is cleaved they leave the active site. The carbonium ion reacts with the hydroxyl of a water molecule which was originally bound to Asp-52 and it leaves the active site, too. The proton of this water molecule serves for regeneration of the neutral state of Glu-35. It is assumed that the concerted protonation of substrate and cleavage of the C1-O4 bond is the rate-determining step of the enzymatic process. Three factors may considerably contribute to the overall catalytic effect: (1) acid catalysis in the protonation of the glycosyl oxygen by Glu-35; (2) strain of the sugar ring at subsite D; (3) stabilization of the positively charged carbonium ion by the negative charge on Asp-52.

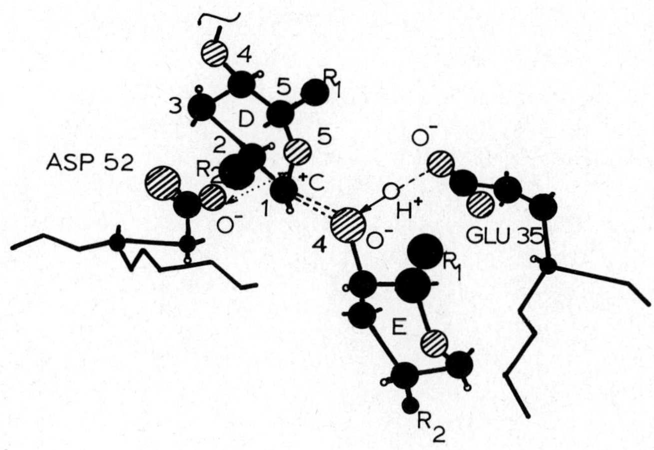

Fig. 16. Steric arrangement of the central machinery in lysozyme. Saccharide units A, B, C and F are omitted. Open circles: hydrogen atoms, full circles: carbon atoms, hatched circles: oxygen atoms. R_1=CH_2OH, R_2=NHAc.

The first step towards understanding the detailed mechanism of action of lysozyme is to gain information on the structure of the enzyme-substrate complex. This problem was investigated by Scheraga's group in a series of papers [65, 69, 71, 136]. They used their molecular mechanics method to find preferred conformations of oligosaccharide inhibitors in their free and enzyme-bound states. In the first publication they state that the interactions which determine conformations for all oligosaccharides containing both N--acetyl-β-D-glucosamine and β-N-acetylmuramic acid are exclusively of intraresidue and of nearest neighbour origin [65]. This means that it is possible to predict all stable conformations which contain these units in any sequence. Low-energy conformations for monomer units are shown in Fig. 17.

Starting from conformations of oligosaccharide inhibitors and substrates in the enzyme-free state, Scheraga and his coworkers minimized the total energy of the enzyme-substrate complex keeping protein coordinates fixed [69, 71]. It was found that the energy of the complex lowers as the number of saccharide units, bound to the active site, raises from one to six. However, no further energy gain is obtained on adding more units. Table IV summarizes their

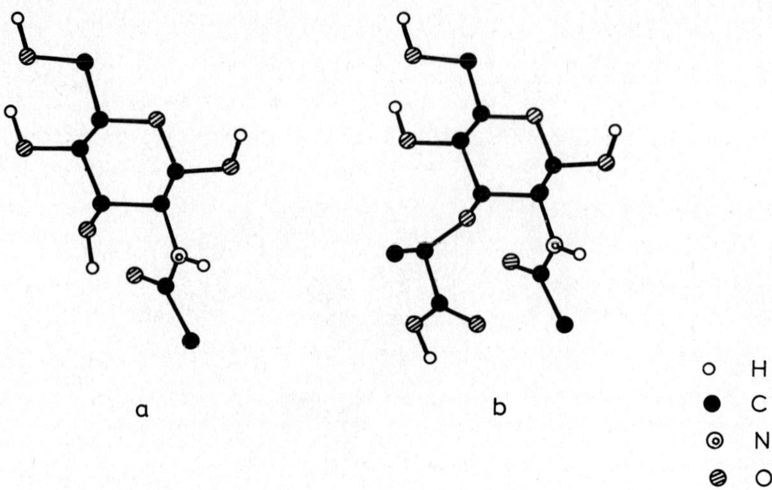

Fig. 17. Low-energy conformations for N-acetyl-β-D-glucosamine (a) and for β-N-acetylmuramic acid (b).

TABLE IV
Positions and binding energies (in kJ/mol) of oligosaccharides at the active site of lysozyme. GlcNAc denotes an N-acetyl-β-D-glucosamine monomer unit. E_{sub} and E_{int} stand for the conformational energy of the oligosaccharide and for the total energy of interaction between the active site of the enzyme and the oligosaccharide, respectively. $E_{tot} = E_{sub} + E_{int}$.

Oligosaccharide	Binding sites[a]	E_{sub}	E_{int}	E_{tot}
$(GlcNAc)_2$	C-D	-78	-190	-268
$(GlcNAc)_3$	B-C-D	-101	-265	-366
$(GlcNAc)_3$ X-ray[b]	A-B-C	-92	1406	1314
$(GlcNAc)_3$ optimized from X-ray disposition	A-B-C	-120	-180	-300
$(GlcNAc)_4$	A-B-C-D	-154	-285	-439

[a] See Fig. 15.
[b] With three-dimensional coordinates taken from X-ray measurements.

results which were obtained for oligomers of N-acetyl-β-D-glucosamine [71]. Scheraga and his coworkers were able to show that subsite C has the highest affinity for a saccharide unit since a stable hydrogen bond between the N-acetyl group of the saccharide and the protein backbone is formed. In addition, other favourable interactions stabilize this arrangement. It is the presence of the N-acetyl group which accounts for the enzyme specificity for binding oligomers of N-acetyl-β-D-glucosamine whereas glucose oligomers are bound with much lower affinity. As it is seen from Table IV preferred binding sites for the di-, tri- and tetrasaccharides are C-D, B-C-D and A-B-C-D. This theoretical result is in agreement with X-ray diffraction studies for the case of the di- and tetrasaccharide. The optimal position of the trisaccharide is B-C-D according to molecular mechanics calculations and it is A-B-C from experiment. It is interesting to notice that for the experimental geometry of $(GlcNAc)_3$ the interaction energy with the enzyme, $\underline{E}_{int}$, is a very large positive value indicating that several unfavourable contacts between the trisaccharide and the enzyme exists. Optimization, starting from this geometry, releases these contacts yielding negative interaction energy. From this we may suppose that it is possible to refine experimental three-dimensional coordinates by molecular mechanics methods.

Let us discuss now the catalytic process treating the three major factors, acid catalysis, strain and electrostatic stabilization, separately. The role of Glu-35 as a general acid catalyst was studied by Loew and Thomas with the IEHT method [137, 138]. Probably this is the first application of quantum chemistry to a well-defined enzymatic reaction. They considered dimethoxymethane, $MeOCH_2OMe$, in a rigid, all--trans conformation as a substrate and a formic acid molecule simulated the action of Glu-35 which represented the whole enzyme. Analyzing changes in charge densities and bond populations they were able to show that the sole action of Glu-35 is an important factor in lowering the energy of activation of the enzymatic process by stabilizing the carbonium ion intermediate. Conformational properties of dimethoxymethane, as a model substrate for lysozyme, were studied by Gorenstein et al. [139] and by one of us [140].

The problem of strain in lysozyme is treated in three theoretical papers [10, 72, 136]. In general, experimentalists stress its importance in the catalytic process [141]. However, molecular mechanics calculations from different laboratories seem to prove that strain does not play an important role as a rate accelerating factor. This problem will be discussed in more detail in Sec. 4.2. According to

Levitt [72]who minimized the total energy of the lysozyme-substrate complex, the D ring of the substrate can be accomodated in the active site in the full-chair conformation even if the enzyme is held completely rigid. This was confirmed by subsequent calculations [10, 69], too. As we have mentioned previously,molecular mechanics calculations may compete with X-ray diffraction measurements in refinement of three-dimensional structure of enzymes. Therefore, we feel, the role of strain in lysozyme catalysis is at least questionable in the light of theoretical calculations.

Electrostatic stabilization of the transition state oxycarbonium ion was extensively studied by several authors [10, 12, 143-145]. The stabilizing effect of the anionic form of Asp-52 was estimated using the CNDO/2 method [143, 144]. It was found that the ability of Glu-35 to donate a proton to the substrate is considerably enhanced by the assistance of the neighbouring aspartate group. Furthermore, proton transfer may be facilitated by additional complexation with a water molecule of the hydration shell [143] and with Trp-108 [144]. The most comprehensive theoretical study of the role of electrostatic stabilization in lysozyme activity was reported in a series of papers by Warshel and his coworkers [10, 12, 145]. The complete enzyme-substrate structure was considered explicitely in the calculations using a new combination of a semi-empirical molecular orbital and of a molecular mechanics method [10]. In two subsequent papers]145] an empirical valence-bond model was developed which is able to reproduce proton-transfer potential energy curves with considerable reliability. Calculations with different methods lead to the same conclusion. The transition state is represented by a quasiplanar carbonium ion (see Fig. 18). It is considerably stabilized by the environment including protein and bulk water. As it is seen from Fig. 19 the amount of stabilization relative to the reaction in aqueous medium is 30 kJ/mol. This corresponds to a rate enhancement of about 10^6. The charged Asp-52 side chain plays a crucial role in stabilization. Here again we have an example for through-space electrostatic effects causing considerable rate enhancements in enzymatic reactions (see also Sec. 3.1).

An interesting proposition for the lysozyme action, the torsional catalysis concept was outlined by Gorenstein and coworkers [139]. They investigated the effect of substrate conformation on enzyme activity and suggested that the proper orbital interaction is important for the cleavage of the carbon-oxygen bond. Based on

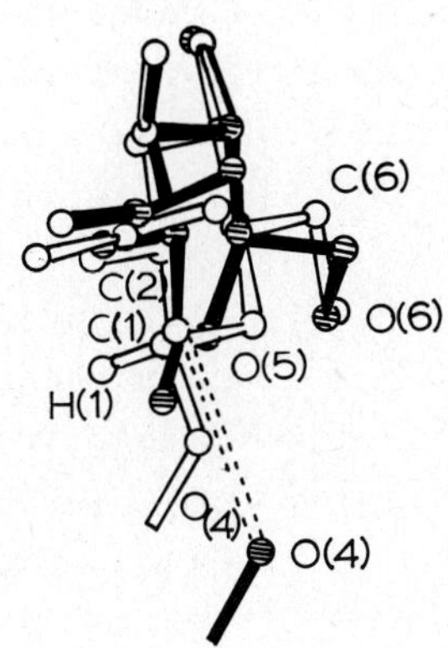

Fig. 18. Calculated equilibrium conformations of the substrate in the ground state (open bonds, full chair) and in the carbonium ion intermediate (solid bonds, quasiplanar). Note how the N-acetyl and C6O6 groups move. After Warshel and Levitt [10].

theoretical studies on dimethoxymethane they state that orientation of neighbouring oxygen lone pairs relative the scissile C-O bond influences overlap population. Trans lone-pair orientation results in a decrease of population which corresponds to bond weakening. Applying this principle to the full-chair form of the sugar residue at subsite D of lysozyme it is the C1-O5 bond that should be disposed to cleavage. On the other hand, if the sugar ring is distorted into a boat, or even sofa, conformation the C1-O4 bond weakens and, in accordance with experiment, this bond is broken. From this Gorenstein et al. concluded that it is the sofa conformation which should be stabilized in the enzyme-substrate complex. Though this hypothesis is appealing and it can be adapted to other enzymatic reactions, too, it is based on the presence of strain for which it does not yield any direct evidence. We feel that it is the electrostatic, rather than this torsional, catalysis which plays the major role in lysozyme action.

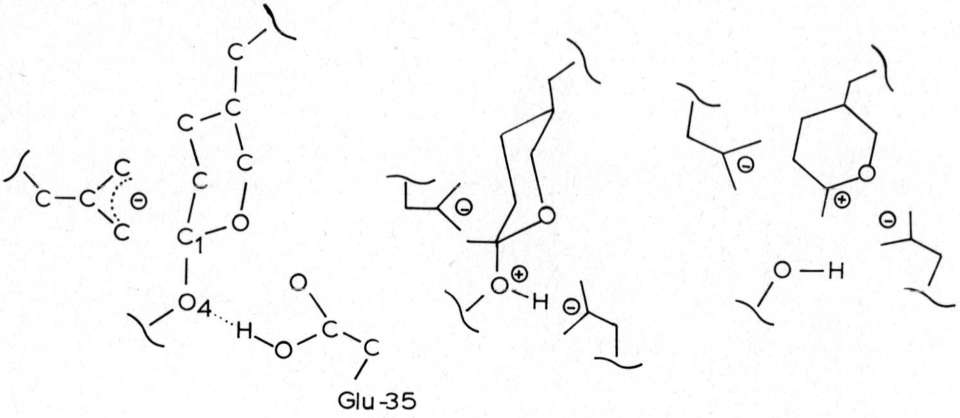

a	b	c
0	42	105
0	25	75

Fig. 19. Schematic representation of the catalytic bond cleavage of the sugar ring in subsite D of lysozyme. (a) Ground state; (b) O4 is protonated by the proton of Glu-35; (c) planar carbonium ion intermediate. Numbers denote relative energies in aqueous solution (upper row) and in the enzyme (lower row)(in kJ/mol). After Warshel [12].

3.4 Carboxypeptidase A

Carboxypeptidase A is a zinc-containing metalloenzyme which cleaves peptide bonds at the C-terminal end of the polypeptide chain. It also catalyses cleavage of esters and zinc can be replaced by other divalent heavy metal atoms (Co^{2+}, Mn^{2+}, Fe^{2+}, Ni^{2+}) without a loss in the peptidase activity. The structure and probable mechanism of action have been studied by Quiocho and Lipscomb [146]. The three-dimensional structure is shown in Fig. 20. Upon binding the substrate, the enzyme undergoes remarkable conformational changes involving Arg-145. Tyr-248 and Glu-270 (see Fig. 21). First, the guanidinium group of Arg-145 moves about 200 pm toward the terminal carboxylate of the substrate forming a salt link. Secondly, the side chain of Tyr-248 rotates to place itself about 200-300 pm from the nitrogen of the peptide bond. Thirdly, the

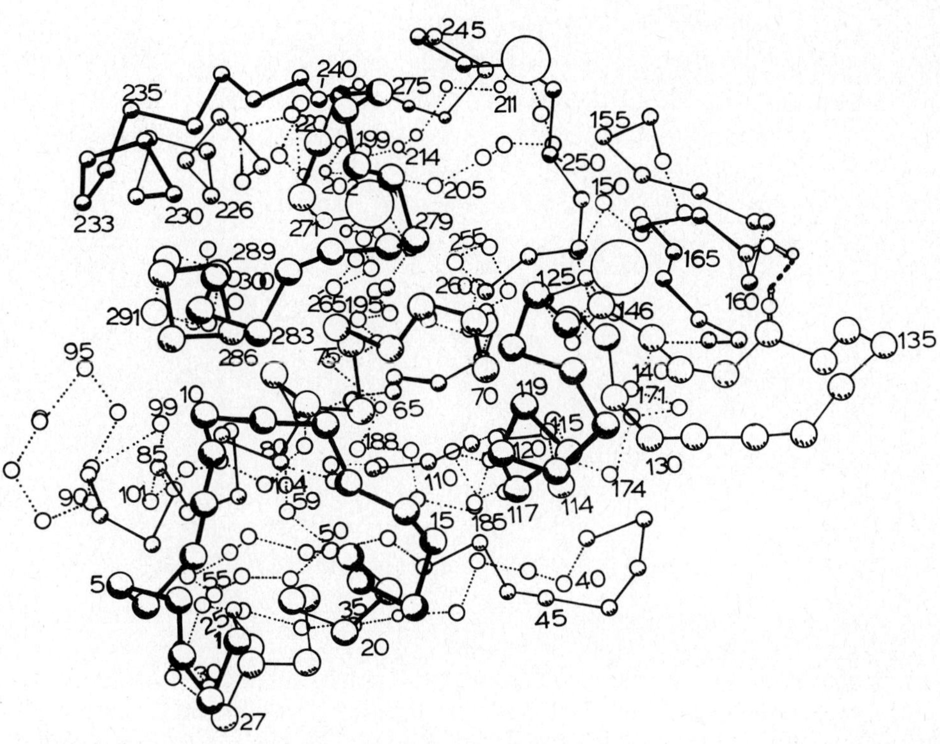

Fig. 20. A view of the carboxypeptidase molecule represented by α-carbon atoms of the polypeptide chain. Side chains of the central machinery are indicated by large open circles. Dashed line represents disulfide bridge. After R.E. Dickerson and I. Geis, The Structure and Action of Proteins Benjamin, Menlo Park, 1969.

carboxylate of Glu-270 comes within 300-350 pm of the carbon atom of the substrate carboxyl group. In addition, the carbonyl oxygen, adjacent to the scissile peptide bond, replaces a water molecule coordinated to Zn^{2+} and itself becomes a ligand to the metal atom.

Two alternate reaction schemes for the action of carboxypeptidase A have been proposed. Quiocho and Lipscomb [146] suggested the "zinc-carbonyl" mechanism. According to this the phenolic OH of Tyr-248 is the likely proton donor to the nitrogen of the susceptible peptide bond. It can form a hydrogen bond to nitrogen and decreases the conjugate stabilization of the peptide linkage. The carbonyl bond may be further polarized by the neighbouring zinc atom which would weaken the C-N bond. Donation of charge from the nearby carboxylate of Glu-270 to the carbonyl carbon is expected to

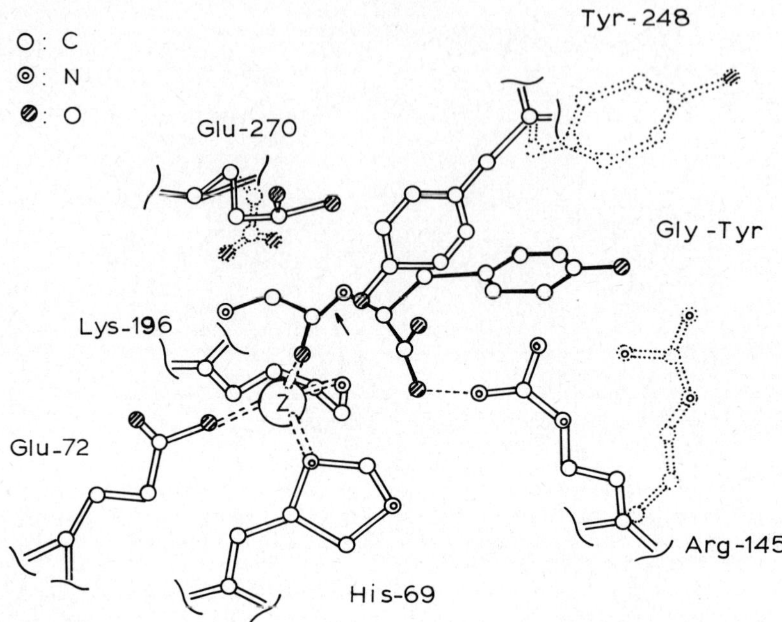

Fig. 21. Schematic representation of the complex of carboxypeptidase A with Gly-Tyr inhibitor. Original and final positions of moving groups are indicated by dotted and full lines. Z denotes metal ion, hydrogen bonds are shown by dashed lines. The site of hydrolytic cleavage of the substrate peptide bond is indicated by the arrow. After Dickerson and Geis (see Fig. 20).

promote rehybridization at this atom and loss of conjugative stabilization. It is thought that the carboxylate group of Glu-270 acts as a nucleophile by attacking the substrate.

According to the "zinc-hydroxide" mechanism [147](see Fig. 22) the substrate is attacked by a nucleophilic water which is activated by Glu-270 as a general base. After formation of the tetrahedral intermediate a proton from this water is released to the phenolic OH of Tyr-248 which transfers it to the leaving amino group.

The first step of the peptide bond cleavage reaction, formation of the carboxypeptidase-substrate complex, was studied by Hayes and Kollman [89]. They considered the dipeptide inhibitor, Gly-Tyr, as a "substrate" since the structure of its complex with carboxypeptidase A has been determined by X-ray diffraction measurements

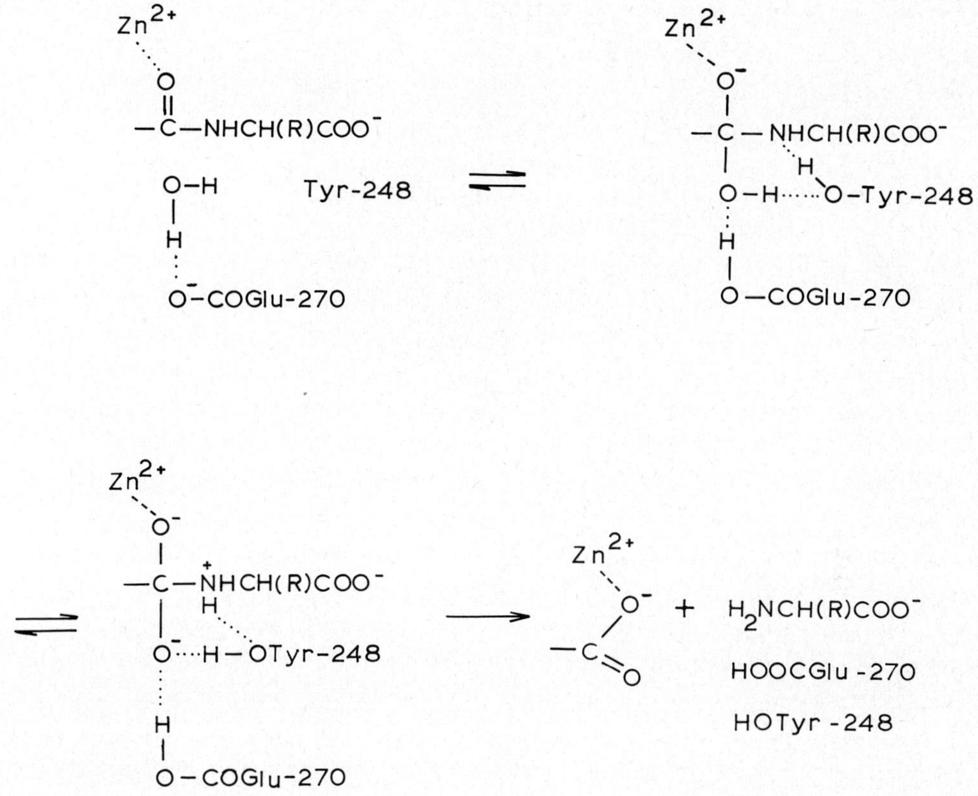

Fig. 22. Schematic mechanism of action of carboxypeptidase A (after Breslow et al. [147].)

[148]. The electrostatic interaction energy between enzyme and "substrate", as calculated using point-charge representations of both carboxypeptidase A and Gly-Tyr, was found to be different if the geometries of the native enzyme or the bound complex were considered. Putting a point charge of +2 on the Zn ion and using STO-3G atomic net charges the interaction energy with the native enzyme was 10 kJ/mol less than in the bound state. This would explain why conformational changes are induced in the enzyme upon binding of a substrate (see Fig. 21). However, if the charge of zinc was put equal to +1.4, trying to mimick delocalization of the positive charge on the metal atom onto neighbouring ligands, an opposite result was obtained: the native enzyme binds the substrate more firmly. Hayes and Kollman also investigated the relative binding abilities of various analogues of the natural Gly-Tyr inhibitor.

They have found that if the phenylalanine side chain is substituted in the _ortho_ instead of _para_ position the binding energy consdierably increases indicating that the _o_-HO-Gly-Tyr analogue would be a more potent inhibitor than Gly-Tyr.

Nakagawa and Umeyama [149] studied the role of arginyl side chains in carboxypeptidase A during substrate binding. There are three Arg residues in the active-site groove: Arg-71 and Arg-127 which are located at the surface and Arg-145 buried in the enzyme. Based on CNDO/2 and _ab initio_ STO-3G basis set calculations on suitable models, a "sliding mechanism" for substrate binding was suggested. It is assumed that the C-terminal carboxylate of the substrate is captured by the outermost Arg-71 and afterwards it may slide along Arg-127 before it fits into the final active site at Arg-145 (see Fig. 23). Nakagawa and Umeyama have shown that their proposed mechanism is considerably favoured energetically over the case where Arg-145 moves to the C-terminal carboxylate only after binding of the substrate.

Once the substrate is bound to the active site, cleavage of the peptide link starts with polarization coming from the close environment. It was supposed in experimental reaction mechanisms (_vide infra_) that polarization weakens the C-N bond and this effect would partly account for the rate acceleration relative to the uncatalyzed process. However, theoretical calculations, coming from two different laboratories, have shown that C-N overlap population in a model substrate, N-methyl-acetamide, gets larger upon placing the molecule from the gas phase to the enzyme environment [90, 150]. In turn, the polarization effect of the environment increases the net charge on the carbonyl carbon atom from 327 to 372 millielectrons (_ab initio_ calculations with STO-3G basis set [90]). Accordingly, the nucleophilic attack either by the carboxylate of Glu-270 or by a water molecule from the hydration shell is facilitated. A similar effect, though with other numerical values, was found by Scheiner and Lipscomb with PRDDO calculations on another model [150]. Both papers agree, that the principal source of rate acceleration seems to be the stabilization of an sp^3 _versus_ an sp^2 nitrogen in the peptide link of N-methyl-acetamide. This results in a stabilization of the tetrahedral intermediate and in a lowering of the activation energy [90, 150].

Scheiner and Lipscomb [150] tried to distinguish betweeen "zinc-carbonyl" and "zinc-hydroxide" mechanisms of the enzymatic reaction.

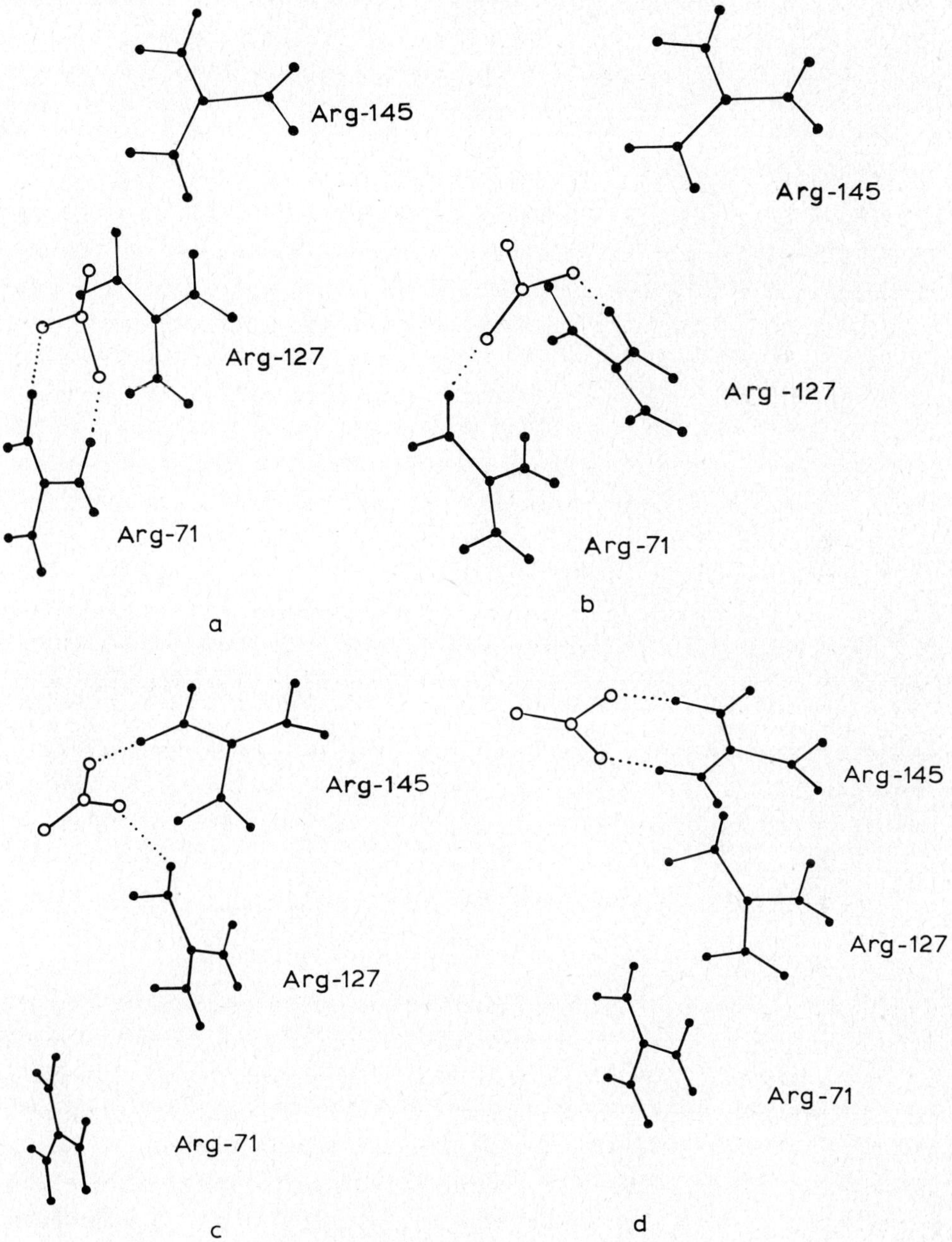

Fig. 23. "Sliding mechanism" of substrate binding by carboxypeptidase A [149]. C-terminal carboxylate is modelled by $HCOO^-$ (open circles) while arginyl side chains are replaced by $(NH_2)_2C{=}NH_2^+$ (full circles). Hydrogen bonds are indicated by dotted lines. The sliding proceeds from the initial state a via intermediates b and c into the final position d which is obtained by X-ray measurements.

They have found that the tetrahedral intermediate is stabilized by 128 kJ/mol relative to the isolated species in the "zinc-carbonyl" and by only 66 kJ/mol in the "zinc-hydroxide" mechanism. From this they concluded that the former is the more probable. However, minimum basis set errors may invalidate this conclusion though one might expect that they approximately cancel in this case. In the light of the criticism on minimal basis set calculations for proton transfer equilibria (see Sec. 2.2) we feel, that this statement should be handled with caution.

The role of metal ion in catalysis cannot be cleared by mere experimental methods, therefore theoretical calculations are of considerable importance here. Hayes and Kollman used a simple point charge model for Zn^{2+} and they have found that this point charge has a strong effect both in increasing the net charge on the carbonyl carbon atom and in stabilizing the tetrahedral intermediate [90]. Scheiner and Lipscomb [150] modelled Zn^{2+} by Li^{+} and by Be^{2+} and they have found a similar, but more pronounced, effect. They state that the stabilization of the tetrahedral adduct occurs through partial bonding between the metal electrophile and the carbonyl oxygen. Further, they have found that metal electrophiles are more efficient in facilitating the hydrolysis of a peptide bond than hydrogen-bonding electrophiles, NH_4^+ or H_2O. According to Osman and Weinstein [20], Zn^{2+} has even more pronounced effect than Be^{2+}. Moreover, an incoming ligand is perturbed more strongly by the metal and directionality constraints are also more pronounced in the Zn-complex than in the Be analogue.

3.5 Other metalloenzymes

Metals represent frequently occuring standard elements of enzymes either as prosthetic groups or as cofactors. The best resolved and characterized metalloenzyme is carboxypeptidase A which was treated above. This section is devoted to quantum chemical studies on other metalloenzymes.

Human carbonic anhydrase contains Zn^{2+} ion. It catalyzes the reversible hydration of carbon dioxide to hydrocarbonate, the hydrolysis of esters and phosphates and the hydration of aldehydes and ketones [151]. The schematic view of the active site, constructed on the basis of X-ray crystallographic studies [152, 153], is given in Fig. 24. Two molecular forms of the enzyme are available: human erythrocyte carbonic anhydrase B (HCAB) and C (HCAC), possessing slightly different biological activity [154]. The mechanism of

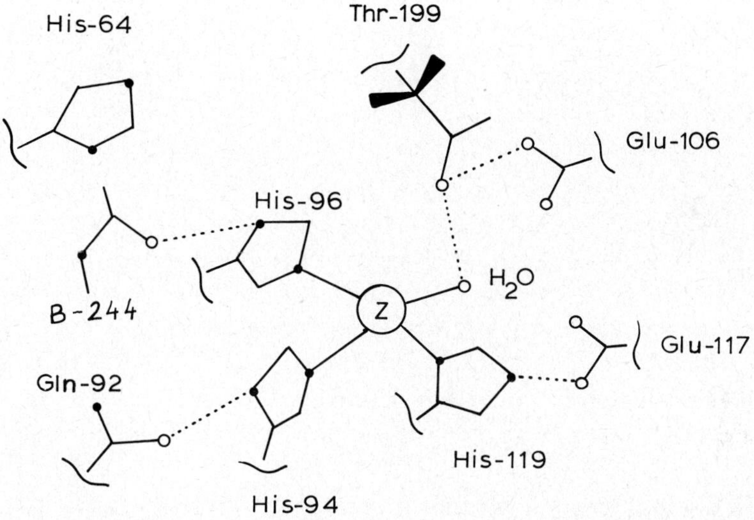

Fig. 24. Active site of human carbonic anhydrase. Z denotes Zn^{2+} ion. Vertices, full and open circles represent carbon, nitrogen and oxygen atoms, respectively B stands for backbone (after Sheridan and Allen [91]).

action of carbonic anhydrase is still under debate, three alternative reaction schemes have been postulated. In the first and second a zinc-bound imidazolate would attack CO_2 either directly [155] (see Fig. 25) or by activating a water molecule [156](see Fig. 26). In the third case (Fig. 27) the zinc-bound hydroxyl ion would be the ultimate nucleophile. In this hypothesis a neutral imidazole acts as a general base to promote the transfer of water from zinc to carbon dioxide.

Quantum chemical studies on carbonic anhydrase are focused primarily on comparison of alternative reaction schemes [157-161]. Pullman and Demoulin performed *ab initio* SCF calculations on different models of the active-site complex [157,158]. They state that though free imidazole deprotonates more easily than free water, tetracoordinated zinc ion (see Fig. 24) lowers pK_a values drastically and reverses the order of the ease of deprotonation. This is true

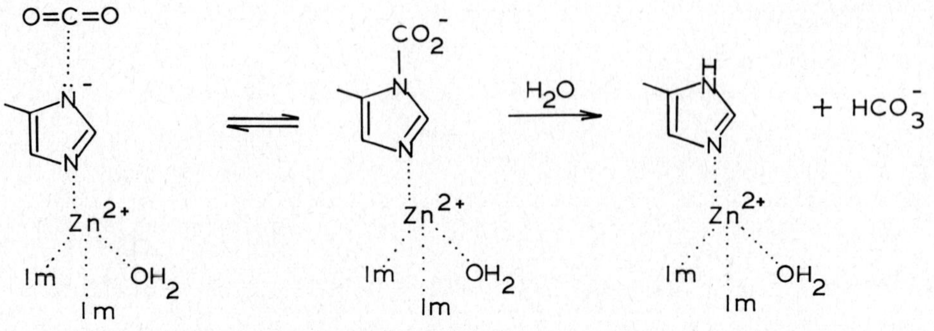

Fig. 25. Nucleophilic "zinc-imidazolate" mechanism for carbonic anhydrase catalyzed hydration of CO_2. Im denotes imidazole.

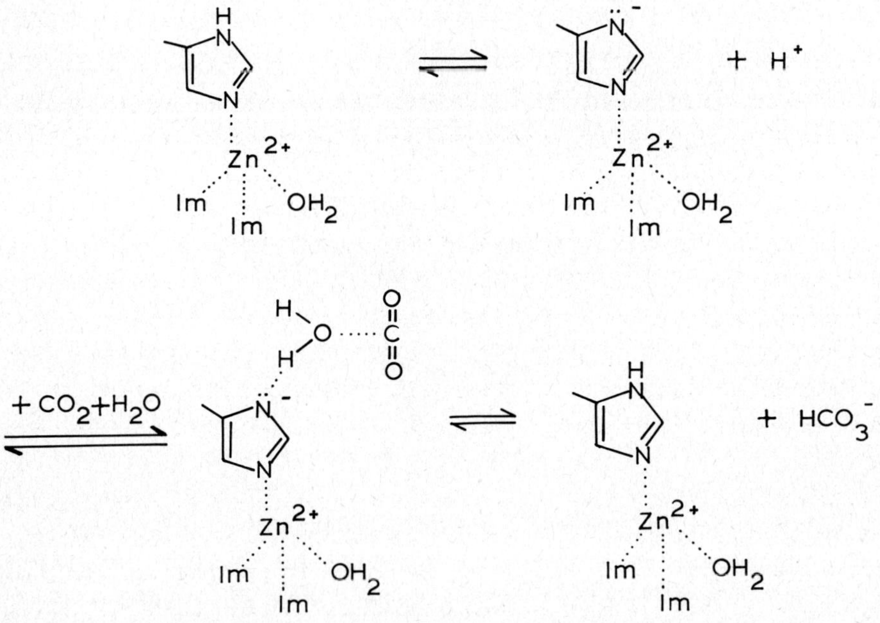

Fig. 26. General base "zinc-imidazolate" mechanism for carbonic anhydrase catalyzed hydration of CO_2. Im denotes imidazole.

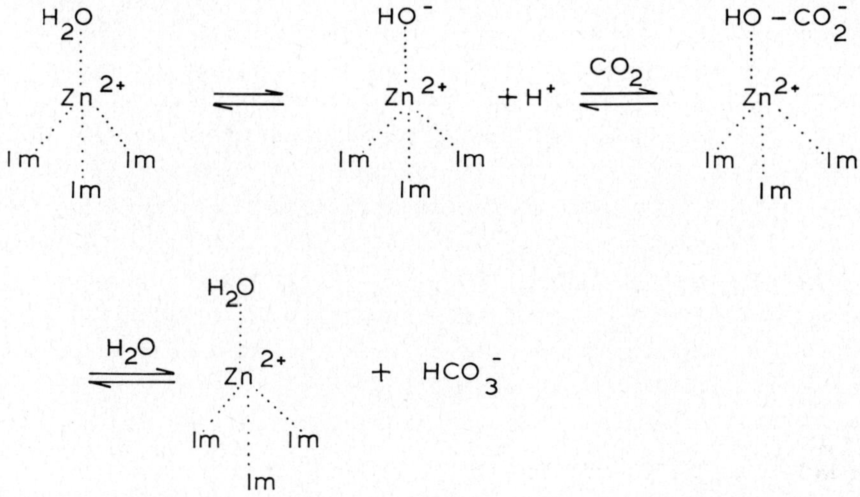

$$\begin{matrix} H_2O \\ \vdots \\ Zn^{2+} \\ Im \quad Im \quad Im \end{matrix} \rightleftharpoons \begin{matrix} HO^- \\ \vdots \\ Zn^{2+} \\ Im \quad Im \quad Im \end{matrix} + H^+ \overset{CO_2}{\rightleftharpoons} \begin{matrix} HO-CO_2^- \\ \vdots \\ Zn^{2+} \\ Im \quad Im \quad Im \end{matrix}$$

$$\overset{H_2O}{\rightleftharpoons} \begin{matrix} H_2O \\ \vdots \\ Zn^{2+} \\ Im \quad Im \quad Im \end{matrix} + HCO_3^-$$

Fig. 27. Nucleophilic "zinc-hydroxide" mechanism for carbonic anhydrase catalyzed hydration of CO_2. Im denotes imidazole.

in the single-coordinated case, too [162]. The above findings suggest that the nucleophilic "zinc-hydroxide" mechanism (Fig. 27) is the more probable since "zinc-imidazolate" mechanisms (Figs. 25-26) start with deprotonation of the imidazole which is unfavoured. Pullman and Demoulin state that as the coordination number of zinc increases pK_a of the bound H_2O and nucleophilicity of the bound OH^- ion also increase. There is thus a balance to be reached between both properties which is reached at a coordination number of four for zinc in carbonic anhydrase. Furthermore, carbon dioxide can approach the active site positioning itself as a fifth ligand to zinc. This statement gains support also from experimental data [163].

Like in several other enzymatic processes (*vide infra*) electrostatics plays a considerable role in carbonic anhydrase action, too. Sheridan and Allen [91] compared the electrostatic potential near the active site in both isozymes of carbonic anhydrase, HCAB and HCAC. Differences in their electrostatic isopotential maps may account for slightly different function of the two forms. For example, only HCAB and not HCAC binds imidazole, the only known competitive inhibitor of CO_2 hydration [163]. Sawaryn and Sokalski have pointed out [159] that the electrostatic field of neighbouring amino-acid side chains considerably stabilizes the transition state of the hydration reaction.

Clementi and his coworkers used the Monte Carlo technique to consider hydration of zinc cation, CO_2 and their complex [161]. The simulation of the hydration was possible using analytical atom-atom potentials by fitting the ab initio binding energies between H_2O and the system under study. On the basis of calculations on Zn plus 27 nearby residues for the active site of HCAB the estimation of the shape of the enzyme cavity and the number of fixed water molecules inside became possible [164].

Another metalloenzyme, which has been studied theoretically, is liver alcohol dehydrogenase. It catalyzes the oxydation of primary and secondary alcohols to aldehydes and ketones, respectively. The enzyme binds the coenzyme, nicotinamide dinucleotide, NAD^+, and two zinc atoms. One of these atoms is participating directly in catalysis and it is liganded to three side chains: Cys-46, His-67 and Cys-174 and to an additional water molecule [165]. The first step of the mechanism is a rapid initial binding of NAD^+. This is followed by an isomerization of the complex between NAD^+ and the enzyme. As a result, the pK_a of the zinc-bound water molecule considerably lowers and a hydroxyl group is formed by releasing a proton. Ionization of the zinc-bound water proceeds through a proton-relay system (see Fig. 28) which is formed by neighbouring Ser-48 and His-51 side chains. The presence of the hydroxyl ion facilitates the formation of a complex between the alcohol to be oxidized and the catalytic zinc. The substrate alcoholate replaces the hydroxyl group and in the last, rate-determining step NADH dissociates and the enzyme izomerizes back to the initial structure with water bound to the catalytic zinc.

Tapia and coworkers modelled the proton-relay system by a complex of imidazole, methanol and water plus a unit positive charge for NAD^+ and a fractional positive charge for the Zn atom [9]. They used their self-consistent reaction-field theory of protein core effects [97] at the CNDO/2 level to introduce inhomogeneous electric field of the enzyme into the calculations. Theoretical energy curves for possible proton transfers from the zinc-bound water to Ser-48 and from Ser-48 to His-51 show that it is the neutral form of the proton-relay system which is most stable (see Fig. 28). The joint effect of static protein field, polarization, presence of Zn

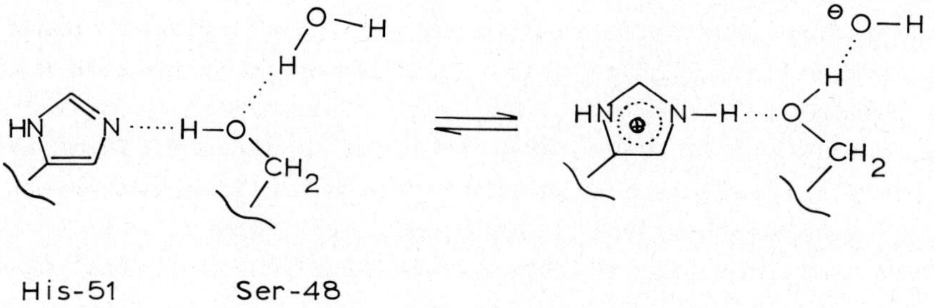

Fig. 28. The proton-relay system in liver alcohol dehydrogenase.

atom and NAD^+ facilitates the formation of each $ImH^+...Ser...OH^-$. $ImH^+...Ser^-...OH_2$ and $Im...SerH^+...OH^-$. The largest stabilization energy, 332 kJ/mol or 480 kJ/mol depending on the value of the reaction field strength, is found for the first structure (see Fig. 28). A smaller fraction of this amount comes from the presence of zinc atom and the positively charged NAD^+. The major stabilizing factor is due to the static protein field and especially to polarization of the proton-relay system by this field. The failure to predict the existence of the $ImH^+...Ser...OH^-$ form of the triad in the active state of the enzyme is most probably due to the minimal basis set error discussed in Sec. 2.2 and especially in Sec. 3.1 for the catalytic His...Asp diad of serine proteinases.

3.6 Miscellaneous topics

Enzymes, discussed previously, occupy a central position in biophysical enzymology. The three-dimensional structure is known with considerable accuracy and detailed mechanisms for their action have been proposed. Theoretical calculations on these systems can often be checked directly thus a relatively firm basis exists on which further discussions can be made. In this section a rather different category is treated. For most cases information on spatial construction of the active site is incomplete or it is even totally lacking. Consequently, discussion of the catalytic mechanism in these systems is often speculative. In spite of the considerable ambiguity in the construction of adequate models quantum chemical calculations may play an important role in interpretation of available experimental data.

First we treat two cases where three-dimensional structure of the enzyme is available. Holmes et al. [166] used a molecular mechanics method to study the action of ribonuclease S on the substrate, uridylyl-(3'-5')-adenosin (UpA). The proposed mechanism is summarized in Fig. 29. In Step 1 the phosphodiester bond is cleaved and a cyclic nucleotide is formed which is hydrolyzed in Step 2 to give a 3'-nucleoside monophosphate. Holmes et al. [166] considered only His-12, His-119 and Lys-41 of the whole enzyme to model the active site. They have compared the minimum energy conformations of the enzyme-bound substrate, UpA, and of the inhibitor, UpcA. This latter compound corresponds to the former except that the O5' atom of the ribose attached to the adenine base is replaced by a methylene group (see Fig. 29). The principal difference between conformations of UpA and UpcA lies in the position of the phosphate moiety. The full reaction path was followed and calculations suggest that His-119 and Lys-41 may move to come in contact with the phosphate group. Transphosphorylation starts by decreasing the O2'...P distance while P-O5 distance simultaneously increases resulting in bond cleavage and in formation of the cyclic intermediate. The phosphorus atom moves about 200 pm bringing the phosphate group within hydrogen-bonding distance to Cys-41. The reaction proceeds by an in-line attack of the O2' atom on phosphorus and passes through a pentacoordinate transition state which is intermediate in geometry between a trigonal bipyramid and a square pyramid. These results support the mechanism proposed by Roberts et al. on the basis of experimental data [168]. Calculations indicate that there are no high energy intermediates along the reaction path for Step 1 of the enzyme action.

Deakyne and Allen [169] used the CNDO/2 method to study transphosphorylation step (Step 1 in Fig. 29) in the catalytic mechanism of ribonuclease A. The role of His-12, Lys-41, Asn-44, His-119, the 119-120 backbone NH and Asp-121 was discussed. The above authors concluded that the 119-120 backbone NH and Gln-11 plus an attached water molecule of the first hydration shell increase electrophilicity of phosphorus facilitating attack of O2'. Furthermore, they found that His-12 is deprotonating O2' but in its absence

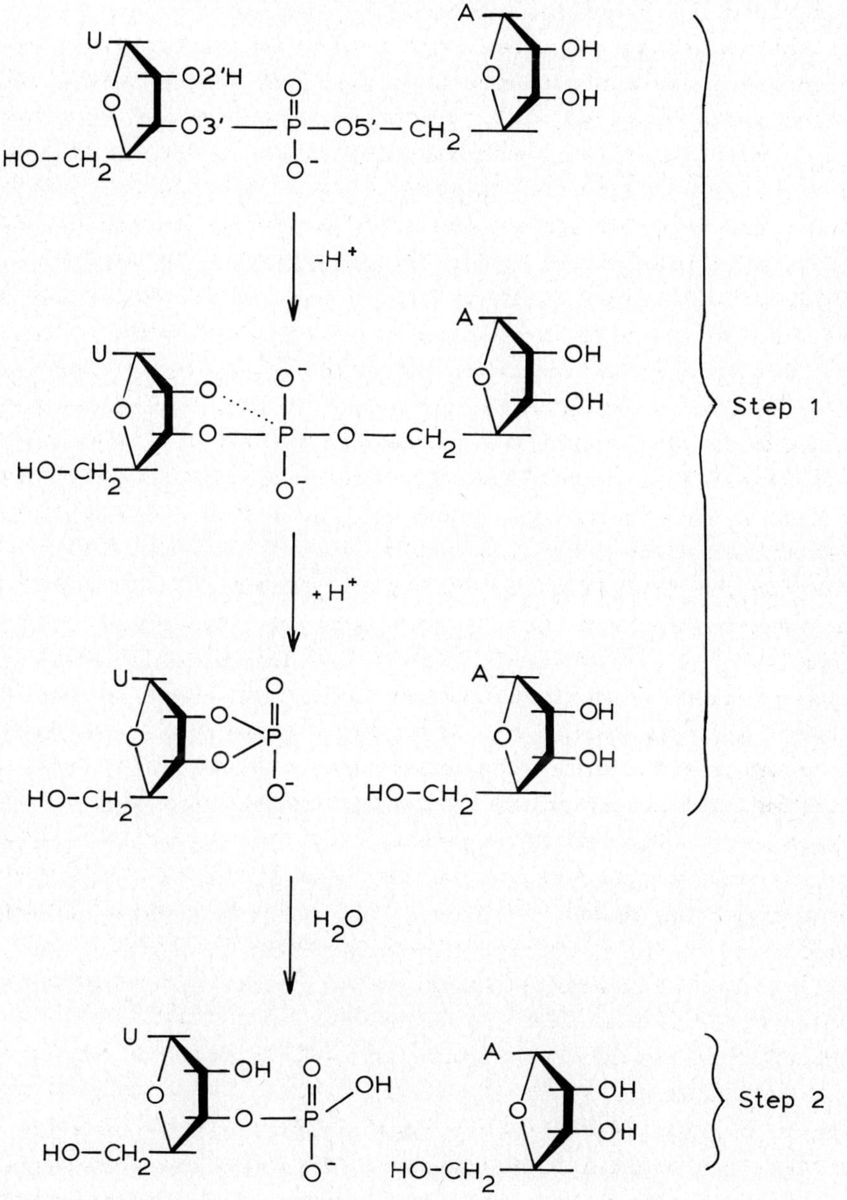

U — O2'H — O3' — P(=O)(O⁻) — O5' — CH2 — A (OH, OH); HO–CH2

-H+

+H+

Step 1

H2O

Step 2

Fig. 29. Mechanism of ribonuclease S action on a dinucleotide substrate, UpA [167]. U = uridyl, A = adenyl. Note 5' oxygen atom on top which is replaced by a methylene group in UpcA.

Asn-44 may fulfil this role. Side chains of His-119 and of Lys-41 play an important role by protonating the leaving group at O5′ and thus facilitating the breakdown of the trigonal-bipyramidal transition state and increasing nucleophilicity of phosphorus.

Gorenstein et *al*. applied their torsional catalysis concept (see Sec. 3.3) to the mechanism of action of ribonuclease A and staphylococcal nuclease [139]. They have found an explanation to the fact that exclusively the conformationally restricted linkage between uridine 33 and O2′-methylguanosine 34 of yeast phenylalanyl transfer RNA is attacked by ribonuclease A. According to X-ray diffraction measurements, only this internucleotide phosphate of the substrate is in a *gauche*-*trans* orientation where, due to interactions with neighbouring lone pairs, the P-O5′ bond weakens and the P-O3′ bond strengthens relative to other, *gauche*-*gauche* or *trans*-*trans*, conformations. For numbering see Fig. 29. Furthermore, it is understood, why only the P-O5′ bond is cleaved by this enzyme. Similarly, selective scission of the P-O5′ bond in thymidine-3′-phosphate, 5′-*p*-nitrophenyl phosphate by staphylococcal nuclease is explained, also.

Several theoretical calculations are performed to clear the mechanism of heme proteins, like cytochromes, catalases and peroxidases [28, 32, 42, 51, 53, 170]. These proteins act as carriers of electrons or of small molecules, like O_2. Especially cytochrome P450 has been extensively studied, which has a primary biological function in the oxidative metabolism of many drugs and endogenous substrates. Speculations are restricted to the heme group with axial ligands (Fig. 30). Kirchner and Loew [170]studied the geometry of the dioxygen ligand in the heme pocket ($X = O_2$, Y = imidazole with $N^{\epsilon 2}$ attached to Fe, see Fig. 30). In agreement with experiments, both *ab initio* and Iterative Extended Hückel calculations favor a bent, end-on structure over a double-coordinated arrangement (see Fig. 31). In another paper [51] the active state of cytochrome P450 has been characterized (X = O, Y = imidazole with $N^{\epsilon 2}$ attached to Fe or Y = methanethiolate, see Fig. 30). Calculations predict negatively charged oxygen and low-energy virtual or half-filled orbitals with substantial oxygen character suggesting that the electrophilic activity of oxygen in the protein is orbital-controlled instead of being charge-controlled. This statement is confirmed by the fact that calculated electric field gradients and quadrupole splittings are in good agreement with experiment. Models for oxy- and carboxy-ferrous cytochrome P450 ($X = O_2$

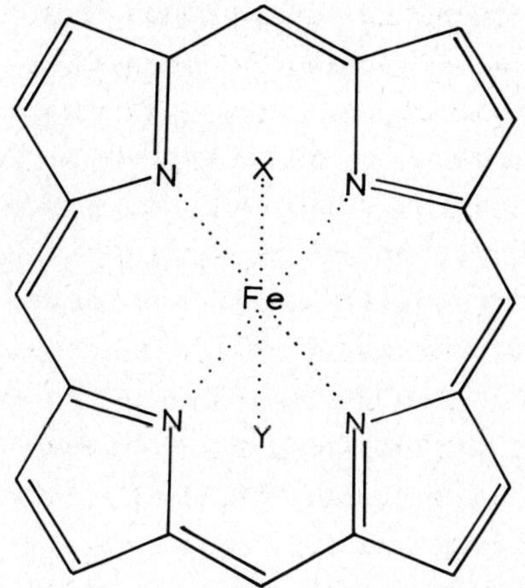

Fig. 30. Heme group of cytochrome P450. X and Y stand for axial ligands.

or CO, Y = methanethiolate, see Fig. 30 and Ref. 53) as well as for iron and manganese porphyrins (X = O_2, Y = imidazole and Fe is replaced by Mn, see Fig. 30 and Ref. 32) were also compared.

Fig. 31. Possible orientations of O_2 in heme proteins. (a) End-on structure; (b) double-coordinated arrangement.

Interaction of cytochrome P450 with aromatic substrates, modelled by quinoline, has been investigated by Nagata et al. [192] It was found that both mechanism of activation and transfer of oxygen to substrate depends on spatial orientation of the enzyme and the substrate. Several models of the heme - quinoline complex were examined by comparing spectral characteristics with observations on isolated cytochrome P450. Pudzianowski and Loew [28] studied hydroxylation and epoxidation pathways for methane and for ethylene, modelling the whole protein by a single oxygen atom. According to MINDO/3 calculations, they predict that for the $CH_4 + O(^1S) \rightarrow CH_3OH$ reaction no ionic intermediates are formed and a retention of configuration occurs. Epoxidation is found to be a concerted process with retention of configuration. In both cases reaction with triplet oxygen involves considerably higher barrier height. It is believed that enzymatic N-hydroxylation by cytochrome P450 is required as a first step in the transformation of aromatic amines to active carcinogens. The mechanism of this reaction has been studied on model compounds by semiempirical MINDO/2 and MINDO/3 methods [42].

Fixation of molecular nitrogen by the enzyme nitrogenase has been studied by semiempirical methods [111, 171, 172]. Since the precise structure of the active site is unknown, only Fe or Mn atoms have been assumed as models. The stereochemical principles of complex formation between N_2 and the metal atom have been rationalized on the basis of mutual orientation of frontier orbitals [111, 171]. A crude model of the active site of nitrogenase has been proposed for which detailed CNDO/2 calculations were done [172].

Recently *ab initio* calculations were reported on a model of glyoxylase [22]. According to Szent-Györgyi's suggestion, methylglyoxal, CH_3COCHO, is involved in regulation of cellular division. According to present knowledge, glyoxylase converts methylglyoxal to lactic acid, $CH_3CH(OH)COOH$, in two steps. Lavery and Pullman studied the first where glutathione acts as a cofactor. Modelling this by methanethiol possible pathways of rearrangement were studied. The involvement of a magnesium cation and of a histidine side chain in the catalytic process, suggested by experiment, has been confirmed by the calculations.

Andrews and coworkers [43, 173] studied the intramolecular isomerization of chorismate to prephenate experimentally and theoretically. This reaction is an aliphatic Claisen rearrangement and it serves as an adequate model for comparison of enzymatic and nonenzymatic processes, since entropy effects do not play a role here. Isomerization can proceed either through a chair-like or through a boat-like transition state. Both states were studied with full geometry optimization using Extended Hückel [43] and

MINDO/3 [173] methods. While for the nonenzymatic process in aqueous solution both pathways are possible, the reaction, catalyzed by the enzymes chorismate mutase and prephenate dehydrogenase, proceeds via a chair-like transition state.

One of the first molecular orbital studies on substrate-enzyme complexes utilized the semiempirical Pariser-Parr-Pople method [174]. Complexation of adenine kinase with adenine analogues was studied on the basis of charge distribution and other reactivity indices of the substrate.

Mechanism of inhibition of alanine racemase and aminotransferase by the antibiotics, cycloserin, has been studied by Nagata and Yamagouchi [175]. It was shown by CNDO/2 calculations that after complexation of the inhibitor with the coenzyme, pyridoxal, its reactivity towards the nucleophilic site in the enzyme is considerably increased. This is in agreement with previously proposed mechanism of inhibitory action which is based on experimental observations.

Interaction of the coenzyme, nicotinamide dinucleotide, NAD^+, with the apoenzyme, modelled by a tryptophan side chain, was studied by Umeyama et al. [176]. Calculations with Pariser-Parr-Pople SCF and CNDO/2 methods implied the existence of a complex between tryptophan and NAD^+ with a charge transfer from the former to the latter. Salem et al. [177] studied another coenzyme, vitamin B12, and they were able to propose a transition-state structure for the coenzyme-catalyzed migration of hydroxyl group from one carbon atom of ethanol to the other.

4 POSSIBLE SOURCES OF RATE ENHANCEMENTS IN ENZYMATIC REACTIONS

One of the most challenging problems in enzymology is to find the source of the extreme catalytic power of enzymes. For example, amides, esters and peptides hydrolyze about 10^{10} times faster in the presence of a serine proteinase than in its absence (see e.g. [5]). The classical approach to understanding the nature of enzymatic catalysis has been from the extrapolation of the study of simple reactions in solution. This approach has several pitfalls as pointed out by Fersht [178] and it seems that the precise explanation for the rate acceleration in enzyme catalysis may come from theoretical calculations.

First of all we want to make clear that rate enhancement is not due to a single factor. However, of the several possible sources of catalytic power of enzymes some seem to be more important than others. Often the same physical or chemical effect is denoted by a different term in the literature. Furthermore, we hawe at present no evidence that the same factor is responsible for catalytic power in all types of enzymes.

In the following we discuss some of the most popular concepts from the theoretical point of view. The charge-relay mechanism was treated in Sec. 3.1 where we pointed out that this hypothesis has been disproved by recent experimental evidence. Therefore we omit this from the present overview.

4.1 Entropy effects

We have shown in preceding sections how an enzyme brings two or more amino-acid side chains together to form an effective catalytic machinery. Since these side chains are fixed in the protein environment they may be oriented to ensure maximum reactivity. For example, the presence of an imidazole as a general base (proton acceptor) accelerates the formation of the tetrahedral intermediate in serine proteinases (Sec. 3.1). Several organic chemical reactions are designed to show that such proximity effects may result in rate enhancement by a factor of 10^{10} or more [178]. An example is given in Fig. 32. It is demonstrated that if reacting groups are fixed in an advantageous relative position the reaction rate is consdierably increased.

Page and Jencks [179] have calculated the entropy change for reactions in which two molecules react to give one molecule of product. Using the Sackur-Tetrode equation [180] they conclude that translational and rotational entropy provides an important driving force for enzymatic and intramolecular rate accelerations. They estimate the rate enhancement, due to entropy factors, to be on the order of 10^8. This calculation is based on an enzyme-substrate model *in vacuo* and it was criticized by Larsen [108] who compared enzymatic rates with the uncatalyzed reaction in water. He found that the loss of translational degrees of freedom, due to the combination of two reactants, is accompanied by the destruction of one solvent cage as the two reactants form a single species. Entropy effects, due to both processes, nearly cancel each other. Delisi and Crothers [109] estimated the rate acceleration in ring cyclizations due to entropy effects at 10^3. Furthermore, Kraut [5] pointed out that, when dealing with serine proteinases, the entropy effect is involved in the acylation step. He argued that, invoking the principle of microscopic reversibility, any interactions which contribute in

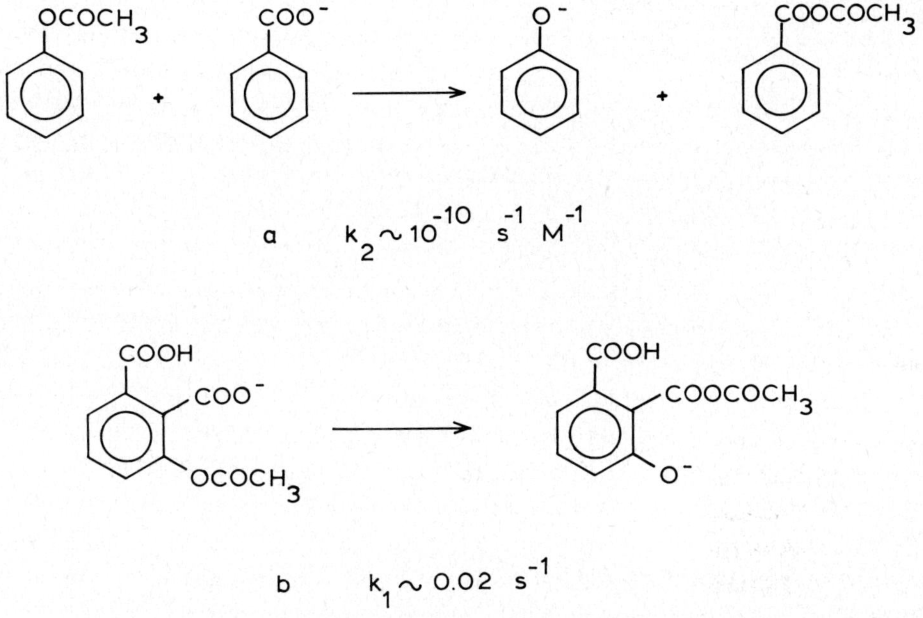

a $k_2 \sim 10^{-10}\ s^{-1}\ M^{-1}$

b $k_1 \sim 0.02\ s^{-1}$

Fig. 32. Relative rates of intermolecular (a) and intramolecular (b) acetyl migration [178].

the acylation step must also contribute to the reverse deacylation reaction. Since this is not the case more than just entropy effects must be involved even for acylation.

Storm and Koshland [181] suggested that the catalytic efficiency of enzymes is due to their ability not only to bring the reacting species together but also to orient them along a path which takes advantage of this directional preference. They called this effect <u>orbital steering</u> and predicted to be as large as 10^4 per reacting atom with an orbital alignment preference. Dafforn and Koshland used two calculation methods to estimate the magnitude of orientational factors. First, they applied collision theory [182] and considered the magnitude of contributions from bending vibrations and the distribution of approach orientations. Second, they used transition-state theory and considered contributions from rotational entropy, bending and stretching vibrations [183].

The concept of orbital steering was heavily criticized. First, Jencks and Page claimed [184] that orbital steering, if carefully defined, may be an adjunct but it is not a satisfactory substitute for the concept of entropy. Lipscomb and coworkers [185] concluded from a PRDDO quantum chemical study of the addition of methanol to formic acid that orientational requirements for this reaction are minimal.

Certain calculations, however, support the orbital steering hypothesis. It was suggested that the energy required to form a misaligned transition state in a solvent cage indicated that precise orientation is important [186]. Umeyama and coworkers [11] also supported this mechanism by their CNDO/2 calculations on α-chymotrypsin.

The effect of geometric restriction on the rates of enzyme catalyzed reactions have been described over the years by several terms [178, 184, 187]. As we have pointed out earlier simple organic reactions are often considered as models for the enzymatic processes. This is not always justified therefore calculations performed on such model systems, even if completely correct, may lead to artefacts if the problem of enzymatic rate enhancement is concerned.

Theoretical calculations of the entropy contribution to the Gibbs free energy of gas-phase reactions are available at present [34, 106]. Extension of such calculations to enzymatic systems is very difficult since in the latter vibrational and rotational terms are coupled due to aperiodic displacements of atoms in the hydrogen--bonded network of the polypeptide chain (see also Sec. 2.5). Another problem is to construct the adequate model since, as we have seen, hydration plays an important role.

4.2 Strain

Several authors believe that, in many cases, enzyme-bound substrates are distorted away from their most stable geometries [4, 6]. This results in an enhanced reactivity of the activated complex which may count for the rate acceleration. For example, the reactivity of an ester or an amide substrate would be substantially increased if it were bound to the enzyme in a non-planar formation. Such deformations would result from interactions between the substrate and the three-dimensional structure of the protein. Since this latter is not rigid it will be deformed also. Strain would be only effective in catalytic rate enhancement where there is sufficient positive binding energy from interaction of the protein with other atoms of the substrate molecule.

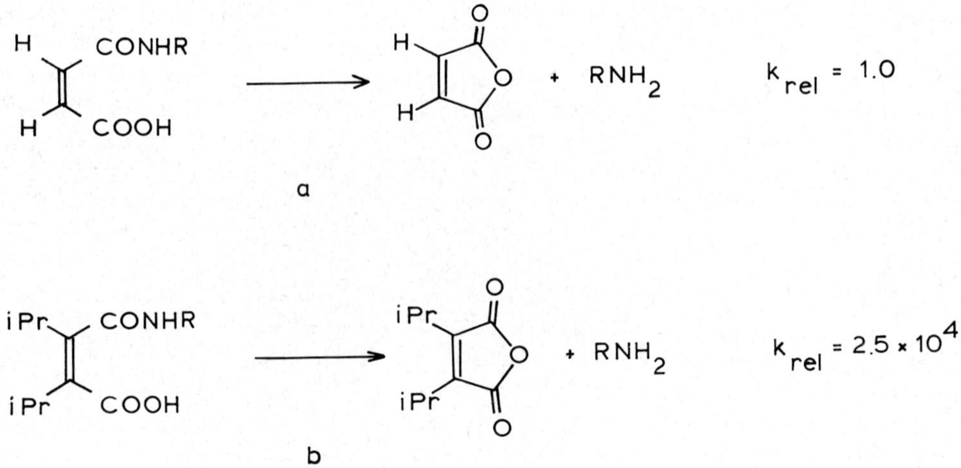

H CONHR / H COOH ⟶ (maleic anhydride, H, H) + RNH_2 $k_{rel} = 1.0$

a

iPr CONHR / iPr COOH ⟶ (diisopropylmaleic anhydride, iPr, iPr) + RNH_2 $k_{rel} = 2.5 \times 10^4$

b

Fig. 33. Effect of strain on ring closure reaction rate. (a) Unstrained; (b) bulky isopropyl (iPr) substituents induce strain in the initial state which leads to rate acceleration [188].

The strain hypothesis was supported by a vast number of model reactions [6], see Fig. 33 for an example. However, as we pointed out in Sec. 4.1, these are not necessarily true models of enzymatic reactions and often they even lead to confusions. Molecular mechanics calculations are available in the literature which deal with the role of geometric strain. In the following we discuss their main conclusions.

Warshel and Levitt [10] treated the mechanism of lysozyme reaction by a combination of quantum chemical and molecular mechanics methods (see Sec. 3.3). They emphasized factors contributing to the stabilization of the carbonium ion intermediate. They found that strain plays a negligible role here. There is no torsional deformation in the initial stage of the reaction. Almost the same geometrical change is obtained if the calculation is performed on the gas-phase reaction or if steric interactions with the enzyme are also considered. Though heme proteins are not enzymes, study of heme-protein interactions in hemoglobin [189, 190] may yield useful informations which can be adapted to enzymatic systems. Molecular mechanics calculations indicate that strain does not account for the entire 15.1 kJ/mol interaction energy.